FRACTURE MECHANICS TESTING METHODS FOR POLYMERS, ADHESIVES AND COMPOSITES

Other titles in the ESIS Series

EGF 1	*The Behaviour of Short Fatigue Cracks* Edited by K.J. Miller and E.R. de los Rios
EGF 2	*The Fracture Mechanics of Welds* Edited by J.G. Blauel and K.-H. Schwalbe
EGF 3	*Biaxial and Multiaxial Fatigue* Edited by M.W. Brown and K.J. Miller
EGF 4	*The Assessment of Cracked Components by Fracture Mechanics* Edited by L.H. Larsson
EGF 5	*Yielding, Damage, and Failure of Anisotropic Solids* Edited by J.P. Boehler
EGF 6	*High Temperature Fracture Mechanisms and Mechanics* Edited by P. Bensussan and J.P. Mascarell
EGF 7	*Environment Assisted Fatigue* Edited by P. Scott and R.A. Cottis
EGF/ESIS 8	*Fracture Mechanics Verification by Large Scale Testing* Edited by K. Kussmaul
ESIS/EGF 9	*Defect Assessment in Components - Fundamentals and Applications* Edited by J.G. Blauel and K.-H. Schwalbe
ESIS 10	*Fatigue under Biaxial and Multiaxial Loading* Edited by K. Kussmaul, D.L. McDiarmid and D.F. Socie
ESIS 11	*Mechanics and Mechanisms of Damage in Composites and Multi-Materials* Edited by D. Baptiste
ESIS 12	*High Temperature Structural Design* Edited by L.H. Larsson
ESIS 13	*Short Fatigue Cracks* Edited by K.J. Miller and E.R. de los Rios
ESIS 14	*Mixed-Mode Fatigue and Fracture* Edited by H.P. Rossmanith and K.J. Miller
ESIS 15	*Behaviour of Defects at High Temperatures* Edited by R.A. Ainsworth and R.P. Skelton
ESIS 16	*Fatigue Design* Edited by J. Solin, G. Marquis, A. siljander and S. Sipilä
ESIS 17	*Mis-Matching of Welds* Edited by K.-H. Schwalbe and M. Koçak
ESIS 18	*Fretting Fatigue* Edited by R.B. Waterhouse and T.C. Lindley
ESIS 19	*Impact of Dynamic Fracture of Polymers and Composites* Edited by J.G. Williams and A. Pavan
ESIS 20	*Evaluating Material Properties by Dynamic Testing* Edited by E. van Walle
ESIS 21	*Multiaxial Fatigue & Design* Edited by A. Pinian, G. Cailletand and T.C. Lindley
ESIS 22	*Fatigue Design of Components.* ISBN 008-043318-9 Edited by G. Marquis and J. Solin
ESIS 23	*Fatigue Design and Reliability.* ISBN 008-043329-4 Edited by G. Marquis and J. Solin
ESIS 24	*Minimum Reinforcement in Concrete Members.* ISBN 008-043022-8 Edited by Alberto Carpinteri
ESIS 25	*Multiaxial Fatigue and Fracture.* ISBN 008-043336-7 Edited by E. Macha, W. Będkowski and T. Łagoda
ESIS 26	*Fracture Mechanic: Applications and Challenges.* ISBN 008-043699-4 Edited by M. Fuentes, M. Elices, A. Martin-Meizoso and J.M. Martinez-Esnaola
ESIS 27	*Fracture of Polymers, Composites and Adhesives.* ISBN 008-043710-9 Edited by J.G. Williams and A. Pavan

For information on how to order titles 1-21, please contact MEP Ltd, Northgate Avenue, Bury St Edmonds, Suffolk, IP32 6BW, UK. Titles 22-27 can be ordered from Elsevier Science (http://www.elsevier.com).

FRACTURE MECHANICS TESTING METHODS FOR POLYMERS, ADHESIVES AND COMPOSITES

Editors: D.R. Moore, A. Pavan and J.G. Williams

ESIS Publication 28

This volume has evolved from the work of the Technical Committee 4
of the European Structural Integrity Society and is
an overview of their activities since 1985.

2001

Elsevier

Amsterdam - London - New York - Oxford - Paris - Shannon - Tokyo

ELSEVIER SCIENCE Ltd.
The Boulevard, Langford Lane
Kidlington, Oxford OX5 IGB, UK

First edition 2001

Library of Congress Cataloging in Publication Data.
A catalog record from the Library of Congress has been applied for.

British Library Cataloguing in Publication Data.
A catalogue record from the Britisch Library has been applied for.

ISBN : 0 08 043689 7
ISSN : 1566-1369

Transferred to digital printing 2006

CONTENTS

Chapter 2 - Elastic-Plastic Fracture Mechanics

Chapter 3 - Adhesion Fracture Mechanics

Chapter 4 - Delamination Fracture Mechanics

INTRODUCTION TO THE WORK OF ESIS TC 4

J.G. WILLIAMS

HISTORICAL INTRODUCTION

Technical Committee 4 of the then European Group on Fracture (now ESIS) started with the decision to form an activity in Polymers and Composites at the ECF conference in Portugal in 1984. Professor Kausch of EPFL (Lausanne) and myself were asked to chair it and we had the opportunity to have a discussion of interested parties at the Churchill College Conference* in April 1985. There was enthusiastic support for the idea and we decided to hold the first meeting in Les Diablerets, Switzerland in October 1985. The venue arose from my involvement with the village and the proximity to Lausanne. The venue and pattern of the meetings, ie 2 1/2 days held in May and October, became established and has continued without interruption.

Two major areas were identified as appropriate for the activity. Firstly there was an urgent need for standard, fracture mechanics based, test methods to be designed for polymers and composites. A good deal of academic work had been done, but the usefulness to industry was limited by the lack of agreed standards. Secondly there was a perceived need to explore the use of such data in the design of plastic parts. Some modest efforts were made in early meetings to explore this, but little progress was made. In contrast things moved along briskly in the standards work and this has dominated the activity for the last fourteen years. The design issue remains a future goal.

The development of standards has a poor reputation in some academic circles. The importance is conceded, but the task is perceived as being of a low academic level. This analysis is quite untrue. Producing a test protocol that gives reliable and meaningful results requires a deep understanding of the physics, and weakness in this regard is soon exposed by poor results. We developed a method, based somewhat on ASTM procedures, of evolving protocols via our regular meeting. An initial version is prepared by the project leader and one or more of the industrial members agreed to supply material. At the next meeting each participant describes their results and experience via presentation. The protocol is then modified in the light of this and the process repeated. About six iterations, ie three years, seems to be necessary to produce a satisfactory result. We have learnt a great deal about topics we felt we understood beforehand by this process. One's experience is multiplied many times by listening to others who have been through the same process. Several PhD students gained a good grounding in their subject via the group.

This book is an overview of our activities over the last fifteen years. A wide range of tests is described and the numerous authors is a reflection of the wide and enthusiastic support we have had. It has been my privilege to act as co-chairman first with Henning Kausch and subsequently with Andrea Pavan, to such a talented and devoted group.

[*Yield, Deformation and Fracture of Polymers, Institute of Materials]

CHAPTER 1

Linear Elastic Fracture Mechanics

INTRODUCTION TO LINEAR ELASTIC FRACTURE MECHANICS

J.G. WILLIAMS

1. INTRODUCTION

Linear Elastic Fracture Mechanics (LEFM) is the basic scheme used for most of the protocols described here. It has a secure theoretical basis in that all energy dissipation is associated with the fracture process and the deformation which occurs is linear elastic. This turns out to be true for many of the situations covered here, i.e. brittle failures in polymers, impact tests, fatigue, delamination of composites and failure of adhesive joints. This is a great benefit since useful and simple methods can be developed in contrast to metals testing, for example, where plasticity and non-linear effects are important in most tests. Such phenomena can be important in polymers and will be described later, but the main emphasis will initially be on LEFM.

2. TOUGHNESS DEFINITIONS

LEFM assumes that a linear elastic body contains a sharp crack and then describes the energy change which occurs when such a body undergoes an increase in crack area. (It should be noted that it is the growth of an already existing crack, or flaw, which is described and nothing is said about the generation of flaws in otherwise perfect bodies.) The parameter of most fundamental importance is the Energy Release Rate, G, which is defined as the rate of energy released by the crack growth,

$$G = \frac{dU}{dA} \tag{1}$$

where dU is the energy change and dA is the area increase. dA is taken as positive for crack growth, and a positive dU and hence G implies a positive energy release. It is this energy release which is available to drive the crack growth and overcomes the fracture resistance, G_c. Therefore, at fracture

$$G = \frac{dU}{Bda} = G_c \tag{2}$$

where a is the crack length for a uniform thickness B. G is determined by the loading and geometry of the cracked body while G_c is a material property and is the energy per unit area necessary to create the new surface area of the crack. As such it may include the effects of many micro-mechanisms occurring in the region of the crack tip. Usually the cracks propagate in the opening, or mode I, in which the crack faces move apart with the displacement being normal to the crack faces. The toughness for this mode is designated G_{IC}. In composites and adhesives it is possible to propagate cracks such that the displacements are parallel to the crack faces giving shear or mode II propagation and a toughness G_{IIC}. Mixed mode tests are combinations of these and loci of G_c for the degree of mode mix are determined. Out of plane sliding, or mode III, is possible, but is not discussed here.

An important aspect of fracture resistance is that it may vary as the crack grows such that G_c is a function of the crack growth, Δa. Thus we may have a curve of G_c versus Δa, which usually rises, and is termed the resistance or 'R' curve as sketched in Figure 1. This curve is a complete description of the fracture toughness of a material and some tests have its determination as the goal (e.g. delamination of composites). Some however, concentrate on the initiation value, i.e. when $\Delta a = 0$. This is usually the lowest value and is thus judged to be most critical. It may also be so regarded on the basis that once fracture has initiated, then a component has failed. Such arguments are valid, but lead to many practical difficulties of definition. Initiation may be defined via direct visual observation but this is difficult to achieve. Indirect, but more reproducible methods are therefore employed, such as the onset of non-linearity in linear load deflection curves, or the occurrence of a specific (5%) reduction in the slope of the load deflection line. These schemes are a good example of where practicalities have required that exact definitions be replaced with something definable, but only indirectly related to the real phenomena. Many 'R' curves tend to level out to give a plateau value which can be a useful upper limit for G_c although the definition is somewhat arbitrary. The special case, as shown by a broken line in Figure 1, of a constant G_c is often observed.

It is worth noting that LEFM conditions require linear load-deflection behaviour and thus very localised deformation at the crack tip. The stress state of this local deformation zone is however, not determined by the LEFM condition. The local nature of the deformation requires that the zone of deformation is small, compared with the in-plane dimensions of the body including the crack length. The stress state however, is determined by the size of the zone compared to the out-of-plane thickness. In many cases the zone is small compared to the thickness and is thus constrained transversely leading to the highly constrained, plane strain condition. In many situations plane strain will occur if the out-of-plane thickness, B, satisfies the condition:

$$B > 2.5\left(\frac{EG_c}{\sigma_c^2}\right) \tag{3}$$

Where E is Young's modulus and σ_c is the local critical stress and is usually taken as the shear yield stress. Testing under such conditions is of practical importance because this highly constrained condition is often assumed to provide a minimum toughness value. The same parameter is taken to define all the in-plane dimensions also,

i.e.
$$a,\ W - a > 2.5\left(\frac{EG_c}{\sigma_c^2}\right) \tag{4}$$

where a is the crack length and W is the width. If both conditions are met we have plane strain and LEFM and this is most commonly sought. However if equation (4) is met, but equation (3) is not then lower degrees of constraint are possible, generally giving higher values of G_c with an upper limit at the plane stress state.

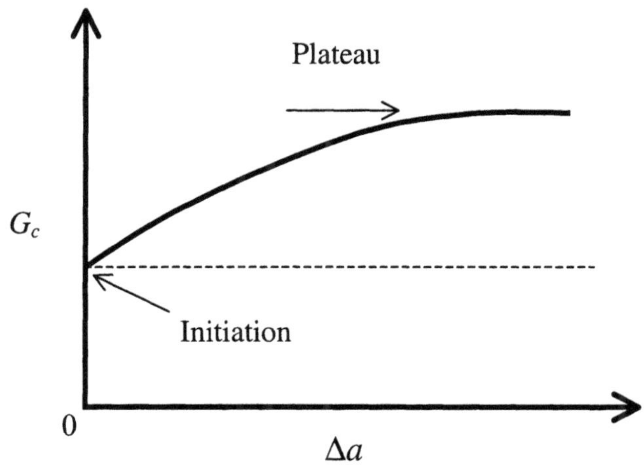

Figure 1. Resistance or 'R' curve

These criteria have been developed for homogeneous materials and will be discussed in the protocols for these later. For delamination in composites and for cracks in adhesive joints the proximity of stiff layers enhances constraint and tends to give plane strain conditions though the situation is often complex giving, for example, toughness variations with adhesive layer thickness.

3. CALIBRATION PROCEDURES

In all the protocols to be described various specimen geometries are used and each must be calibrated so that load or energy measurements at fracture may be converted to G_c. Two schemes are used for effecting this calibration. For many specimens, which are beams in one form or another, it is possible to measure their stiffness, or more conveniently compliance C (= $(\text{stiffness})^{-1}$) as a function of crack length. For all loading systems, G may be defined as

$$G = \frac{dU_{ext}}{dA} - \frac{dU_s}{dA} - \frac{dU_k}{dA} - \frac{dU_d}{dA} \tag{5}$$

where U_{ext} is the external work
U_s is the strain energy
U_k is the kinetic energy
U_d is the dissipated energy
and $dA = Bda$, the change in crack area for a uniform thickness, B.

For low rate testing $U_k = 0$ and if all the energy dissipation is local to the crack tip then $U_d = 0$. For LEFM the load deflection lines are as shown in Figure 2 in which the compliance increases as a increases to $a + da$. The energy changes are,

$$dU_{ext} = P\ du$$

and
$$U_s = \frac{1}{2} Pu$$

ie
$$dU_s = \frac{1}{2}(Pdu + udP).$$

Thus
$$G = \frac{1}{2B}\left(\frac{Pdu}{da} - \frac{udP}{da}\right) \tag{6}$$

and G is the energy change represented by the shaded area in Figure 2. We may now invoke compliance,

i.e. $u = C.P$ and $du = CdP + PdC$

and substituting in equation (6) we have,

$$G = \frac{P^2}{2B} \cdot \frac{dC}{da} = \frac{Pu}{2B} \cdot \frac{1}{C} \cdot \frac{dC}{da} = \frac{u^2}{2B} \cdot \frac{1}{C^2} \cdot \frac{dC}{da} \tag{7}$$

Thus if $C(a)$ is known dC/da may be found and hence G calculated from either load, load and displacement or energy, and displacement alone. These forms are all used in the various protocols described later.

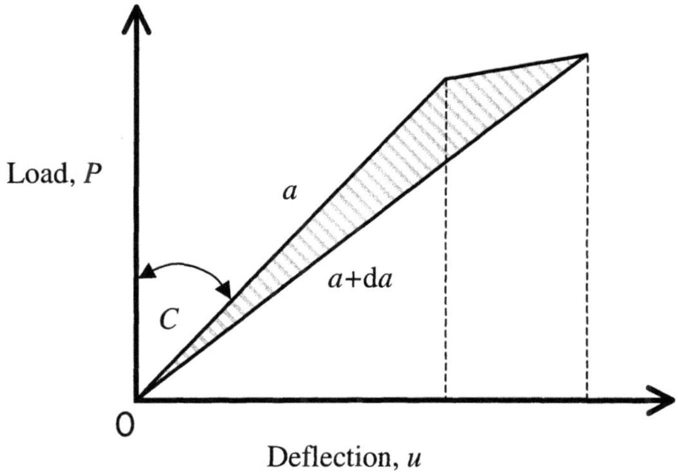

Figure 2. Load-deflection curves for LEFM

The delamination tests on composites generally give stable crack growth using double cantilever beam (DCB) specimens so that P and u are recorded as a increases thus giving $C(a)$. This can then be empirically curve fitted by a power law.

$$C = A.a^n$$

which is termed the Berry Method resulting in,

$$G = \frac{n}{2} \cdot \left(\frac{Pu}{Ba} \right) \tag{8}$$

from the second of equations (7). This form is generally preferable because P, u and a may be used directly and only n is required. The protocols also employ beam theory to determine C which has the advantage of giving a value of Young's modulus which serves as a useful cross check. In most cases simple beam theory has to be corrected for shear deformation and rotation at the end of the crack. The corrected results given are from this corrected beam theory (CBT). The adhesives tests also employ DCB specimens, but in addition use contoured beams which are designed to give a constant dC/da so that a constant load gives a constant G.

The peel test protocol also uses this approach for analysis though it is somewhat more direct. For peeling a strip with a force P at an angle θ the rate of external work may be found directly and is given by:

$$\frac{dU_{ext}}{Bda} = \frac{P}{B}(1 - cos\theta)$$

There are only minor changes in U_s but plastic bending can give significant U_d values. These can be computed and must be deducted from the external work to give G.

Some geometries of practical interest do not lend themselves well to analysis via compliance measurements. Plates in tension and bending are examples and although equations (7) are still correct it is very difficult to find dC/da experimentally. A more accurate method is to consider the local stress field around the crack tip which has the form

$$\sigma = \frac{K}{\sqrt{2\pi r}} . f(\theta), \quad f(0) = 1 \tag{9}$$

where r is the distance from the crack tip and θ is the angle measured from the crack line. The stresses are singular as $r \to 0$ but the product $\sigma\sqrt{r}$ remains finite and is characterised by K, the Stress Intensity Factor. Two relationships for K are important in calibrating specimens. Firstly, K is related to G via,

$$K^2 = EG \tag{10}$$

and for the generic case of a large plate containing a central crack of length $2a$ subjected to a uniform stress σ,

$$K^2 = \sigma^2 \pi a \tag{11}$$

i.e.

$$G = \frac{\sigma^2 \pi a}{E} \tag{12}$$

π is the calibration factor for this case and noting that $\sigma = P/BW$, where W is the width, then from equation (7) we have,

$$\frac{dC}{d\alpha} = \frac{2\pi}{EB} \cdot \alpha, \qquad \alpha = \frac{a}{W}$$

For other geometries the calibration factor π is a function of α and these factors have been derived in great detail both via analysis and computation. The general form is expressed as,

$$K = f(\alpha) \cdot \frac{P}{B\sqrt{W}} \tag{13}$$

and for the large plate case $f(\alpha) = \sqrt{\pi\alpha}$. This form is used in several of the protocols to give the critical K at fracture, K_c. This is used in engineering design because it requires no knowledge of E to determine loads at fracture since, from equation (13), if $f(\alpha)$ is known by measuring P at fracture, then K_c can be found. Thus in any other geometry if $f(\alpha)$ is known, a critical P value may be found without resort to G_c.

In general we are more concerned with characterisation here and hence finding G_c. This may be done via K_c and use of equation (10) when E is determined. This process may be included in the same test by finding G_c via the energy route using the second of equations (7). Thus,

$$G_c = \frac{U_s}{BW} \cdot \left(\frac{1}{C} \frac{dC}{d\alpha} \right) = \frac{U_s}{BW\phi(\alpha)} \tag{14}$$

The calibration factor $\phi(\alpha)$ can be deduced from $f(\alpha)$ if the compliance at $\alpha = 0$, C_o, can be estimated since,

$$\frac{dC}{d\alpha} = \frac{2}{EB} \cdot f^2 \text{ and } \phi(\alpha) = \frac{1}{f^2} \left[\frac{EB}{2} C_o + \int_0^\alpha f^2 d\alpha \right]$$

The $f(\alpha)$ and $\phi(\alpha)$ values and functions are given in the protocols for three point bend and compact tension specimens, which are used for slow rate and 1m/s impact tests to determine K_c and G_c. The value of E from equation (10) is compared with that deduced from compliance measurements since,

$$E = \frac{2}{B} \frac{f^2}{\frac{dC}{d\alpha}} = \frac{2}{BC} \cdot f^2 \phi \tag{15}$$

and is used as a cross check on accuracy.

All the protocols are quasi-static in that $U_k = 0$ is assumed except for the higher rate impact test. Here the loads cannot be measured and the test is conducted at a constant speed, V and timed to give the displacement at fracture. The static value of K_c is then found by deducing the load from,

$$P = \frac{u}{C} = \frac{Vt_f}{C}$$

Where t_f is the fracture time giving a static value of K.

$$K_s = \frac{E}{\sqrt{W}} \cdot \frac{(Vt_f)}{2\phi . f} \tag{16}$$

A correction is made for kinetic energy effects via an experimentally or computed correction factor, k_d, such that:

$$K_c = K_s . k_d . \tag{17}$$

K_C AND G_C AT SLOW SPEEDS FOR POLYMERS

J.G. WILLIAMS

1. INTRODUCTION

A linear elastic fracture mechanics (LEFM) protocol for determining K_c and G_c for plastics is reproduced as the appendix to this paper. This was the first protocol developed by TC4 and was chosen as a starting point because many members had experience of the test method and it was felt to be of practical importance. The basic method was that developed by ASTM for K_c testing of metals [1] but with significant changes to make it suitable for polymers and to include G_c determination. The version in the appendix is the final form produced by TC4 and was the basis used for the ASTM version [2] and subsequently the ISO version [3]. These latter contain changes made to conform to the style and practices of those bodies, but none of substance occurred.

2. BACKGROUND TO THE TECHNICAL ISSUES

The major technical issues addressed in the protocol are notching and the definition of initiation. The method requires that a natural sharp crack is first grown and then the conditions for its re-initiation used to define K_c and G_c. Great skill and care is required to produce these initial cracks and the results are critically dependent on their quality. Different techniques are required for different materials ranging from razor blade tapping in hard materials to blade sliding for soft materials. Initiation is defined as either the maximum load or the load which gives a 5% increase in compliance. Neither is true initiation but represents a reproducible value for a small amount of crack growth.

The size criteria for validity are designed to ensure both LEFM and plane strain and a further restriction, that the maximum load should be no more than 10% greater than that for the 5% compliance change, is a guarantee on linearity and hence LEFM conditions. It is also worth noting that the energy result used to find G_c requires a compliance correction for load point indentation, a notion which arises in several protocols.

3. RESULTS OBTAINED USING THE PROTOCOL

An example of a set of data, in this case a nylon, is given below. Nine groups performed the tests and it can be seen that the average standard deviation are 5% for K_{IC} and 12% for G_{IC}. The agreement between the two values of E is generally good with differences of less than 1% for five sets of data and only one of up to 10%. The data sets do show the common characteristic of such exercises in that some values are wildly out suggesting an error which is usually difficult to identify. Nylon is given as an example because it is not among those materials which are eas[illegible] notch (e.g. epoxies, PMMA), nor is it amongst those which are rather dif (e.g. PE, PP). However, with perseverance, good results can be achieved as demonstrated in Table 1.

TABLE 1. Results of the measurements of K_{IC} and G_{IC} performed on a Nylon by nine groups of ESIS TC4.

Group No	Specimen type	Notching	K_{IC} (mean) $MPa.\sqrt{m}$	G_{IC} (mean) kJ/m^2	E_{stiff} GPa	E_{fract} GPa
1	SENB	RS	4.14 ± 0.17	4.76 ± 0.98	3.65	3.65
		RT	4.03 ± 0.10	3.92 ± 0.15	4.14	4.14
2	SENB	RT	3.79 ± 0.08	4.01 ± 0.17	2.24[a]	3.58
3	SENB	RS	3.84 ± 0.17	4.48 ± 0.70	3.32	3.33
		RT	4.21 ± 0.26	4.82 ± 0.73	3.64	3.71
4	SENB	RT	4.10 ± 0.35	5.14 ± 0.67	3.30	3.28
5	SENB	RS	3.82 ± 0.21	4.20 ± 0.41	3.63	3.32
6	CT	RT	4.46 ± 0.13	5.82 ± 0.24	3.57	3.42
7	SENB	RT	3.99 ± 0.10	4.80 ± 0.46	-	3.32
8	SENB	RT	3.9 ± 0.3	4.7 ± 0.8	3.22	3.21
9	SENB	RT	4.10 ± 0.22	6.40 ± 0.8[b]	-	2.63[b]
		Mean	4.03 ± 0.19	4.82 ± 0.56		

a Error suspected $E_{stiff} = \psi / BC$, $E_{fract} = K_{IC}^2 / G_{IC}$

b Without indentation correction

RS Razor sliding

RT Razor tapping

4. CONCLUDING COMMENTS

In general this protocol has worked well and has now been adopted as an ISO standard for plastics. It does incidentally work well for other materials, which are reasonably stiff and linear in their loading response, as would be expected. With some additional procedures it has been used to measure toughness values in injection moulded discontinuous fibre composites [4] and more recently it has been successfully applied to foods [5].

5. ACKNOWLEDGEMENTS

The protocol was developed over about six years and many groups contributed results and insights, which gave rise to the final version. Below is a list of those contributors with their affiliation in 1990.

Professor JG Williams:	Dept of Mechanical Engineering, Imperial College, UK
Professor HH Kausch:	Lab de Polymere, Dept de Materiaux, Ecole Polytechnique Federal de Lausanne, Switzerland.
Dr P Czarnocki:	Warsaw University of Technology, Poland.
Dr G C Adams:	EI Du Pont de Nemours & Co Inc, USA
Professor W Bradley:	Texas A & M University, USA.
Dr MJ Cawood:	BP Chemicals, UK
Dr ML Clerbois:	Solvay et Compagnie, Belgium
Dr MH Daeniker,	Du Pont de Nemours Int. SA, Switzerland
Dr P Davies:	Universite de Compiegne, France
Dr GE Hale:	The Welding Institute, Cambridge, UK
Dr Ing W Doll:	Fraunhofer-Institute fur Werkstoff-mechanik, FRG.
Mr B Echalier:	Atochem, France.
Mr M P Flueler:	Swiss Federal Laboratories for Materials Testing (EMPA), Switzerland
Professor K Friedrich:	Technische Universitat Hamburg, FRG
Mr E Reese:	Technische Universitat Hamburg, FRG
Mr H Wittich:	Ciba-Geigy, AG, Switzerland
Dr KP Jud:	Ciba-Geigy, AG, Switzerland
Mr M Fischer:	Ciba-Geigy, AG, Switzerland
Professor AJ Kinloch:	Imperial College, UK
Mr I Malkin:	Instron Ltd, UK
Dr B Melve:	Sintef, Norway.
Dr DR Moore:	ICI Petrochemical & Plastics Division, UK.
Professor A Pavan:	Politecnico di Milano, Italy.
Dr F Ramsteiner,	BASF, AG, FRG.
Dr A Roulin-Molony:	Ecole Polytechnique Federal de Lausanne, Switzerland
Dr R Schirrer:	EAHP, France
Dr SD Sjoerdsma:	DSM, The Netherlands
Dr C Wrotecki,	CdF Chimie SA, France
Dr Wutthrich:	BBC, Baden-Datwil, Switzerland

6. REFERENCES

[1] ASTM E399-90: Standard Test Method for Plane-Strain Fracture Toughness of Metallic Materials, 1990.

[2] ASTM D5045-99: Standard Test Methods for Plane-Strain Fracture Toughness and Strain Energy Release Rate of Plastic Materials, 1999.

[3] ISO 13586-1: Standard Test Method for "Determination of Fracture Toughness (G_C and K_C) -Linear Elastic Fracture Mechanics (LEFM) Approach." (2000).

[4] Moore, D.R., K_C and G_C at "Slow Speeds for Discontinuous Fibre Composites", in "Fracture Mechanics Testing Methods for Polymers, Adhesives and Composites" Elsevier Science, 2001.

[5] Kamyab, I., Chakrabarti, S. & Williams, J.G. Cutting Cheese with Wire. Journal of Materials Science, 33, 2763-2770, 1998.

7. THE TEST PROTOCOL

A Linear Elastic Fracture Mechanics (LEFM) Standard
for Determining K_{IC} and G_{IC} for Plastics.

Testing Protocol – March 1990

This protocol has been created by the activities of the ESIS TC4 Task Group on Polymers and Composites and is the result of a series of Round-Robin tests. It is intended to form the basis of national and international standards. It has been drafted by Professor JG Williams, Mechanical Engineering Department, Imperial College London, UK.

1. Introduction

These tests are designed to characterise the toughness of plastics in terms of the critical stress intensity factor K_c, and the energy per unit area of crack G_c, at fracture initiation. The scheme used assumes liner elastic behaviour of the cracked sample, so certain restrictions on linearity of the load – displacement diagram and specimen width must be imposed to ensure validity. In addition, a state of plane strain at the crack tip is required so that thickness normal to the crack front must be sufficient to ensure this state. Finally the crack must be sufficiently sharp to ensure that a minimum value of toughness is obtained. These requirements are common to the ASTM metals standard E399 and much of what follows is drawn from this source. There are, however, special problems associated with plastics and these are accommodated in what follows. Items not covered here will be found in E399. It should also be noted that G_c is of particular importance for plastics and this protocol covers its determination, while E399 does not.

2. Specimen Preparation

Three point bend (also called single edge notched bend, SENB) and compact tension (CT) geometries are recommended, because these have predominantly bending stress states which require smaller sizes to achieve plane strain than other configurations. It is usually helpful to maximise the thickness used from a sheet sample and this is best achieved by making the sheet thickness that of the specimen i.e. B in Figure 1 where the two configurations are shown. In both tests the crack length range should be; $0.45 < a/W < 0.55$ and it is usually convenient to make $W = 2B$ initially.

3. Notching

The ideal case is when a natural crack is re-initiated and this is embodied in the metals test by requiring that an initial machined notched sample is fatigued to give some growth. This method may be used for plastics, but often it is difficult to do, because of unstable fatigue crack growth and the necessity of using low frequencies (<4Hz in some plastics) to avoid hysteretic heating. In plastics it is possible to produce sufficiently sharp initial cracks by other methods and particularly by first machining a sharp notch and then further sharpening it by using a razor blade. This is generally a

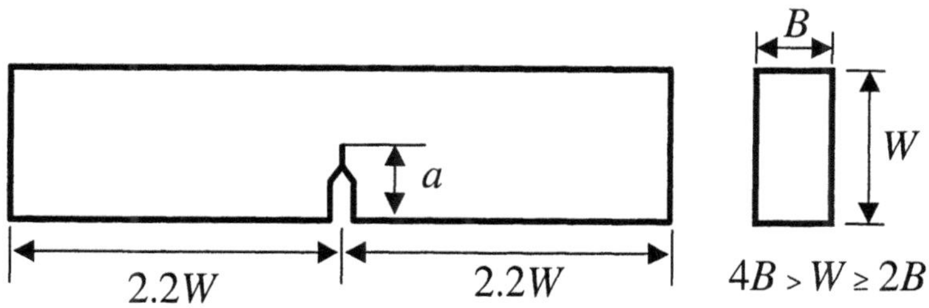

a) Three point bend specimen (SENB)

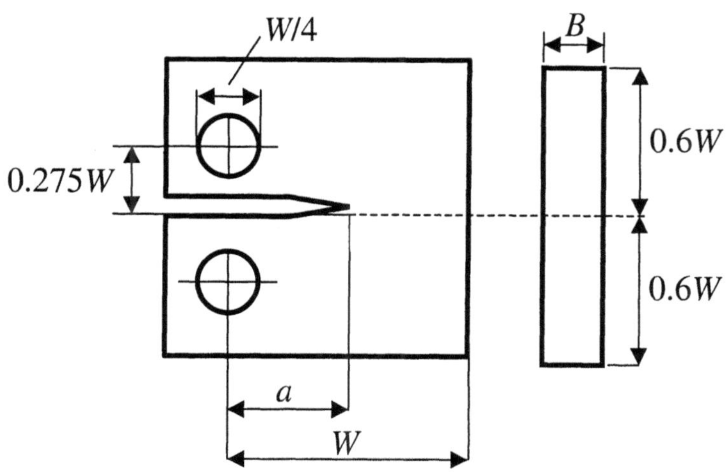

b) Compact tension configuration (CT)

Figure 1. Specimen configurations (as in document E399)

much simpler technique than growing cracks in fatigue. The procedure to be followed is:

i) Machine or saw a sharp notch in the specimen. Then generate a natural crack by tapping on a new razor blade placed in the notch. It is essential to persevere with this since in brittle specimens a natural crack can be generated by this process, but some skill is required in avoiding too long a crack or local damage. (Some precompression of the specimen may be helpful). The cracks grown should be several times longer than the pre-notch tip radius. Failure to generate a natural crack will result in too high values.

ii) If a natural crack cannot be generated, as in some tough specimens, then the notch can be sharpened by sliding or sawing a new razor blade across the notch. Again the depth of this extension should be greater than the original notch tip radius. Pressing the blade into the notch is not recommended because of induced residual stresses.

4. Test Conditions

Since plastics are viscoelastic materials it is necessary to specify both the temperature and time scale under which the result was obtained. As a basic test condition, it is recommended that 23°C and a crosshead rate of 10mm/min be used. In all cases the loading time should be quoted.

If it is not possible to obtain valid results at 23°C it is often possible to do so by decreasing the temperature, which usually does not change K_c greatly, but increases the yield stress rendering the fractures more brittle. If this procedure is used then again both temperature and loading time must be stated.

It is recommended that speeds of greater than 1m/s or loading times of less than 1ms should be avoided, because of the danger of dynamic effects causing errors.

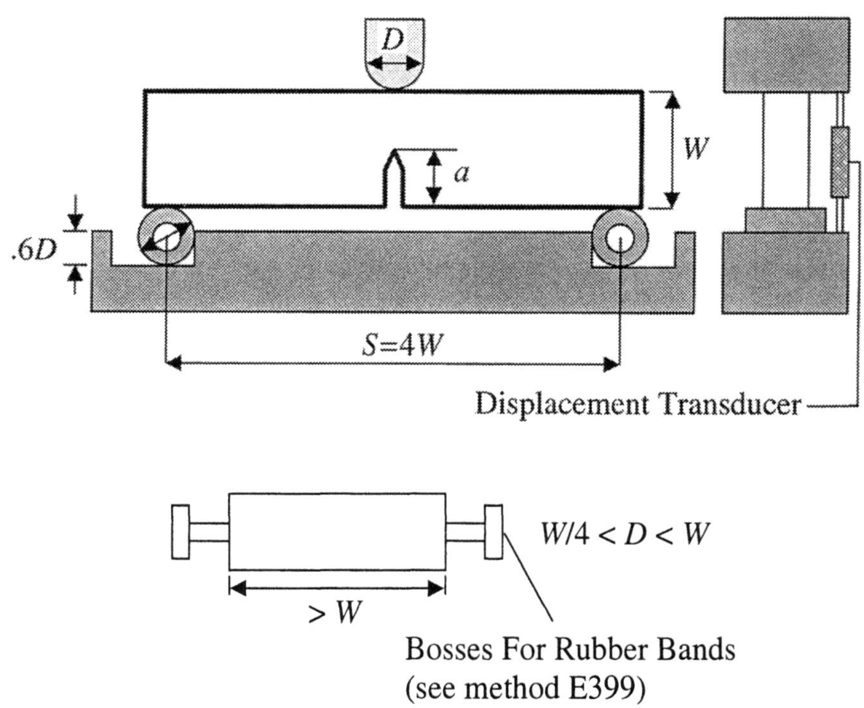

Figure 2. Bending rig used for SENB testing

5. Loading Rigs

For SENB a rig with moving rollers[1] of sufficiently large diameter to avoid plastic indentation is recommended. That shown in Figure 2 is based on E399. For the CT test the loading is via pins in the holes.

For either sample configurations, the displacement measurement can be performed using the loading machine's internally provided stroke (position) transducer. The fracture test displacement data must then be corrected for total system compliance, loading pin penetration and sample compression. This can be performed by a simple calibration of the testing system. The procedure is as follows.

A test configuration as shown in Figures 3a or 3d using identically prepared, but unnotched, samples is used to generate a load-displacement correction curve. This correction curve is then "subtracted" from the load-displacement curve obtained during the actual fracture test with notched samples. This subtraction is performed by subtracting the correction curve displacement from the fracture test displacement at corresponding loads.

In practice, a linear correction curve can usually be obtained (up to the maximum loads recorded in the fracture test). Use of a linear correction simplifies the displacement correction. Any initial non-linearity due to penetration of the loading pins into the sample is observed during both the calibration test and the actual fracture test, so a linearisation of the near-zero correction data and the fracture test data can effectively correct for this initial non-linearity.

This displacement correction must be performed for each material and at each different temperature or rate. Polymers are generally temperature and rate sensitive and the degree of loading pin penetration and sample compression can vary with changes in these variables.

If the internally provided displacement transducer is not available, then externally applied displacement measuring devices may be used. For this case, displacement should be taken at the load point.

For CT samples, this is preferred to crack mouth opening since the load point displacement is required for the energy calculation used in finding G_c. For CT, a clip gauge near the pins will be satisfactory. (If a stiff metals gauge is used it may be necessary to correct the loads in a plastics test.)

For SENB a displacement transducer can be placed between the load point and the base as indicated in Figure 2. In the G_c tests it is necessary to correct the measured displacement for indentation effects and machine compliance. This can be done by two methods:

---

[1] It has subsequently become more common to use fixed supports of the same radius as specified for the rollers. No significant errors have been observed by this simplification of the test rig.

First method: A load displacement curve from that in Figure 3a may be subtracted from that in the fracture test to obtain the true displacement. The load-displacement curve is usually linear and its slope determines the compliance due to indentation and machine stiffness, C_{cor}. The value here would be slightly high because of flexing, so a more precise result can be obtained from the second method described below.

Second method: Using the arrangement shown in Figure 3c the compliance of the machine, C_m, is determined. This is subtracted from the compliance obtained from the arrangement in Figure 3b, to give the compliance due to the indentation at the centre point of the sample, C_i. Therefore, indention compliance due to both the loading striker and the rollers is given by $3/2\ C_i$. Thus C_{cor} is: $C_m + 3/2\ C_i$.

The indentation tests should be performed such that the loading times are the same as the fracture tests. Since the indentations are stiffer, this will involve lower rates to reach the same load; in many cases about half the speed. (More details on energy calculations are given in section 8).

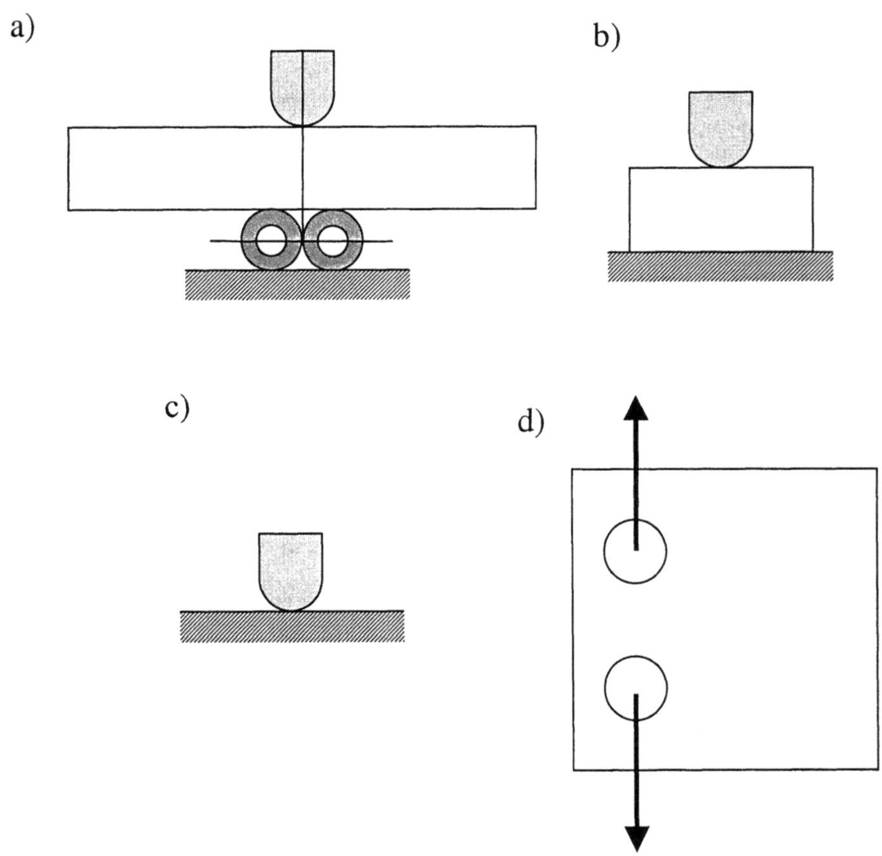

Figure 3. Arrangements for finding indentation displacement

6. Test Procedure

It is recommended that three replicates be used. The test is performed and the load versus load-point displacement curve obtained. In the ideal case this is a linear diagram with an abrupt drop of load to zero at the instant of crack growth initiation. In some cases this occurs and K_Q can be found from the maximum load. (In such cases a natural crack will be required, see section 3). In most cases there is some non-linearity in the diagram and this can be due to plastic deformation at the crack tip, non-linear elasticity, general visco-elasticity and stable crack growth after initiation, but prior to instability. The first three effects violate the LEFM assumption and the fourth one means that the true initiation load is not defined by the maximum. Indeed it is doubtful if an exact definition of initiation could be made and with this, and a need for simplicity in mind, the arbitrary rule of E399 is used here. The diagram is shown (exaggerated) in Figure 4, and a best straight line is drawn to determine the initial compliance C as shown. This is then increased by 5% and a further line drawn. If P_{max} falls within these two lines then P_{max} is used to find K_Q. If the $C+5\%$ intersects the load curve then $P_{5\%}$ is found and this is taken as the load at crack initiation. In fact if all the non-linearity is due to crack growth, then it corresponds to a particular amount of crack growth given by:

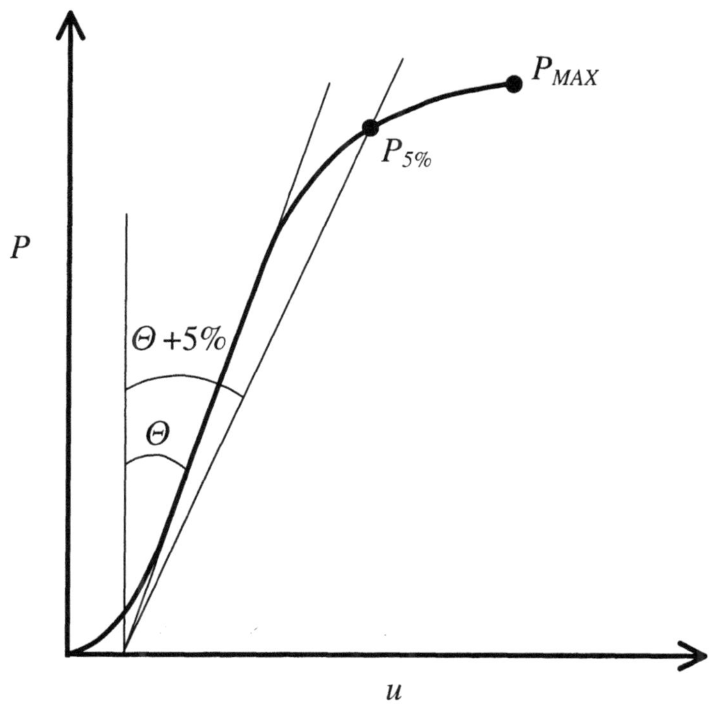

Figure 4. Determination of $P_{5\%}$ and C

$$\frac{\Delta a}{a}=\frac{\phi}{(a/W)}\cdot\frac{\Delta C}{C}$$

Where $\phi(a/W)$ is the calibration factor discussed in section 8. For the configuration in SENB used here, $\phi/(a/W) \approx 0.5$ so we have $\Delta a/a \approx 2.5\%$ i.e. a 2.5% increase in crack length.

To stay within the LEFM condition it is further specified that:

$$\frac{P_{max}}{P_{5\%}}<1.1$$

i.e. a 10% non-linearity is allowed. If $P_{max}/P_{5\%}>1.1$ then the test is invalid. If $P_{max}/P_{5\%}<1.1$ then $P_{5\%}$ is used in calculation of K_Q or P_{max} it if falls within the two lines. (It should also be noted that crack 'pop-in' can occur in which the crack jumps forward a small distance and then arrests. This results in a short drop in the curve and then a continued rise. This value of load can be used and quoted as a 'pop-in' value.)

Values of K_Q are computed from the original crack length a which is best determined from the fracture surface after testing. An average value may be used but the difference between the shortest and longest length should not exceed 10%. Care should be taken that it is the original crack which is being observed since slow growth can occur. K_Q is then calculated from the following relationship:

$$K_Q=f\frac{P_Q}{BW^{\frac{1}{2}}}$$

Tabulated values of f ϕ, ψ and η_e are presented in Table 1 for the SENB specimen and Table 2 for the CT specimen.

7. Size Criteria

The validity of K_Q should now be checked via the size criteria;

$$B,\ a,\ (W-a)>2.5\left(\frac{K_Q}{\sigma_y}\right)^2$$

Since specimen dimensions require that $W=2B$ initially (see section 2) and $a/W \approx 0.5$ then usually all are satisfied if one is. In fact the criteria covers two limitations in that B must be sufficient to ensure plane strain but $(W-a)$ has to be sufficient to avoid excessive plasticity in the ligament. If $(W-a)$ is too small the test will usually violate the linearity criteria but not necessarily so. If the linearity criteria is violated a possible option is to increase W for the same B. Values of W/B of up to 4 are permitted. It should also be noted that if the specimen is too small B will result in K_Q being high whilst $(W-a)$ will result in it being low. The nett effect may be close to correct, but unfortunately in an unpredictable way, since the dependence on B cannot be quantified.

σ_y is the uniaxial tensile yield stress and for polymers this is conventionally taken at the maximum load. Because of visco-elastic effects the 0.2% offset value as used for metals is not a yield stress and gives too low a value. Shear yielding in tensile tests in most polymers can be achieved by carefully polishing the specimen edges, but if brittle fracture does occur then, since yielding is at a larger load, the stress at fracture may be used in the criteria to give a conservative size value. An alternative is to use 0.7 times the compressive yield stress. In all cases the time to yielding should be within ± 20% of the fracture loading time and the method of finding σ_y given.

If these criteria are met then, $K_Q = K_{IC}$, the plane strain value.

8. G_c Calculations

G_{IC} can, in principle, be obtained from,

$$G_{IC} = \frac{(1-\nu^2)K_{IC}^2}{E}, \text{ (for plane strain)}$$

but for plastics E must be obtained at the same time and temperature conditions because of visco-elastic effects. Many uncertainties are introduced by this procedure and it is considered preferable to determine G_{IC} directly from the energy derived from integrating the load versus load-point displacement diagram. The procedure to be followed is via K_Q for validity testing and then to determine the energy U_Q up to the same load point as used for K_Q, as shown in Figure 5a. The correction curve, as sketched in Figure 5b, is usually quite linear and the energy from indentation and machine compliance, U_{cor} can be estimated from C_{cor} from,

$$U_{cor} = \frac{1}{2}P^2 C_{cor}$$

where $P = P_{5\%}$ or P_{max}

The true fracture energy is,

$$U = U_Q - U_{cor}$$

It is considered easier to correct for initial curvature by extrapolation as shown, but subtracting the total curves is permitted. Total energy corrections are usually <20%.

G_c may be calculated from this energy U via,[+]

$$G = \frac{U}{BW\phi} \quad \text{or} \quad \frac{\eta_e U}{B(W-a)}; \quad \eta_e = \frac{1-a/W}{\phi}$$

(The η_e form is of the same form to that used for J tests). Values of η_e are given in Table 1 for SENB and Table 2 for CT.

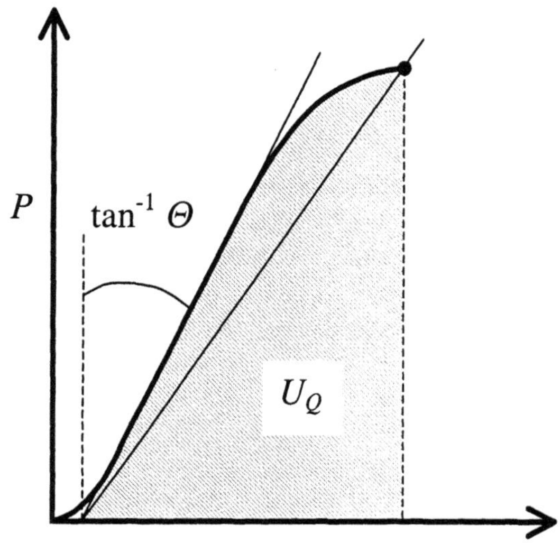

a) Load-deflection; Fracture test

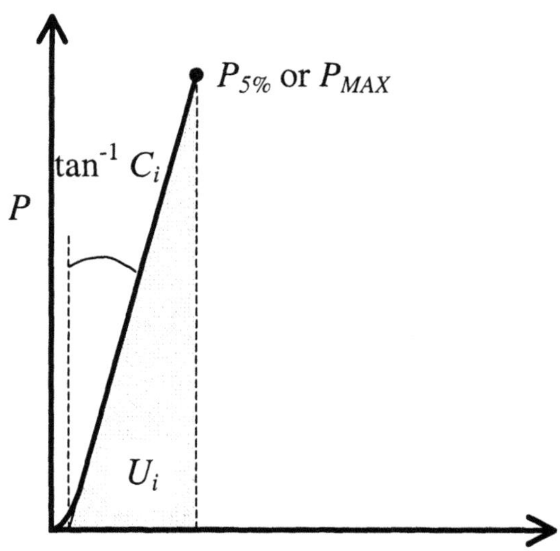

b) Load-deflection; Indentation

Figure 5. Method of correcting for indentation

The energy calibration factor ϕ is defined as

$$\phi = \frac{C}{\dfrac{dC}{d\left(\dfrac{a}{W}\right)}}$$

and may be computed as shown in the Appendix . Values for the test geometries used are also given in Tables 1 and 2.

A useful cross check on accuracy may be made since $E/(1-\nu^2)$ can be found from the true compliance C, ie, $C = C_Q - C_{cor}$ from:

$$\left(\frac{E}{1-\nu^2}\right).BC = 2f^2\phi = \psi$$

and the factor ψ is given in Tables 1 and 2. This value of $E/(1-\nu^2)$ should be compared with that obtained from K_{IC}^2/G_{IC} and the former value should be the larger but the difference should be <15%. If the difference is greater, then the results should be examined for possible errors.

+ JG Williams, "Fracture Mechanics of Polymers", Ellis Horwood/Wiley, 1985.

9. Reporting

The following format for reporting results is suggested;

i) Specimen Configuration and Dimensions
ii) Notching method
iii) Temperature and Loading Rate
iv) One example of Load – Displacement diagram
v) P_{max} and/or $P_{5\%}$ values for all (3) specimens plus loading times
vi) $P_{max}/P_{5\%}$ < or > 1.1?
vii) K_Q value
viii) σ_y value at maximum load and loading time
ix) $2.5(K_Q/\sigma_y)^2$; < or > B, $W-a$?
x) Energy value (indentation and machine compliance corrected?)
xi) G_{IC} value via ϕ or η_e
xii) $E/(1-\nu^2)$ via C_c
xiii) $E/(1-\nu^2)$ via K_{IC}^2/G_{IC}

Appendix: Calibration factors for SENB and CT specimens

a/W	f	ϕ	ψ	η_e
0.05	2.50	1.502	18.7	0.63
0.10	3.39	0.857	19.7	1.05
0.15	4.07	0.641	21.2	1.33
0.20	4.70	0.526	23.2	1.52
0.25	5.36	0.449	25.7	1.67
0.30	6.09	0.391	28.9	1.79
0.35	6.93	0.345	33.1	1.89
0.40	7.93	0.307	38.5	1.96
0.45	9.14	0.275	45.9	2.00
0.50	10.65	0.246	55.9	2.03
0.55	12.57	0.220	69.6	2.04
0.60	15.09	0.195	88.7	2.05
0.65	18.51	0.170	116.4	2.06
0.70	23.40	0.145	158.5	2.07
0.75	30.84	0.120	228.2	2.08
0.80	43.21	0.096	357.7	2.09
0.85	66.76	0.072	643.7	2.08
0.90	123.30	0.049	1484.6	2.05
0.95	351.62	0.025	6148.3	2.01

Table 1 Calibration Factors for SENB geometry with S/W=4.

Note: Values calculated using:

Bakker, A. Compatible Compliance and Stress Intensity Expressions for the Standard three-point Bend Specimen. *International Journal of Fatigue and Fracture of Engineering Materials and Structures* 13 (2), 145, 1990.

a/W	f	ϕ	ψ	η_e
0.25	4.92	0.199	9.6	3.77
0.30	5.62	0.208	13.2	3.36
0.35	6.39	0.213	17.4	3.05
0.40	7.28	0.213	22.5	2.82
0.45	8.34	0.208	28.9	2.64
0.50	9.66	0.199	37.1	2.51
0.55	11.36	0.186	48.1	2.42
0.60	13.65	0.170	63.6	2.35
0.65	16.86	0.152	86.6	2.30
0.70	21.55	0.133	123.2	2.26
0.75	28.86	0.112	186.3	2.23

Table 2 Calibration factors for CT geometry

Note: Values calculated using:

J.A. Kapp, G.S. Leger and B. Gross;
Fracture Mechanics Sixteenth Symposium, ASTM, STP 868, pp 27-44.

Appendix

Compact Tension Specimen $(0.2 < a/w < 0.8)$

$$f = \frac{(2+\alpha)}{(1-\alpha)^{\frac{3}{2}}}\left[0.886 + 4.64\alpha - 13.32\alpha^2 + 14.72\alpha^3 - 5.6\alpha^4\right]$$

$$\phi = \frac{(1.9118 + 19.118\alpha - 2.5122\alpha^2 - 23.226\alpha^3 + 20.54\alpha^4)(1-\alpha)}{(19.118 - 5.0244\alpha - 69.678\alpha^2 + 82.16\alpha^3)(1-\alpha) + 2(1.9118 + 19.118\alpha - 2.5122\alpha^2 - 23.226\alpha^3 + 20.54\alpha^4)}$$

Where $\alpha = a/w$

Single Edge Notched Bend Specimen $(0 < a/w < 1)$

$$f = 6\alpha^{\frac{1}{2}} \frac{\left[1.99 - \alpha(1-\alpha)\left(2.15 - 3.93\alpha + 2.7\alpha^2\right)\right]}{(1+2\alpha)(1-\alpha)^{\frac{3}{2}}}$$

$$\phi = \frac{A + 18.64}{dA/d\alpha}$$

$$A = \frac{16\alpha^2}{(1-\alpha)^2}\left[8.9 - 33.717\alpha + 79.616\alpha^2 - 112.952\alpha^3 + 84.815\alpha^4 - 25.672\alpha^5\right]$$

$$dA/d\alpha = \frac{16\alpha^2}{(1-\alpha)^2}\left[-33.717 + 159.232\alpha - 338.856\alpha^2 + 339.26\alpha^3 - 128.36\alpha^4\right] +$$

$$16\left[8.9 - 33.717\alpha + 79.616\alpha^2 - 112.952\alpha^3 + 84.815\alpha^4 - 25.672\alpha^5\right]\left[\frac{2\alpha(1-\alpha) + 2\alpha^2}{(1-\alpha)^3}\right]$$

DETERMINATION OF FRACTURE TOUGHNESS (G_{IC} AND K_{IC}) AT MODERATELY HIGH LOADING RATES

A. PAVAN

1. INTRODUCTION

Characterisation of the fracture resistance of plastics under high loading rate conditions, such as those encountered in impact tests, has been in use for a long time. Traditional impact testing methods, such as pendulum impact tests (in both versions, Charpy and Izod) and falling weight impact tests, however, provide only conventional measurements of toughness which depend to a great degree on test set up and specimen, which are arbitrarily set out.

Data obtained by tests of that kind, once standardised, can be useful for some purposes (e.g.: material quality control and purchase specifications) but have very limited or only a general relationship to performance. Since the fastest growing applications area for plastics is today in engineering parts, where performance under a variety of mechanical stresses is critical, the engineering design of plastic products is necessary. For that reason, intrinsic material property data are needed.

As previously pointed out, Fracture Mechanics now provides a logical framework and a rational scheme for determining intrinsic fracture resistance values [e.g. *1*].

The test method outlined in the preceding paper to determine fracture resistance in the two variants, G_{IC} (the fracture energy) and K_{IC} (the critical stress intensity factor) in plastics, applies at quasi-static loading rates, where dynamic effects are absent. The method dealt with in this paper is meant to extend the applicability of that method to rates of up to around 1 m/s, where dynamic effects are present but can still be 'controlled'.

The method is based on the experience gathered within ESIS Technical Committee 4 as the result of a series of round-robin exercises, which involved some thirty industrial and academic laboratories from twelve countries during an eight years' period. The author bore the responsibility of drafting the testing protocol, co-ordinating the inter-laboratory round robins, analysing the data generated thereby, from which an improved draft of the testing protocol was proposed, again and again, up to the final version which was agreed by the Committee and issued in September 1997 [*2*].

The protocol reported in section 3 below is the version of the final ESIS TC4 document issued in September 1997 adapted for the (mostly stylistic) specifications set out by the International Standard Organisation (ISO), in order to have the document accepted as an official international standard [*3*].

2. BACKGROUND TO THE PROCEDURE AND ISSUES CONSIDERED IN DEVELOPING THE PROTOCOL

In this section the background for the extension of the scope of the protocol for quasi-static testing to higher rates is illustrated, the problems encountered are highlighted and the solutions found and adopted in the final draft of the protocol are reviewed.

Origin of dynamic effects

In order to take the dynamic effects occurring at high loading rates into proper account and define the scope of applicability of the testing protocol precisely, it is necessary to understand the nature of those effects. The dynamic phenomena observed when a test-piece is loaded rapidly have two possible origins. One is the finite, though great, speed of the stress wave propagation in the material under test, which prevents stresses from attaining equilibrium during the short period of the impact event, and is inherent in fast loading. The second, commonly termed 'inertial effect', is the high acceleration imparted to the specimen initially, which excites inertial forces and complex motions in both specimen and striker and is mainly instrumental, depending largely on the characteristics of the latter. The relative importance of the two varies with the rate of loading.

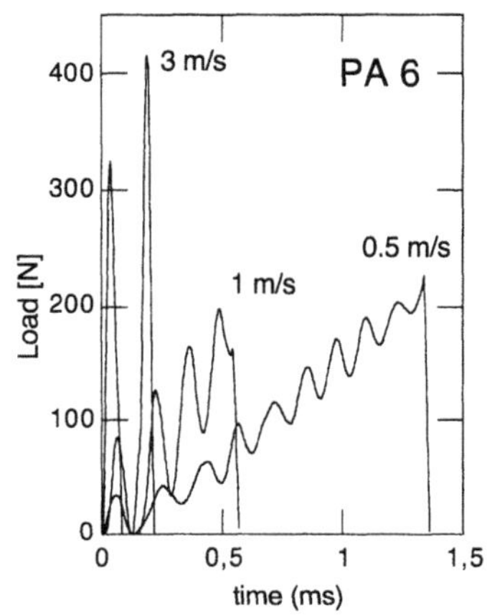

Figure 1 — Typical force/time curves recorded from impact test at different speeds. Tester: Fractovis dart drop by CEAST, Turin, Italy, Test configuration: SE(B)

- At *very high loading rates* the time-scale of the fracture event is comparable with the time taken by the stress waves to travel through the test-piece. The subsequent stress wave reflections and their interference with the crack may give some effect.

- At *moderately high loading rates* (load-point displacement rates of the order of 1 m/s, loading times of the order of 1 ms), it is the dynamic effects related to the specimen motion which predominate. The inertial forces caused by the acceleration imparted to the specimen produce vibrations in the test system, oscillations in the recorded signal and forces on the test specimen, which are different from the forces sensed by the test fixture. Possible loss (and regaining) of contact between the specimen and the tup of the moving arm of the testing machine and also between the specimen and the shoulders of the mounting vice, may also occur during the test.

- At *lower loading rates* these effects become negligible and the fracture mechanics methods used for quasi-static test conditions (preceding paper) can be applied as they stand.

It is evident from the examples shown in Figure 1 that at high rates the amplitude of the oscillations may become comparable to or even exceed the total load, and the interpretation of the test record becomes difficult.

Control of the dynamic effects

Considerable work has been published, dealing with the assessment, analysis, modelling and control of the dynamic effects manifested by fluctuations of the measured force signal such as those shown in Figure 1. As a result, several alternative remedies could be thought out.

• *Reducing the test speed* to contain those effects was not, of course, considered as a sensible solution, because the high speed condition is just the point of interest here, where the materials may show significant decrease in toughness.

• Since a part of the observed dynamic effects is instrumental in origin, there is room for controlling these effects by *improving machine design and test equipment*. However, one aim set out in developing this protocol was to keep the procedure within the reach of laboratories having standard equipment. Therefore, deliberately, no restriction was placed *a priori* on the type of loading machine and test equipment, but, nevertheless, their performance must meet certain minimum requirements.

Developments in instrumentation in recent years, on the other hand, now offer the possibility of visualising the high speed loading processes precisely at a fairly low cost. Instrumented impact testers (either swinging pendulums or falling weights) and high-speed hydraulic testing machines are becoming generally available, which justifies the attempt of working out a testing method based on force measurement which can be adopted as a standard usable for routine testing in reasonably equipped laboratories.

• Previous studies have shown that the force oscillations recorded by force transducers mounted in the moving arm of the test instrument are considerably greater than the ones actually experienced by the specimen at its crack tip (see for example [*4*, *5*]). It is thus tempting to reduce the disturbing oscillations in the recorded force signal *a posteriori*, by *electronic filtering or attenuation*. Electronic filtering or attenuation, however, may wipe out some real features of the mechanical response of the test specimen, especially at the higher speeds (when the period of the oscillations becomes comparable with test duration). To ward off this danger, electronic filtering or attenuation was banned in developing the protocol and another solution was sought.

• Previous studies have shown that the amplitude of the force oscillations depends largely on the 'contact stiffness' of the tup-specimen interface (see for example [*6*, *7*]). Some reduction of these effects by proper *control of the 'contact stiffness'* can thus be envisaged as possible and this solution was adopted in the proposed protocol.

• *A full analysis of the dynamics* of the impact event (e.g. based on lumped mass-spring-dashpot models, on dynamic finite element modelling, on comparison with a dynamic response calibration curve, etc.) can also offer possible alternative routes to the characterisation of fracture properties at high loading rates. An approach of that type generally requires a higher degree of technical sophistication and expertise than it is in the scope of the present protocol. One such procedure is considered in the paper by Böhme to deal with testing speeds of > 1 m/s.

Viability and side effects of the mechanical damping method

With pendulum and falling weight impact testers, the impact may be cushioned by means of a soft pad, placed where the tup strikes the specimen. With servo-hydraulic testing machines, initial acceleration of the specimen can be controlled by means of a damper applied in the motion transmission unit. The degree of damping can be varied by changing consistency and thickness of the damping material used.

Ample evidence of the effectiveness of this expedient has been gathered within ESIS TC4. As the examples in Figure 2 show, signal oscillations can be drastically reduced or even suppressed at 1 m/s testing speed.

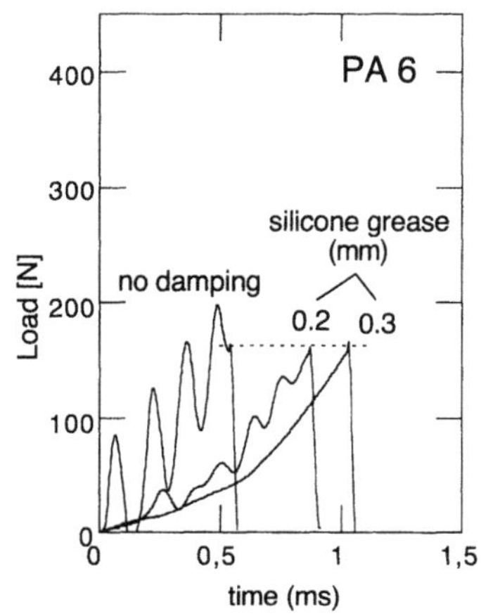

Figure 2 — Effect of placing a layer of silicone grease on a SE(B) specimen struck at 1 m/s

No adverse effects are observed if damping is contained. The value of the load at fracture is not affected (Fig. 3 shows an example) and the load-point displacement rate (in a displacement control mode of testing) can be kept essentially constant during the test, provided the testing machine is of sufficient capacity. Time to fracture is somewhat increased due to damping, so the testing speed needs to be adjusted to maintain the load-point displacement rate or the time-to-fracture fixed (see below for this alternative).

Overdamping may induce some initial non-linearity in the load trace, as can be seen in Fig.2. That effect must be balanced against the effect on load fluctuations: to this end the protocol requires that damping is contained to a minimum sufficient to confine load fluctuations within the allowed limits of ±5% of the load at fracture initiation.

In view of the energy measurements, the degree of mechanical damping must be strictly controlled so as to have similar effect in each test when a series of similar specimens is tested. Preparation of the damping device requires some skill to obtain reproducibility. Some suggestions as to the preparation are provided in the protocol, but its effectiveness is to be assessed each time, from the performance obtained during the test.

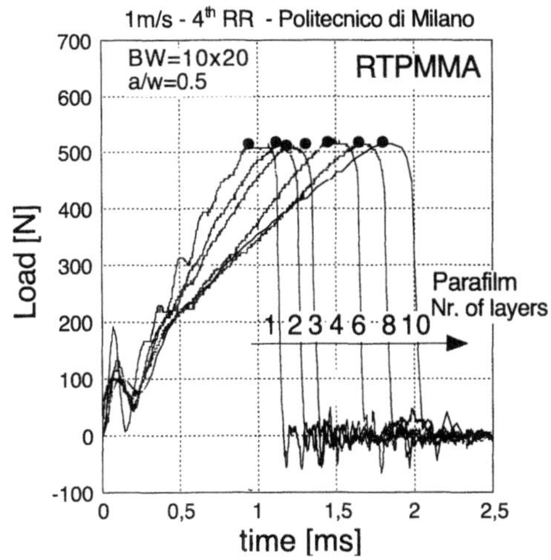

Figure 3 — Effect of placing a damping pad of varying thickness on a SE(B) specimen of rubber-toughened polymethylmethacrylate (RTPMMA) struck at 1 m/s. The point of fracture initiation is indicated on each curve.

Identification of the point of fracture initiation

The tests dealt with in this and in the paper on quasi-static testing, are designed to characterise toughness *at fracture initiation*. There exist several possible experimental techniques to detect fracture initiation, but they generally require some sophisticated instrumentation and complex calibration procedures, especially so in the case of high speed testing. For the sake of simplicity, the same approach used in the protocol for quasi-static testing was maintained in this protocol, namely the point of fracture initiation is to be deduced from the load diagram.

In order to make that approach possible, the occurrence of force peaks and major fluctuations in the initial part of the load/time record is tolerated, but in the portion of the force/time record close to the point of fracture initiation force fluctuations must be contained within a prescribed limit. This is to be obtained by means of the mechanical damping expedient.

In order to allow for some non-linearity in the fracture behaviour and apply the construction based on a 5% increase in specimen compliance to determine the 5% offset load, as set out in the protocol for quasi-static testing, a further smoothing of the load diagram is suggested. This is obtained by a curve-fitting procedure. The regression analysis is applied to the higher portion of the recorded load trace, where the remaining force fluctuations are minor, and its result is extrapolated back to the origin. The procedure can be easily carried out with the aid of a computer.

Energy measurements and corrections

As in the low-rate case covered in the protocol for quasi-static testing, G_{IC} should be determined directly from the energy derived from integrating the load diagram and the method must include careful measurements and corrections for machine compliance and specimen indentation, unless an external displacement measuring device is used, (e.g. optical), which, however, would be impractical at high testing speeds.

Under impact testing conditions and with a mechanical damping device in place, however, the area under the measured load/displacement curve, U_Q, contains additional spurious contributions which need to be removed before G_{IC} can be calculated. As in the low-rate case, a portion of the correction can be estimated from a separate test, to be performed on an un-notched specimen. Damping rather complicates the energy analysis, since the damper absorbs a large fraction of the energy applied. Even more problematic is the evaluation of the kinetic energy of the moving test specimen and of the energy associated with the inertial loads.

Since inertial loads are essentially independent of crack length [*11*] and the same is true for the kinetic energy term, the protocol suggests an alternative, multispecimen procedure, which circumvents the need of evaluating those two terms at all. This procedure is explained in detail in the protocol.

Test speed

The effectiveness of the mechanical damping technique and the applicability of the data handling procedures provided for in the protocol were tested within the ESIS TC4 group at a testing speed of 1 m/s and thereabout, i.e. in the intermediate ('*moderately high*') loading rate range identified above. The upper test speed limit at which the method may fail has not been assessed precisely and may differ with the material evaluated.

The test is to be performed under controlled load-point displacement and, as a standard test condition, it is recommended that a constant load-point displacement rate of 1 m/s be used. However - for the sake of comparing different materials - testing at a fixed time-to-fracture (e.g. 1 ms) is also contemplated in the protocol as a possible alternative. Both alternatives are considered since it is still debated whether it is the load-point displacement speed or the current rate of loading, as expressed by the time derivative of the stress intensity factor, dK/dt (see e.g. [*8, 9*]), or the total failure time [*10*], which is important for rate-sensitive materials such as plastics.

Note – In the copy of the protocol reported in the following section the original numbering of subsections and figures is maintained. To avoid confusion with subsection and figure numbers in other sections, a letter P (for protocol) is prefixed to subsection and figure numbers in this section.

3. THE PROTOCOL

A LINEAR ELASTIC FRACTURE MECHANICS (LEFM) STANDARD FOR DETERMINING K_{IC} AND G_{IC} FOR PLASTICS AT MODERATELY HIGH LOADING RATES

P.1. Scope

This protocol provides guidelines for determining the fracture toughness of plastics in the crack-opening mode (Mode I) by a linear elastic fracture mechanics (LEMF) approach, at load-point displacement rates of 1 m/s or thereabouts. It supplements the document entitled "A Linear Elastic Fracture Mechanics (LEFM) Standard for Determining K_{IC} and G_{IC} for Plastics" (hereafter referred to as the 'parent protocol') so as to extend its applicability to loading rates somewhat higher than it is in the scope of the latter (see also Note 1 below).

The general principles, methods and rules given in the 'parent protocol' for fracture testing at low loading rates remain valid and should be followed except where expressly stated otherwise in the present protocol.

The methods are suitable for use with the same range of materials as covered in the 'parent protocol' (see also Note 2 below).

The same restrictions as to linearity of the load/displacement diagram, specimen size and notch tip sharpness apply as for the 'parent protocol' (see also Note 3 below).

NOTE 1 - Fracture testing at high loading rates presents special problems because of the presence of dynamic effects: vibrations in the test system producing oscillations in the recorded quantities, and inertial loads producing forces on the test specimen different from the forces sensed by the text fixture. These effects need to be either controlled (and, if possible, reduced by appropriate action) or else taken into account through proper analysis of the measured data.

The relative importance of such effects increases with increasing testing rate (decreasing test duration). At speeds of less than 0.1 m/s (loading times of greater than 10 ms) the dynamic effects may be negligible and the testing procedure given in the 'parent protocol' can be applied as it stands. At speeds approaching 1 m/s (loading times of the order of 1 ms) the dynamic effects may become significant but still controllable: the procedure given in the 'parent protocol' can still be used though with some provisos and these are contemplated in the present protocol. At speeds of several meters per second and higher (loading times quite shorter than 1 ms) the dynamic effects become dominant, and different approaches to fracture toughness determination are required, which are out of the scope of this protocol.

NOTE 2 - Although the dynamic effects occurring at high loading rates are largely dependent on the material tested as well as on the test equipment and test geometry used, the guidelines given here are valid in general, irrespective of test equipment, test geometry and material tested.

NOTE 3 - The linearity requirements referred to in the 'parent protocol', clause 6, are to be verified here on the "smoothed" load/displacement curve, to be obtained as specified in P.7.1 below.

P.2. Terms and definitions

For the purposes of this protocol, the same terms and definitions given in the 'parent protocol' apply.

P.3. Test Specimens

P.3.1. Specimen geometry and preparation

As for the low-rate testing case covered by the 'parent protocol', two test configurations are recommended, namely the three point bending (also called single edge notch bend and denoted SE(B)) and the compact tension (denoted C(T)) (see Figure P.1).

Shape and size, preparation, notching and conditioning of test specimens shall comply with the requirements set out in the 'parent protocol', clause 2.

P.3.2. Crack length and number of test replicates

P.3.2.1. Determination of K_{IC}

As in the low-rate testing case covered by the 'parent protocol', measuring test specimens having the same crack length is adequate for determining K_{IC}. The initial crack length a should be in the range $0{,}45 \le a/W \le 0{,}55$. In view of the lower degree of accuracy to be expected with measurements at high rates of loading as compared with low-rate testing, however, it is recommended that at least five replicates, with crack length in the range specified above, be used to determine K_{IC}, and the results averaged.

P.3.2.2. Determination of G_{IC}

At variance with the low-rate testing case covered by the 'parent protocol', a multispecimen procedure, using a series of test specimens with identical dimensions but varying crack-length as specified below, shall be applied for determining G_{IC}.

At least fifteen valid determinations should be made, with initial crack length varying over the range $0{,}20 \le a/W \le 0{,}70$ for the SE(B) configuration and $0{,}40 \le a/W \le 0{,}75$ for the C(T) configuration. They may include the five determinations made on test specimens having initial crack length in the range $0{,}45 \le a/W \le 0{,}55$ to obtain K_{IC}. It is then suggested that, of the remaining ten test specimens to be used, six have initial crack length in the range $0{,}20 \le a/W \le 0{,}45$ and four in the range $0{,}55 \le a/W \le 0{,}70$ in the case of the SE(B) configuration and three have initial crack length in the range $0{,}40 \le a/W \le 0{,}45$ and seven in the range $0{,}55 \le a/W \le 0{,}70$ in the case of the C(T) configuration.

P.3.3. Measurement of test specimen dimensions

As per the 'parent protocol'.

P.4. Test Conditions

P.4.1. Loading mode

The test is to be performed at constant load-point displacement rate. A maximum variation of 10% in the load-point displacement rate during the test is allowed (see P.5.1).

P.4.2. Test speed

As a basic test condition, it is recommended that a load-point displacement rate of 1 m/s be used. If a different rate is applied, it should be quoted in the test report.

With rate-sensitive materials such as plastics, a more significant measure of the rate of the experiment is probably its duration, i.e. the time required to bring the test specimen to fracture. The time to fracture, t_f, is understood here as the time interval between the moment when the load starts acting on the test specimen and the point of fracture initiation as defined in P.7.1.

With a fixed load-point displacement rate the time to fracture varies with material and specimen geometry. If results at a given time to fracture (e.g. 1 ms) are wanted, it is necessary to adapt the load-point displacement rate of the test to each material and specimen geometry (type and dimensions). For this purpose it is expedient to run preliminarily some trial tests at different testing speeds (i.e. load-point displacement rates) to determine the testing speed required to obtain the assigned time to fracture under the given test conditions.

In any case, the time to fracture, t_f, should also be quoted in the test report.

P.4.3. Test atmosphere and temperature

As per the 'parent protocol', clause 4.

P.5. Test Equipment

P.5.1. Loading Machine

Any type of loading machine (impact pendulums, falling-weight towers, servohydraulic universal testing machines, etc.) is allowed, provided it is capable of applying an adequate load to bring the testpiece to fracture at the required load-point displacement rate and of maintaining this rate constant throughout the test up to fracture initiation. With testing machines of limited capacity, this requirement may need to be verified by preliminary tests, especially when new materials are tested or when new test conditions (e.g. specimen size) are used.

Any variation in the load-point displacement rate during the test should be determined and quoted if it exceeds 10% of the rate at fracture initiation.

P.5.2. Loading Rigs

Unlike for low-rate testing, the use of fixed anvils rather than moving rollers is preferred for conducting three point bend (SE(B)) fracture tests under high rate conditions, as it is normally the case with standard impact pendulums. The span between the supports is to be adjustable, however, so that they can accommodate specimens of different size as required in the 'parent protocol', clause 5.

NOTE - In the case of three point bend testing (SE(B) specimens), improved results can be obtained if the testpiece is held in contact with the anvils by light springs (e.g. rubber bands). These will assist in maintaining the testpiece in position during the sudden load transmission from the machine to the test specimen, and ensure more reproducible records.

P.5.3. Instrumentation

Acquisition of a complete record of the load/time response of the material sample under test is essential for the determination of K_{IC}. In addition, a means of evaluating the displacement of the moving load-point during the test is necessary for the independent determination of G_{IC}. Instrumentation of the testing machine should thus comprise, basically, a force sensing and recording system and a displacement measuring and recording system or devices to measure and record quantities from which the load and the load-point displacement can also be indirectly determined.

The adequacy of the response of this equipment to the dynamic events occurring in the relevant determinations should be checked. It can be considered satisfactory if a plain plastic specimen - without any mechanical damping device in place - shows an inertial peak (Figure P.2) larger than 100 N at 1 m/s test speed. The response time should be <20% of the input signal rise time.

If a digital recording system is used, the sampling time should be less than 1/200 of the time to fracture, i.e. at least 200 data points should be collected over the time interval from the first increase of the signal to the point of fracture initiation in order to define the required data curve with sufficient accuracy.

P.6. Control of Dynamic Effects

P.6.1. Electronic filtering

The first manifestation of dynamic effects is the presence of oscillations in the load recording signal. They may complicate the interpretation of the test records up to the point of obscuring the basic response of the specimen under test. It is thus desirable that these effects be contained. Reducing these oscillations artificially, *a posteriori*, by electronic filtering or attenuation can be fallacious, however, since it may wipe out some real features of the specimen response. Therefore, electronic filtering or attenuation is not allowed unless the source of the removed "noise" is known and the effect on the data is understood.

P.6.2. Mechanical damping

Some control of the effects of inertial loads can be achieved by proper mechanical damping of the load transmission. With impact testing machines the impact may be cushioned by means of a soft pad, placed where the tup strikes the specimen. The pad should reduce the inertial effects by reducing the "contact stiffness". With high-speed testing machines (e.g. servohydraulic), initial acceleration of the specimen can be controlled by means of a damper applied in the motion transmission unit.

With impact testing machines and SE(B) test specimens the damping pad can be made by spreading the contact surface either of the tup of the striking hammer or of the testpiece with a layer of a paste or a highly viscous grease. For the sake of reproducibility it is important that the grease be homogeneous and evenly applied, with thickness constant to ±0,05 mm. This can be obtained by delivering the grease with a spatula through an aluminium stencil having the required thickness, normally a few tenths of a millimetre, as shown in Figure P.3.

With high-speed testing machines and C(T) test specimens the damping pad can be more conveniently made of a viscoelastic rubber-like material with a low coefficient of restitution. The rubber-like character should ensure a more or less complete recovery of the pad deformation after each test, thus allowing the same pad to be used repeatedly.

P.6.3. Damping level

If mechanical damping is applied, it should be kept to a minimum, sufficient to contain the fluctuations in the force/time trace within the 10% envelope defined in P.7.1. To obtain this optimal result it is advisable to run some preliminary trial tests to gauge the performance of the damper. This can be varied by changing consistency and thickness of the damping material used.

If the test specimens are in short supply, it is advisable to use an unnotched specimen to assay the performance of the damper. The dynamic effects which are to be controlled by mechanical damping are in fact largely independent of crack length and the use of an unnotched specimen offers the advantage that it can stand repeated strokes without breaking.

In order to determine the level of damping needed to meet the requirement stated in P.7.1, however, reference should be made to the worst case to be expected in the testing programme, that is the case of the specimen with the deepest notch, which will present the lowest fracture resistance and thus the largest force oscillations to fracture load ratio.

P.6.4. Check on speed

Because of damping, some deviations from the pre-set load-point displacement rate may ensue. Thus, if mechanical damping is applied, the instrument should be reset to the desired load-point displacement rate and its constancy should be checked (as requested under clause P.4.2) under the actual test conditions, i.e. with the damping device in place.

If mechanical damping is applied, it shall be recorded in the test report.

P.7. Data handling

P.7.1. Analysis of the test the records and identification of fracture initiation

These tests, as well as the low speed tests contemplated in the 'parent protocol', are designed to characterise the toughness at fracture initiation. Once a fracture test has been performed and the load/time or load/load-point displacement curve has been obtained, the question arises of identifying the point of fracture initiation. Several techniques are possible, but in this standard it should be deduced from the load diagram.

The same rules as those stated in the 'parent protocol' to determine P_Q are used here, but in the case of high-rate testing some preliminary analysis of the load/time record is required to make sure that dynamic effects do not obscure the basic response of the specimen under test.

Firstly, in the case of high-rate testing, a load drop before maximum load should not be assumed to be an arrested crack extension ("pop-in"), unless borne out by examination of the fracture surface.

Secondly, the occurrence of force peaks and fluctuations in the initial part of the load/time record is tolerated, but a limit is placed on force fluctuations in the portion of the force/time record where the force exceeds 1/2 of its value at fracture initiation and the curve is smoothed. The procedure is as follows.

Draw a smooth mean force/time curve through the experimental load/time record, $P(t)$, and determine P_{max} and $P_{max}/3$ on that curve (see Figure P.4). Then improve the determination of the mean load/time curve by a computer-aided curve-fitting procedure. The following empirical fitting equation is suggested:

$$\overline{P}(t) = m(t - t_0) - b(t - t_0)^n \tag{1}$$

where t_0, m, b and n are (positive) fitting parameters, with n preferably ≥ 5. Use the curve drawn previously to obtain a first estimate of these parameters (see annex A) and use this set of values at the start of the regression analysis. The regression analysis should be confined to the portion of the experimental curve comprised in the time interval defined by $P_{max}/3$ and P_{max}. The value of the initial time, t_0, should be derived from the regression analysis too. However, if that value turns out to be smaller than the time when the force signal first rises, take the latter one as initial time t_0 and repeat the curve fitting by forcing the new curve $\overline{P}(t)$ to pass through the point $t = t_0$, $P = 0$. Finally, determine $\overline{P}_Q$ on the curve $\overline{P}(t)$ (Figure P.4), as indicated in the 'parent protocol', clause 6 (see also the Note below). To this end, the "maximum load" – to be denoted $\overline{P}_{max}$ – is defined here as the value of the fitted force, $\overline{P}(t)$, at time $t = t_{max}$ corresponding to the maximum of the experimental curve (see Figure P.4)

The curve $\overline{P}(t)$ so obtained is assumed to be a good representation of what the load/time response of the system would be in the absence of dynamic effects, provided it meets the following requirement (refer to Figure P.5): the force $P(t)$ recorded experimentally must not deviate from the mean current value $\overline{P}(t)$ by more than 5% of the critical value $\overline{P}_Q$ over the time interval defined by $\overline{P}_Q/2$ and $\overline{P}_Q$. To check this draw two lines parallel to the curve $\overline{P}(t)$ at a distance of 5% of $\overline{P}_Q$ on either side of it, over the time interval defined by $\overline{P}_Q/2$ and $\overline{P}_Q$. All parts of the experimental curve $P(t)$ in that interval should fall within this 10% envelope. If the experimental curve $P(t)$ fails this requirement, then the determination must be deemed invalid. Before abandoning any determination, however, action should be taken to try and reduce the dynamic effects further, as stated in clause P.6.

NOTE - Once the parameters of the best fit have been determined, the two straight lines to be used in order to identify $\overline{P}_Q$ (clause 6 of the 'parent protocol') can be simply obtained as given by the equations $\overline{P} = m(t - t_0)$ and $\overline{P} = (m/0.05)(t - t_0)$, and the value of $\overline{P}_5$ can be readily calculated as $\overline{P}_5 = (m/1.05)(0.05\,a/(1.05\,b))^{1/(n-1)}$. Furthermore, if $\overline{P}_Q = \overline{P}_5$ then the time to fracture can be calculated as $t_f = t_5 - t_0 = (0.05\,m/(1.05\,b))^{1/(n-1)}$.

P.7.2. Energy correction

As in the low-rate testing case (see the 'parent protocol'), G_{IC} should be determined directly from the energy derived from integrating the load versus load-point displacement diagram. As in the low-rate case, however, the area U_Q under the measured load vs load-point displacement curve (Figure P.6) contains spurious contributions in excess of the true fracture energy, U_B, and some corrections are required before G_{IC} can be calculated from that energy. As a matter of fact, unless an external displacement measuring device is used (e.g. optical), the apparent load-point displacements are in excess of the specimen deformation. Besides indentation of the testpiece and compliance of the testing machine, the compression of the mechanical damping device (if used) also contributes to this excess. Correction for these effects is covered in P.7.2.1. Moreover, in the case of high-rate testing the area U_Q under the measured load vs load-point displacement curve also contains some contributions from the kinetic energy (U_{kin}) of the moving test specimen and from inertial loads (U_{inert}) produced by test piece acceleration. A procedure to get rid of these parasitic energy terms is given in P.7.2.2.

P.7.2.1. Testpiece indentation, machine compliance and damper compression

The correction for testpiece indentation, machine compliance and damper compression can be estimated from a separate test, to be performed on an unnotched specimen, as specified in the 'parent protocol', clause 5. Suggested unnotched specimen arrangements for the correction test are shown in Figures P.7 (a) and (b), for SE(B) and C(T) configurations, respectively.

It is advisable to carry out two or three replicates of the correction test to check repeatability and, in case of large variations, to check for possible errors.

The force/displacement correlation obtained in the correction test is integrated up to the initiation load $\overline{P}_Q$ determined in the fracture test (Figure P.8) and the obtained energy U_{cor} is subtracted from the energy U_Q obtained by integrating the force/displacement curve measured in the fracture test (Figure P.6).

The magnitude of this correction, U_{cor}, depends on the magnitude of $\overline{P}_Q$, which may vary substantially from specimen to specimen, especially if the initial crack length to width ratio, a/W, is varying (clause P.7.2.2). The correction should therefore be computed for each specimen subjected to the fracture tests, and applied to its respective total energy to fracture, U_Q.

As specified in the 'parent protocol', clause 5, the correction test should be performed such that the loading time (up to load $\overline{P}_Q$) is the same as in the fracture test, i.e. t_f. This will involve lower test speeds to reach the same load in the same time, e.g. about half the speed of the fracture test. Furthermore, with specimen of varying crack length (clause P.7.2.2) this requirement would imply performing different correction tests at different speeds. This is deemed unnecessary provided time-to-fracture variations among the given set of specimens are less than 50% of the mean time-to-fracture: it is then sufficient to perform the correction test at a mean testing speed.

NOTE - Because of the damper compression contribution the correction can be substantially larger than it is in the low-rate tests. Moreover, because of dynamic effects and the effect of mechanical damping the load vs. load-point displacement curve obtained in the correction experiment is seldom linear, and the practice of linearizing the near-zero data before evaluating displacement or compliance corrections, as suggested in the 'parent protocol', is not advisable. It is preferable to follow the alternative way of correcting energies, as stated above.

P.7.2.2. Kinetic energy and inertial loads

The corrected energy $U_{Q,cor} = (U_Q - U_{cor})$ should be further diminished of the two aforementioned parasitic energy terms, U_{kin} and U_{inert}, to obtain the true fracture energy U_B from which G_{IC} can be determined.

An alternative route that circumvents the need of evaluating that correction at all consists in determining G_{IC} from the slope of a plot of fracture energy versus the energy calibration factor ϕ [see the 'parent protocol' clause 8 and Figure P.9(a) here] obtained by testing a series of specimens with equal dimensions but varying crack length. Since U_{kin} and U_{inert} are essentially independent of crack length, their addition to the fracture energy U_B in an energy-versus-ϕ graph will not alter the slope, and no correction is necessary [see Figure P.9(b)].

NOTE - Subtraction of U_{cor} from U_Q does not get rid of the kinetic and inertial contributions contained in U_Q. As a matter of fact, when U_{cor} is measured (correction test) specimen's motion is suppressed and inertial effects are substantially reduced compared with the inertial effects occurring in the fracture test, as a result of the reduced speed used in the correction test (P.7.2.1).

P.8. Expression of results

P.8.1. Determination of K_{IC}

The value of $\overline{P}_Q$ determined as specified in P.7.1 is used to calculate K_Q as specified in the 'parent protocol', clause 6.

The provisional value, K_Q, is to be checked for linearity and size requirements according to the criteria stated in the 'parent protocol', clause 6 and 7, before it can be assumed as a valid K_{IC} value. For the linearity criterion, the "maximum load" which $\overline{P}_Q$ is to be confronted with, is the value $\overline{P}_{max}$ defined in P.7.1 above.

The time to fracture, t_f, is then evaluated as the difference $t_f = t_Q - t_0$ between the time at the instant when the load is $\overline{P}_Q$ and the initial time t_0 as determined above.

P.8.2. Determination of σ_y

The uniaxial tensile yield stress, σ_y, to be used in the size validity criteria should be determined under loading rate conditions comparable to those in the fracture test: the tensile test can be performed at a constant stroke-rate such that the loading time to yield, t_y, is within ±20% of the actual loading time observed in the fracture test, t_f.

Since σ_y is a decreasing function of time, a low-rate value may be used in the first instance to give a conservative size value. If the result is valid, it is then unnecessary to measure σ_y under high-rate conditions. If the result is invalid, determine and use the high-rate σ_y value.

If a high-rate testing machine is not available for the tensile test, σ_y may be determined by extrapolation of values obtained from low-rate tests covering a range of times to yield, on a logarithmic time-scale.

The method of finding σ_y shall be quoted in the test report.

P.8.3. Determination of G_{IC}

Produce a series of test specimens with equal dimensions but varying crack length, test them under equal conditions (including damper characteristics and testing rate), determine $\overline{P}_Q$ for each individual test specimen as specified in P.7.1 and check for its validity (linearity and size criteria) as specified in P.8.1.

Determine U_Q for each individual test specimen by integrating the respective load versus load-point displacement diagram up to the load point ($\overline{P}_Q$) defining fracture initiation (Figure P.6).

Determine the energy correction, U_{cor} for each individual test specimen by integrating the load versus load-point displacement diagram of the correction test up to $\overline{P}_Q$ (Figure P.8).

Plot the corrected energies, $U_{Q,\,cor} = (U_Q - U_{cor})$, as a function of $BW\phi$ and best fit a straight line through the data points (Figure P.9(b)). From the slope of this line the value of G_{IC} is determined.

NOTE 1 - The parasitic energy contributions ($U_{kin} + U_{inert}$) mentioned above will appear on this plot as a positive intercept of the regression line on the energy axis. If a negative intercept is obtained then the results should be examined for possible errors.

NOTE 2 - If results at a fixed time to fracture are wanted, specimens of varying crack length should be tested under different testing speeds (i.e. load-point displacement rates) in order to obtain the same time to fracture.

If the same test speed is used and the effect of varying time to fracture is neglected, the resulting G_{IC} value should be quoted in association with the mean time to fracture obtained in the K_{IC} determination.

NOTE 3 - In view of the difficulty in determining the correct specimen compliance under high-rate conditions, the cross check on accuracy via $E/(1-\nu^2)$ suggested in the 'parent protocol', clause 8, shall not be applied here. The value of $E/(1-\nu^2)$ obtained from K_{IC}^2/G_{IC} should still be reported for information's sake.

P.9. Test report

The test report shall contain the following information:

a) a reference to this protocol;
b) all details necessary for complete identification of the material tested, including source and history;
c) the test specimen shape (SE(B) or C(T)) and dimensions;
d) the notching method used;
e) the test temperature and speed;
f) type of test apparatus used;
g) type of mechanical damping device used (if any);
h) the maximum test speed variation during the tests (if in excess of 10%);
i) one example of load/time or load/displacement curve, showing the 10% envelope and the $\overline{P}_Q$ determination;
j) the number of specimens tested and the ranges of crack length used for determining K_{IC} and G_{IC} respectively;

k) the kind of initiation point (pop-in, 5% offset or maximum load) and the ratio $\overline{P}_{max}/\overline{P}_5$, if relevant;
l) the time to fracture;
m) the yield stress determination procedure used and the loading time;
n) the results of the size criteria assessment;
o) the diagram of energies U_Q and $U_{Q,\,cor}$ vs. $BW\phi$;
p) the critical stress intensity factor K_{IC} and the critical energy release rate G_{IC};
q) the value of K_{IC}^2/G_{IC}.

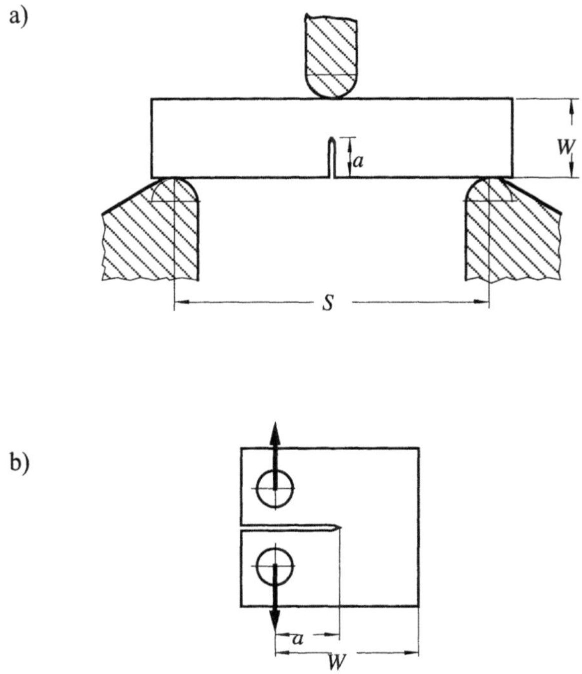

Figure P.1 — Test configurations as specified in P.3.1 and P.5.2: (a) SE(B); (b) C(T)

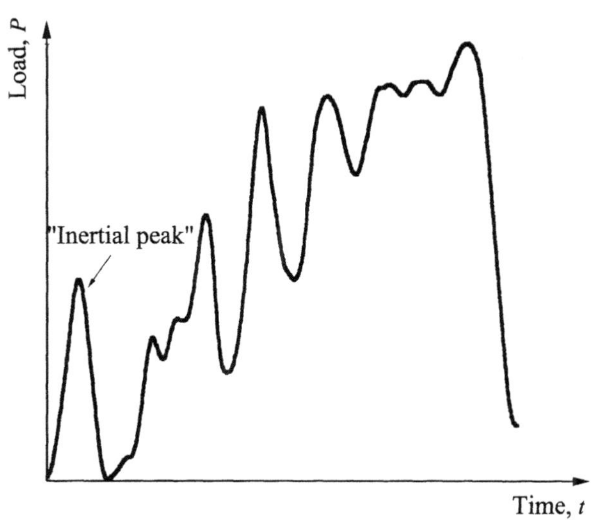

Figure P.2 — Typical load/time record in the absence of signal attenuation and mechanical damping

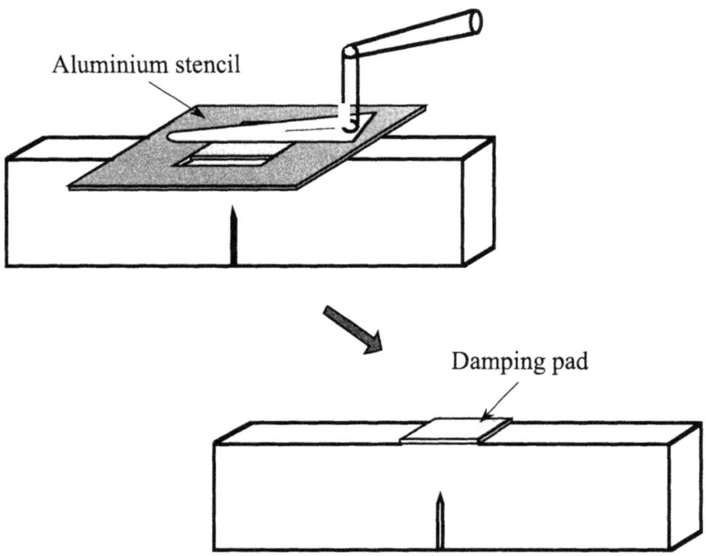

Figure P.3 — Deposition of damping pad on SE(B) test specimen (see P.6.2)

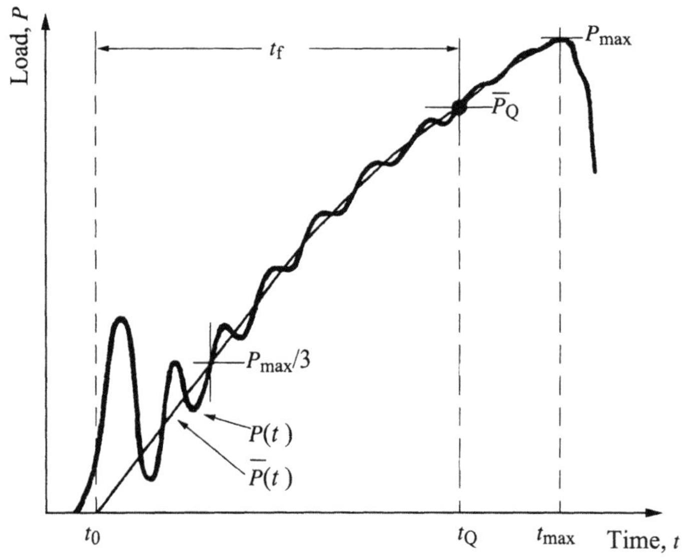

Figure P.4 — Curve fitting and determination of $\overline{P}_Q$ and t_f as specified in P.7.1 (schematic)

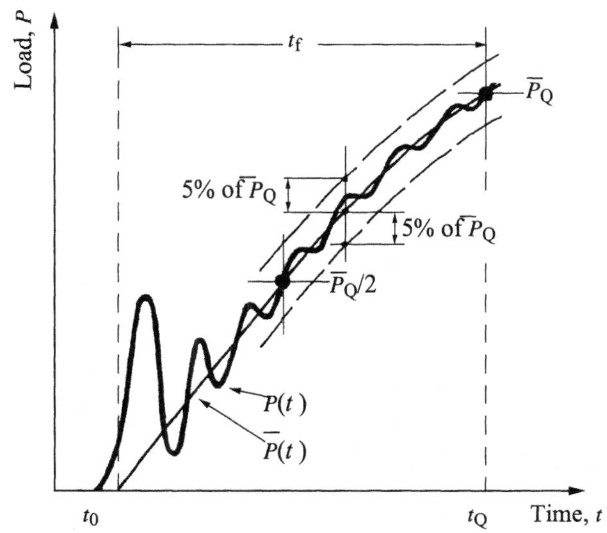

Figure P.5 — Limits of allowable force fluctuations in the fracture test as specified in P.7.1 (schematic)

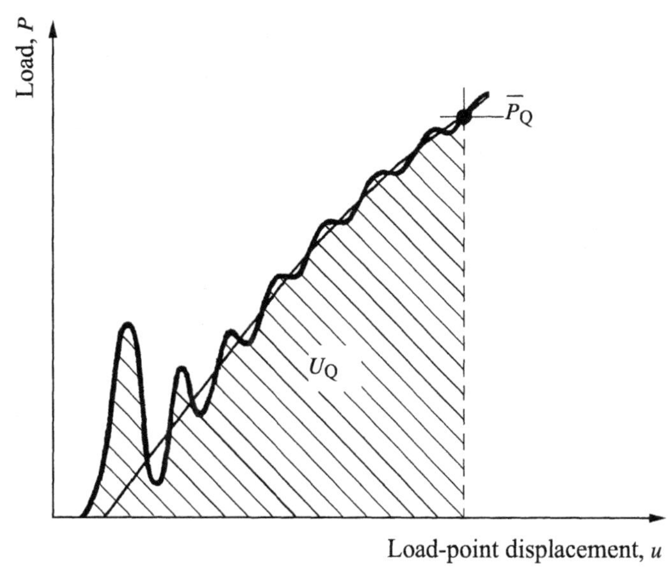

Figure P.6 — Evaluation of energy U_Q from the fracture test (see P.7.2)

a)

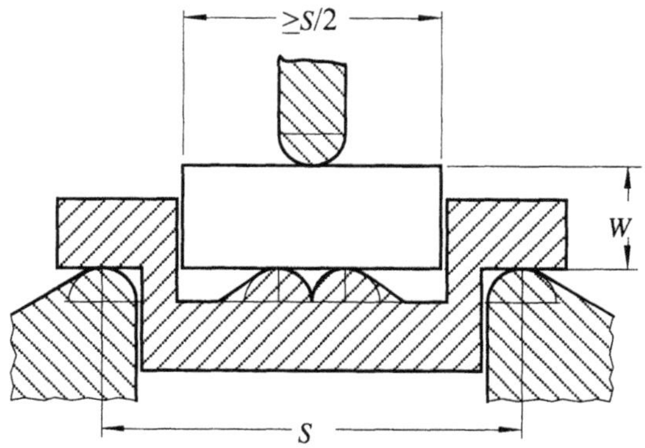

b)

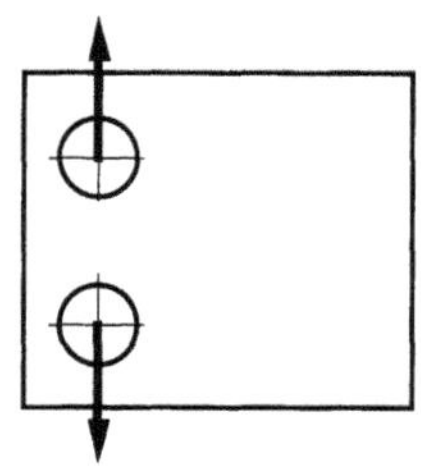

Figure P.7 — Arrangements for the energy correction test: (a) SE(B) configuration; (b) C(T) configuration (see P.7.2.1)

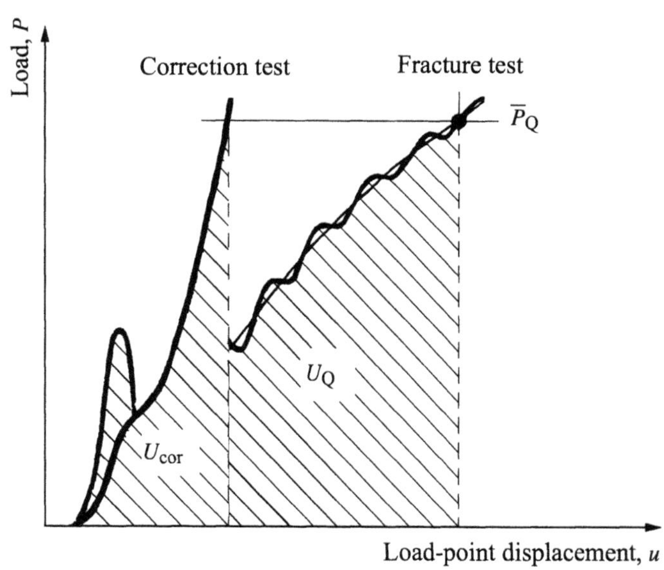

Figure P.8 — Evaluation of energy U_{cor} from the correction test as specified in P.7.2.1 (the plots obtained from the correction test and the fracture test are shown superposed)

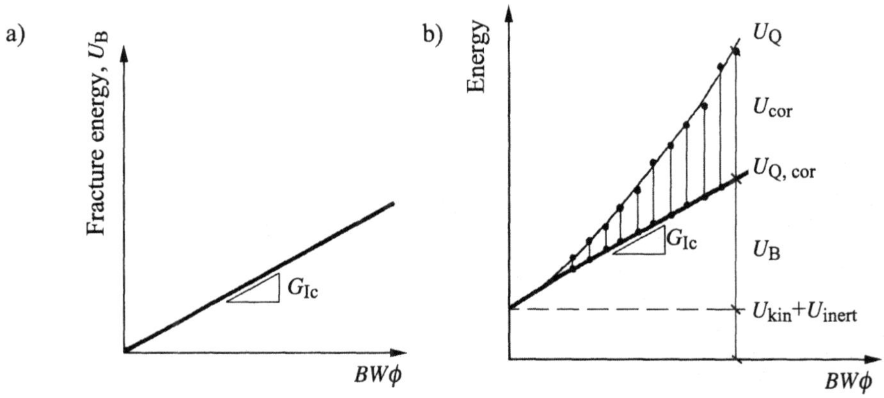

Figure P.9 — Determination of G_{IC} from (a) fracture energy, U_B, and (b) corrected energy, $U_{Q,cor} = (U_Q - U_{cor})$, plotted vs. $BW\phi$ as specified in P.7.2.2

Annex A

Estimation of curve fit parameters

Once a smooth mean force/time curve has been drawn by guesswork, the values of the two parameters characterizing the (initial) linear portion of the curve, i.e. t_0 and the initial slope m, can be evaluated from the initial tangent $P'(t)$ (see Figure A.1).

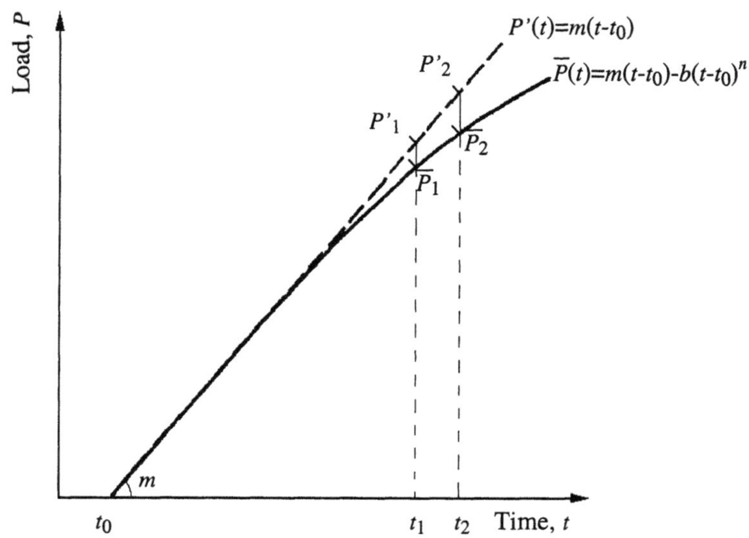

Figure A.1 — Construction for the estimation of the curve fitting parameters t_0, m, b, n (see text).

The values of the two parameters, b and n, which characterize the deviation from linearity can then be estimated as follows:

draw two vertical lines through the curved portion of the $\overline{P}(t)$ curve (e.g. at times t_1 and t_2) and measure the two segments $\overline{P}_1P_1'$ and $\overline{P}_2P_2'$ (see Figure A.1), then n is calculated from

$$n = \frac{ln\left[\left(P_2' - \overline{P}_2\right) / \left(P_1' - \overline{P}_1\right)\right]}{ln\left[P_2' / P_1'\right]} \tag{A.1}$$

and b is obtained from

$$b = m(t_1 - t_0)^{1-n} - \overline{P}_1(t_1 - t_0)^{-n} \tag{A.2}$$

Annex B

Recommended test report form, page 1 of 5 **Form (a)**

Name :	**Date of testing :**
Organization :	**Protocol :**
Material :	Temperature [°C] :

1. TEST EQUIPMENT CHARACTERISTICS

1.1. Type of testing apparatus :

1.2. Test fixture (if different from that stated in the protocol) :

1.3. Instrumentation :

1.4. Quantities monitored :

1.5. Sampling time t_s [μs] :

2. TEST PERFORMANCE

2.1. Inertial peak height (without damping) : > 100 N ?

2.2. Mechanical damping device used (if any) :

2.3. Load-point displacement rate variation during the test : < 10%?

2.4. Minimum time to fracture recorded, $t_{f\ min}$ [ms] :

2.5. Minimum number of data points between t_0 and t_Q (i.e. $t_{f\ min} / t_s$) : > 200 ?

3. DATA HANDLING

3.1. Determination of $\overline{P}_Q$: curve regression analysis applied successfully ?

3.2. Yield stress determination procedure used:

4. REMARKS (Any deviation from procedure and conditions stated in the protocol) :

Recommended test report form, page 2 of 5 — Form (b)

Name :	**Date of testing :**
Organization :	**Protocol :**
Material :	Temperature [°C] :

K_{Ic} DETERMINATION

Notching method: root radius: [μm]

									Mean	St.dev.
1	Specimen N°								Mean	St.dev.
2	Specimen type [SE(B) or C(T) ?]	dimensions	B [mm]							
3			W [mm]							
4			a [mm]							
5			a/W							
6	Damping applied ?		(Y/N)							
7	Speed		[m/s]							
8	Speed variation: < 10% ?		(Y/N)							
9	Load	Fluctuations: within limits ?	(Y/N)							
10		$\overline{P}_{max}$	[N]							
11		$\overline{P}_Q$	[N]							
12		Q = <pop-in>, <5%> or <max> ?								
13		If Q = <5%> :	$\overline{P}_{max} / \overline{P}_5$							
14			< 1.1 ? (Y/N)							
15	Time to fracture, t_f		[ms]							
16	Geometry calibration factor $f(a/W)$									
17	$K_Q = f\overline{P}_Q/BW^{1/2}$		[MPa m$^{1/2}$]							
18	σ_y	measured or calculated ?	(M/C)							
19		σ_y	[MPa]							
20		Time to yield, t_y	[ms]							
21	Size	$2{,}5\,(K_Q/\sigma_y)^2$	[mm]							
22		$< B$?	(Y/N)							
23		$< a$?	(Y/N)							
24		$< W-a$?	(Y/N)							
25	K_{IC}		[MPa m$^{1/2}$]							
26	Time to fracture, t_f		[ms]							
27	$\dot{K}_{IC} = K_{IC} / t_f$		[GPa m$^{1/2}$s^{-1}]							
28	Energy	U_Q, uncorrected	[mJ]							
29		Energy correction, U_{cor}	[mJ]							
30		$U_{Q,cor} = U_Q - U_{cor}$	[mJ]							
31		Energy calibration factor, $\phi(a/W)$								
32		$BW\phi$	[mm^2]							
33	G_{IC} (from $U_{Q,cor}$ slope)		[kJ m^{-2}]							
34	K_{IC}^2 / G_{IC}		[GPa]							

NOTE: white cells = to be filled in with experimental data, light gray cells = to be calculated

Recommended test report form, page 3 of 5 — Form (c)

Name :	**Date of testing :**
Organization :	**Protocol :**
Material :	Temperature [°C] :

K_{Ic} DETERMINATION (cont.d)

EXAMPLE OF LOAD/TIME DIAGRAM
(showing the 10% envelope and $\overline{P}_Q$ determination)

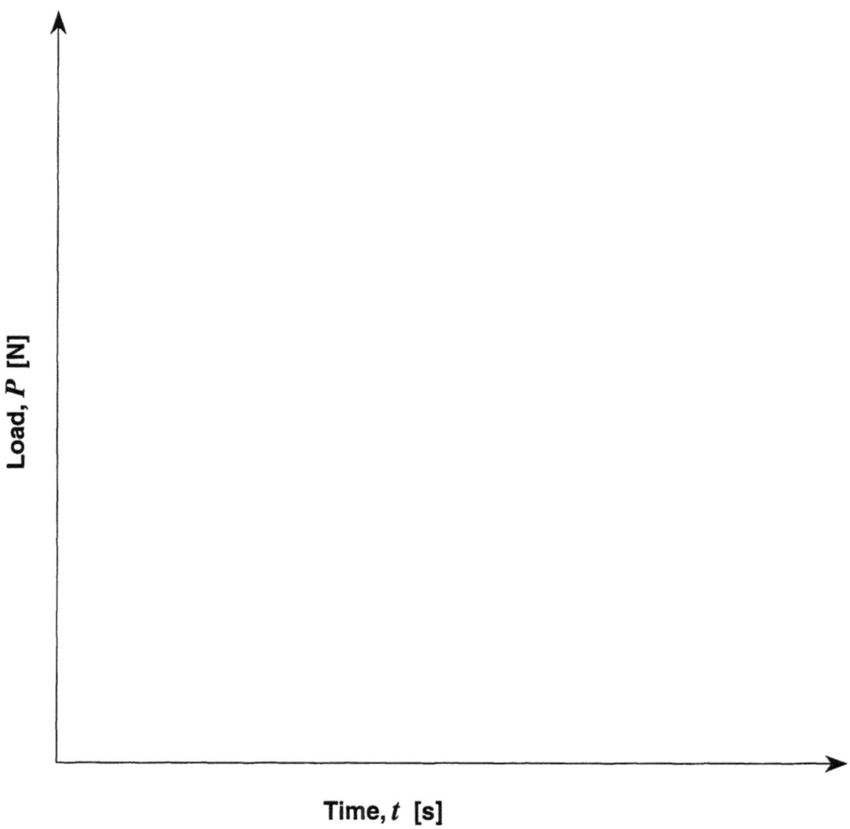

Recommended test report form, page 4 of 5 — Form (d)

Name :	Date of testing :
Organization :	Protocol :

Material :	Temperature [°C] :

G_{Ic} DETERMINATION

(see Form (b) for entries)

1																Mean
2																
3																
4																
5																
6																
7																
8																
9																
10																
11																
12																
13																
14																
15																
16																
17																
18																
19																
20																
21																
22																
23																
24																
25																
26																
27																
28																
29																
30																
31																
32																
33																
34																

NOTE: white cells = to be filled in with experimental data, light gray cells = to be calculated

Recommended test report form, page 5 of 5 — Form (e)

Name :	**Date of testing :**
Organization :	**Protocol :**
Material :	Temperature [°C] :

G_{Ic} DETERMINATION (cont.d)

U_Q and $U_{Q,cor}$ vs. $BW\phi$ DIAGRAM

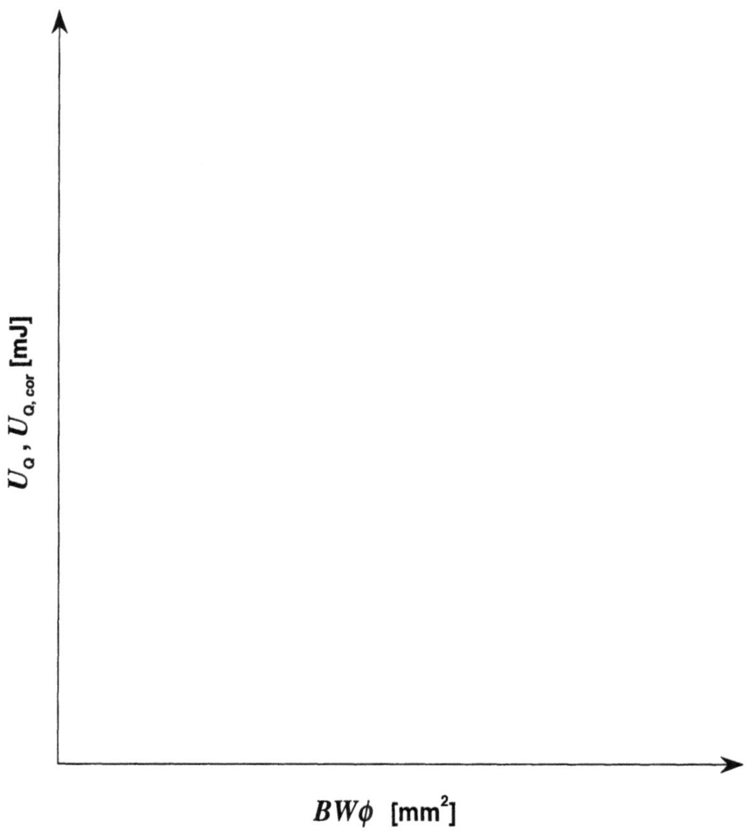

$$U_{Q,cor} = (U_{kin} + U_{inert}) + G_{Ic} \cdot BW\phi$$

4. USING THE PROCEDURE: RESULTS OF INTER-LABORATORY ROUND ROBIN TESTING

In this section a summary of the test results obtained in the series of round-robin tests carried out within ESIS TC4 is reported and the significance of inter-laboratory consistency of the collected data is pointed out with the view of validating the protocol.

Seven materials were tested, varying in stiffness, toughness and ductility. For each material, all the samples were prepared (normally in the form of plane sheets) at one source, but the individual specimens were generally machined and notched at the laboratories which tested them, though the use of specimens prepared from the same workshop was also tried to assess reproducibility. Use of different test equipment, test geometries, damping materials and, of course, operators of varying expertise from the different laboratories was included, in the philosophy of round robin exercises.

The last two columns of Tab. 1 show the improvement in data scattering obtained from the successive round robins as a result of progressive refining of the protocol used.

Table 1 — Summary of ESIS TC4 round robins on G_{IC} and K_{IC} testing at 1 m/s

RR No.	Number of participants	Protocol	Material[a]	K_{IC}		G_{IC}	
				Sets of data	St.dev. (% of mean)	Sets of data	St.dev. (% of mean)
1	7	none	mPVC	7	± 40	7	± 800
2	11	1st draft	mPVC	12	± 23	11	± 37
3	9 9	1st draft	HDPE RTPA/PPO	8 5	± 19 ± 31	6 4	± 55 ± 52
4	15	2nd-3rd draft	PMMA RTPMMA	9 19	± 24 ± 34	7 14	± 19 ± 33
5	10	4th draft	PMMA[b] PMMA	9 13	± 6 ± 15	8 14	± 53 ± 51
6	16	5th draft	PVC	14	± 12	8	± 20
7	6 12	7th (final) draft	ABS PVC[b]	6 12	± 14 ± 5	5 12	± 41 ± 16
8	10	7th (final) draft	RTPMMA	12	± 8	12	± 8

[a] see section 6 (Acknowledgements) for material description and suppliers
[b] notched in one lab

The following (Fig. 4 and Tab. 2 and 3) are results from the last two round robins, which were conducted to validate the final draft of the protocol.

Figure 4 shows examples of the kind of loading curves observed. The 7th round robin used a sample of polyvinylchloride (PVC) producing a loading curve (mean line) nearly linear up to the maximum load, followed by unstable (brittle) fracture (Fig.4a), while for the 8th round robin a sample of rubber-toughened polymethylmethacrylate (RTPMMA) showing some ductility and limited stable crack propagation before the load drop (Fig. 4b) was used to assess the applicability of the protocol to moderately non-linear fractures. Figure 4 also shows the application of the curve fitting, the check on force fluctuations and the construction to determine $\overline{P}_Q$.

All the data from the 7th round robin (Tab. 2) was obtained in SE(B) testing and three types

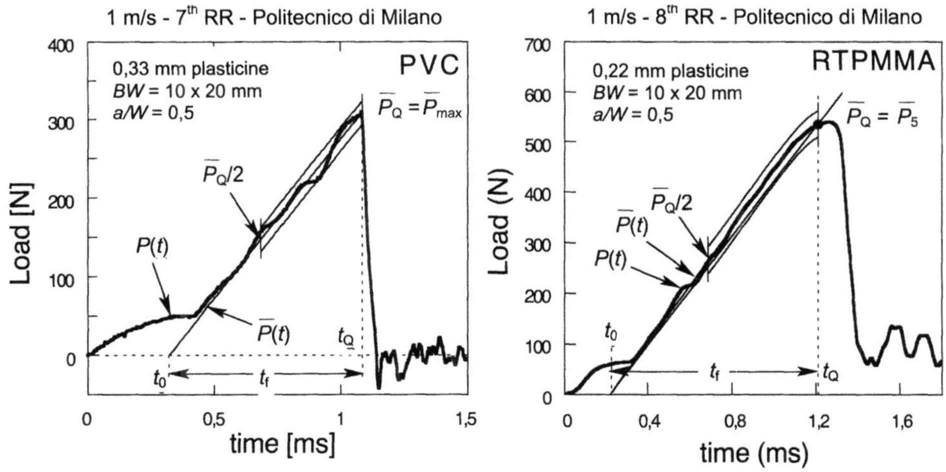

Figure 4 — Examples of loading diagrams obtained in the 7th (a, left) and 8th (b, right) round robins. Damping was obtained by a layer of plasticine (thickness shown) on SE(B) specimens. Material: (a) PVC, (b) RTPMMA. *BW* = specimen cross-section, *a*/*W* = relative crack length.

Table 2 — K_{IC} and G_{IC} measurements on PVC (7th round robin)

Lab. No.	Testing machine	Specimen type	P_5 or P_{max}	Mean t_f (ms)	K_{IC} determination: # Valid tests	K_{IC} determination: a/W	K_{IC} determination: **K_{IC} (MPa $\sqrt{m}$)**	G_{IC} determination: # Valid tests	G_{IC} determination: a/W range	G_{IC} determination: **G_{IC} (kJ/m²)**	K_{IC}^2/G_{IC} (GPa)
1	Falling weight	SE(B)	max	0,73	5	0,54	**2,70 ± 0,26**	9	0,20~0,71	**1,47**	4,96
2	Servohydraulic	SE(B)	max	0,78	5	0,50	**2,65 ± 0,10**	15	0,20~0,70	**2,19**	3,21
3	Pendulum	SE(B)	max	0,60	6	0,50	**2,60 ± 0,17**	13	0,19~0,69	**1,63**	4,15
4	Pendulum	SE(B)	max	0,56	3	0,49	**2,45 ± 0,07**	12	0,20~0,70	**1,50**	4,00
5	Servohydraulic	SE(B)	max	0,70	5	0,50	**2,76 ± 0,13**	15	0,20~0,71	**2,00**	3,81
6	Falling weight	SE(B)	max	0,63	5	0,51	**2,53 ± 0,02**	15	0,20~0,70	**1,90**	3,37
7	Falling weight	SE(B)	max	0,92	5	0,51	**2,51 ± 0,13**	14	0,20~0,65	**1,45**	4,34
8	Falling weight	SE(B)	5%	0,63	4	0,50	**2,82 ± 0,27**	13	0,25~0,70	**1,85**	4,30
9	Falling weight	SE(B)	max	0,82	3	0,50	**2,59 ± 0,08**	11	0,20~0,65	**1,25**	5,37
10	Servohydraulic	SE(B)	max	0,81	4	0,50	**2,56 ± 0,12**	11	0,20~0,72	**[3,23]**[a]	–
11	Servohydraulic	SE(B)	max	1,27	5	0,52	**2,82 ± 0,28**	15	0,21~0,71	**[5,08]**[b]	–
12	Servohydraulic	SE(B)	max	0,50	1	0,52	**2,88**	8	0,21~0,62	**[5,00]**[b]	–
						mean:	**2,66 ± 0,15**		mean:	**1,69**	
						st.dev.:	**0,14 (5%)**		st.dev.:	**0,31 (18%)**	

[a] error suspected
[b] without energy correction

of testing machine were used: impact pendulum, falling weight and servohydraulic testing instrument. Fracture initiation was generally identified with the point of maximum force (fourth column). The means of the K_{IC} values obtained from valid tests at $a/W{\sim}0.5$ are given together with the partial standard deviations (eighth column) and the slope of the linear regression through the corrected energy values obtained from valid tests covering a range of a/W is given for G_{IC} (eleventh column). The standard deviation from the mean values of all participating laboratories (bottom lines) is 5% for K_{IC} and 18% for G_{IC}.

Most data from the 8th round robin (Tab. 3) were obtained in SE(B) testing and three types of testing machine were used: impact pendulum, falling weight and servohydraulic testing instrument. Fracture initiation was mostly identified with the 5% offset (fourth column). The means of the K_{IC} values obtained from valid tests at $a/W{\sim}0.5$ are given together with the partial standard deviations (eighth column) and the slope of the linear regression through the corrected energy values obtained from valid tests covering a range of a/W is given for G_{IC} (eleventh column). The standard deviation from the mean values of all participating laboratories (bottom lines) is 8% for both K_{IC} and G_{IC}.

Table 3 — K_{IC} and G_{IC} measurements on RTPMMA (8th round robin)

Lab. No.	Testing machine	Specimen type	P_5 or P_{max}	Mean t_f (ms)	K_{IC} determination			G_{IC} determination			K_{IC}^2/G_{IC} (GPa)
					# Valid tests	a/W	**K_{IC} (MPa √m)**	# Valid tests	a/W range	**G_{IC} (kJ/m²)**	
1	Falling weight	SE(B)	max-5%	0,98	5	0,45	**4,16 ± 0,46**	7	0,36~0,70	**4,25**	3,58
2	Servohydraulic	SE(B)	5%	1,21	5	0,47	**3,70 ± 0,04**	10	0,16~0,65	**5,03**	2,73
3	Pendulum	SE(B)	5%	1,12	5	0,49	**3,42 ± 0,09**	14	0,25~0,69	**4,08**	2,87
4	Servohydraulic	SE(B)	5%	1,20	5	0,51	**3,71 ± 0,08**	11	0,36~0,66	**4,23**	3,19
5	Falling weight	SE(B)	5%	0,96	5	0,50	**4,12 ± 0,05**	11	0,35~0,65	**4,20**	3,90
	Falling weight	SE(B)	5%	1,00	5	0,50	**4,13 ± 0,07**	11	0,35~0,65	**4,70**	3,43
	Falling weight	SE(B)	5%	1,00	5	0,50	**3,92 ± 0,02**	12	0,30~0,65	**4,30**	3,48
6	Servohydraulic	SE(B)	5%	1,10	5	0,50	**4,20 ± 0,45**	10	0,36~0,60	**4,67**	3,78
7	Servohydraulic	SE(B)	max-5%	1,03	9	0,49	**4,21 ± 0,16**	23	0,19~0,69	**4,95**	3,58
8	Servohydraulic	SE(B)	max	1,32	5	0,49	**4,49 ± 0,04**	15	0,17~0,70	**4,71**	4,21
9	Servohydraulic	SE(B)	max	0,66	3	0,50	**3,71 ± 0,23**	5	0,28~0,50	**[5,54]**[b]	–
10	Servohydraulic	C(T)	5%	1,06	5	0,51	**3,71 ± 0,06**	15	0,31~0,71	**[8,44]**[a]	–
						mean:	**3,96 ± 0,15**		mean:	**4,51**	
						st.dev.:	**0,31 (8%)**		st.dev.:	**0,34 (8%)**	

[a] error suspected
[b] without energy correction

The significance of inter-laboratory consistency of the collected data can be appreciated by comparison of the results obtained here with data reported in [*12*] and [*13*] for inter-laboratory measurements of K_{IC} and G_{IC} at quasi-static testing rates. Precision of K_{IC} determinations at low testing rates [*12*, *13*] and at 1 m/s (Tab. 2 and 3) are similar and, rather unexpectedly, also the reproducibility of the determinations of G_{IC} at 1 m/s (Tab. 2 and 3) compares well with that obtained at lower rates [*12*, *13*] in spite of the more elaborate procedure adopted and the larger corrections involved in the high rate testing determinations as compared to the low rate case.

5. CONCLUDING REMARKS

Application of Fracture Mechanics to characterise toughness at high loading rates deserves special attention because of dynamic effects inherent in the test. With plastics tested at speeds around 1 m/s these effects can be contained or circumvented, and G_{IC} and K_{IC} can be determined with sufficient accuracy. The protocol developed within ESIS TC4 has been validated by a large amount of experimental data and can now be taken into consideration as a basis for national and international standards.

6. ACKNOWLEDGEMENTS

The following organisations and experts took part in several of the eight round-robins carried out within ESIS Technical Committee 4, to establish the 1 m/s testing protocol:

Organisation	Experts
BASF, Ludwigshafen, D	F.Ramsteiner
BP Chemicals, Grangemouth, UK	M.J.Cawood, Q.Clutton, A.Gray, A.J.Hemingway
Ciba-Geigy, Marly, CH	M.Fischer
Cranfield University, Bedford, UK	D.Ayre, A.Karpodinis, I.Partridge
Delft University of Technology, NL	A.Bakker, J.Dekker
DSM, Geleen, NL	H.Bos, P.Habets, G.Struyk
Du Pont, Wilmington, DE, USA	R.G.Bender, B.A.Crouch
EAHP-ICS, Strasbourg, F	Ch.Fond, Ch.Goett, R.Schirrer
Eastman Kodak, Kingsport, Tenn., USA	E.J.Moskala
Elf-Atochem, Serquigny, F	N.Chedozeau, F.Fernagut, L.Tézé
EPFL, Lausanne, CH	Ph.Béguelin
ICI, Wilton, UK	R.Moore, R S Hardy
Imperial College, London, UK	L.Braga, M.Chong, H.MacGillivray, J.G.Williams
INSA Lyon, Villeurbanne, F	H.Sautereau
IWM, Freiburg, D	W.Böhme
LMPL-ENSMA, Futuroscope, F	J.Parisot
Martin-Luther University, Merseburg, D	H.W.Grellmann, S.Seidler
Montan University, Leoben, A	Z.Major
Pipeline Developments, Manchester, UK	P.Marshall, P.Harry, K.Morley
Politecnico di Milano, I	R.Frassine, A.Pavan, M.Rink
Rhône-Poulenc, Aubervilliers, F	G.Orange
Solvay, Brussels, B	A.Goldberg
Technical University, Budapest, H	B.Pukánszky
The Welding Institute, Abington, UK	G.E.Hale
Universidad de Oviedo, Gijón, E	J.Belzunce
University of Kaiserlautern, D	J.Karger-Kocsis
University of Twente, NL	P.E. Reed, L.Warnet
VTT, Espoo, SF	A.Sivola

The materials used for the round robin exercises were kindly supplied by the following manufacturers: rubber-toughened polymethylmethacrylate (RTPMMA) from Atohaas Italia, Rho, Italy; butadiene rubber-toughened acrylonitrile-styrene copolymer (ABS) and rubber-toughened polyamide / poly(phenylene oxide) blend (RTPA/PPO)) from BASF, Ludwigshafen, Germany; high density polyethylene (HDPE) from BP Chemicals, Grangemouth, UK; polyvinylchloride (PVC) from EVC Int.l SA/NV, Brussels, Belgium; polymethylmethacrylate (PMMA) from ICI, Wilton, UK; and modified polyvinylchloride (mPVC) from Pipeline Developments, Manchester, UK. The sample of polyamide 6 (PA6) was kindly supplied to Politecnico di Milano by Radici Novacips, Villa D'Ogna, Italy.

7. REFERENCES

[1] Williams, J. G., "*Fracture Mechanics of Polymers*", Ellis Horwood, Chichester, UK, 1984.

[2] "*A Linear Elastic Fracture Mechanics (LEFM) Standard for Determining K_{IC} and G_{IC} for Plastics at High Loading Rates*", Testing Protocol prepared for ESIS TC4 by A. Pavan (final draft: Sept. 1997).

[3] ISO/CD 17281 "*Plastics - Determination of fracture toughness (G_{IC} and K_{IC}) at moderately high loading rates*",

[4] Venzi, S., Priest, H. A., May, M. J., "Influence of Inertial Load in Instrumented Impact Tests", *Impact Testing of Metals, ASTM STP 466*, American Society for Testing and Materials, 1970, pp. 165-180.

[5] Kalthoff, J. F., Böhme, W., Winkler, S., and Klemm, W., in Proc. CSNI Specialists Meeting on Instrumented Precracked Charpy Testing, Palo Alto, Cal., USA, 1980.

[6] Williams, J. G., *International Journal of Fracture*, Vol. 33, 1987, pp. 47-59.

[7] Williams, J. G., and Adams, G. C., *International Journal of Fracture*, Vol. 33, 1987, pp. 209-222.

[8] Béguelin, Ph., and Kausch, H. H., *Journal of Material Science*, Vol. 29, 1994, pp. 91-98.

[9] Béguelin, Ph., Fond, C., and Kausch, H. H., *International Journal of Fracture*, Vol. 89, 1998, pp. 85-102.

[10] Frassine, R., Rink, M., and Pavan, A., "On the Viscoelastic Time-Dependence of Fracture Toughness at High Loading Rates in Polymers", *Impact and Dynamic Fracture of Polymers and Composites*, ESIS 19, J. G. Williams and A. Pavan, Eds., MEP, London, UK, 1995, pp. 103-111.

[11] Zanichelli, C., Rink, M., Riccò, T., and Pavan, A., *Polymer Engineering and Science*, Vol. 30, 1990, pp. 1117-1124.

[12] Williams, J. G., K_C and G_C at Slow Speeds for Polymers, this book

[13] ASTM D 5045-93 "Standard Test Methods for Plane-Strain Fracture Toughness and Strain Energy Release Rate of Plastic Materials", American Society for Testing and Materials, West Conshohocken, PA, 1993

THE MEASUREMENT OF K_C AND G_C AT SLOW SPEEDS FOR DISCONTINUOUS FIBRE COMPOSITES

D.R. MOORE

1. AIM OF THE FRACTURE PROCEDURE

Discontinuous fibre reinforced composites can be prepared by a number of different processing technologies. These may include:-

(i) Extrusion compounding of the resin and glass to generate pellets that can be subsequently fabricated by processes such as injection moulding.

(ii) Pultrusion of a continuous fibre and resin lace which can be pelletised prior to fabrication by processes such as injection moulding.

(iii) The formation of sheets by laying fibres randomly within a mould into which the resin is added.

These and other processes generate discontinuous fibre reinforced composites but the fibre orientation and morphology of the materials is quite different. The aim of the fracture procedure is to be able to measure fracture toughness (K_c and G_c) for discontinuous fibre reinforced composites. In the first instance, this will apply to only injection moulded samples where the morphology is to some extent predictable.

The procedure for determination of the toughness parameters K_c and G_c for these composites will follow the general approaches of linear elastic fracture mechanics procedures which has already been described for plastics at relatively slow test speeds. (see "Introduction to Linear Elastic Fracture Mechanics" and "K_c and G_c at Slow Speeds for Polymers").

In conducting its work, the ESIS task team made various measurements of both long and short discontinuous fibre composites whilst focusing on what it eventually saw as its most fruitful approach. In developing the final protocol, all our measurements were conducted on a 50% w/w short glass fibre reinforced polyamide composite (Solvay IXEF) which was injection moulded into plaque mouldings of thickness 2 mm and 5 mm. All tests were conducted at 23°C at relatively slow test speeds i.e. in the range 1 to 10 mm/min on various types of Universal test machines.

2. THE FRACTURE PROBLEM

In contemplating a K_C or G_C value for a discontinuous fibre reinforced composite it has to be recognised that although the crack will initiate and propagate in the resin, it is nevertheless influenced by the nature of the fibres embedded in the resin. The orientation, size and interfacial properties of the fibres will influence the toughness of the composite. It is apparent that the composite is not a homogenous material but will act more like a structure. It is therefore unlikely that a single toughness value can exist for such a composite and that it will be important to qualify the type of toughness value that may be determined for this material. However, it is also helpful to attempt to do this within the experience and framework of fracture mechanics theory rather than to create a totally arbitrary framework for obtaining the toughness.

The measurement of fracture properties of discontinuous fibre reinforced composites presents three particular problems that are not encountered when testing unfilled plastics. The first of these relates to the sample morphology stemming from the method of fabrication (injection moulding). The second relates to a feature involved with the detection of crack initiation. The

third problem relates to the recognition that LEFM was developed to analyse sharp cracks in homogeneous materials.
The problem associated with sample fabrication is fundamental to injection moulding technology which presents an issue that is highlighted for discontinuous fibre reinforced composites but also intrinsic to this type of fabrication for any material. During the injection moulding process a melt is delivered into a mould tool under a shear stress field. The first portion of melt strikes a relatively cold mould wall and solidifies; this causes a majority of the fibres to be aligned in the direction of mould-fill. The melt that enters the central or core region of the mould tool is then subjected to a stress field where the deformations are extensional i.e. a diverging stress field [1]. This aligns the fibres in this core region at approximately right angles to the direction of the mould-fill. In simplistic terms, a skin-core-skin structure is established through the thickness of the moulding. The mould thickness decides which "layer" will be dominant. For example, for thin mouldings the size of the total skin layers will be larger than the size of the core; whilst for thick mouldings the core may be larger than the skin.

In generating and growing cracks in such fibre reinforced composites it is apparent that the relative direction of stress, fibre orientation and crack will be fundamental to the measurement of toughness or strength. In addition, this simplistic in-plane description of the fibres in the moulding may often be required to accommodate out-of-plane fibre orientations. The second issue concerns detection of crack initiation. Many types of discontinuous fibre reinforced composite generate force-displacement signals from notched LEFM specimens which can be interpreted in the same manner as unfilled plastics, namely, the maximum force corresponding to crack initiation. However, there is a group of discontinuous fibre reinforced composites that exhibit a force-displacement signal in which the initiation of the crack occurs well before the region of maximum force [2]. Traditional analysis of this type of signal is therefore inaccurate unless the specimen is instrumented in such a manner that positively identifies crack initiation. Such procedures for instrumentation are available and in fact form the subject of other parts of the ESIS Task Committee's activity but are beyond the scope of this section. The third issue in contemplating the application of LEFM to discontinuous fibre reinforced composites is to establish whether it is applicable. There are many ways to articulate such applicability but we will address the issue simply and non-rigorously now. For example, to establish whether the test specimen is of adequate size for generating a plane strain geometry value for K_C and G_C the existing protocol (see "Introduction to Linear Elastic Fracture Mechanics" and "K_c and G_c at Slow Speeds for Polymers") requires that the specimen thickness (B) shall be larger than a minimum value , B_{min}:-

$$B_{min} = 2.5\left(\frac{K_C}{\sigma_Y}\right)^2 \tag{1}$$

Where σ_Y is a tensile yield strength.

It is of considerable value when conducting LEFM measurements to include specimen size criteria checks in order to eliminate invalid data. But there is now an issue as to whether discontinuous fibre reinforced composites can exhibit the usual yielding process. In fact, whether there is any meaning to the term yield strength for these composite materials is questionable. This simple question has far reaching ramifications, since it will influence notions of plastic zone sizes for these composites and related issues concerning damage zones. These three issues generally summarise many of the aspects that need to be accommodated in establishing a protocol for toughness measurement in these materials. Without pre-empting the discussion to follow, it will become apparent that a rigorous solution will not be possible. There are many who will therefore dismiss the problem and abandon a solution that invokes fracture mechanics ideas. However, it is our contention that this is irresponsible on two counts. First, there are many dreadful ad-hoc toughness tests waiting in the wings which scientifically are worst than our most dreaded nightmares! Second, it is our view that where rigour cannot rule, pragmatism can practice.

3. A TEST PROTOCOL

3.1 Aim of the fracture measurements

The purpose of the protocol is to provide a means of measuring K_c and G_c for discontinuous fibre reinforced composites at test speeds in the range 1 to 10 mm/min. This method will be supplementary to the existing protocol document for the measurement of K_c and G_c (see "Introduction to Linear Elastic Fracture Mechanics" and "K_c and G_c at Slow Speeds for Polymers") for plastics and in that section are included the necessary LEFM background. Therefore, in this paper we will only discuss those aspects of the test that are additional for composites.

3.2 Specimen definition and preparation

Two particular specimen geometries are accommodated in the main protocol; a single edge notched (SEN) beam and a compact tension (CT) specimen. An injection moulding will have in-plane anisotropy and through thickness heterogeneity. There are six different directions of toughness for an anisotropic sheet as shown in Figure 1. In addition, there are three mutually perpendicular directions that can be defined in the sheet; these are (according to ASTM E616-81):-

(i) A processing direction e.g. mould-fill, designated the **longitudinal** (L) direction.

(ii) There are two directions which are **transverse** to the longitudinal, namely a long transverse (T) and a short or thickness-transverse (S).

Although Figure 1 includes all the specimen configurations, only two are relevant if the requirements of the main test protocol are met since the moulding thickness "*B*" limits how cracks can be introduced in practice. So, only specimens "T-L" and "L-T" may be used. This two letter code also follows the ASTM method. The first letter designates the direction normal to the crack plane i.e. the direction of applied stress for generating a colinear crack propagation. The second letter is the expected direction of crack propagation.

In common with the preparation of specimens from injection moulded plaques it is helpful to avoid the edges of the moulding where the conditions and circumstances of melt flow are complex. The notching of the specimens is less critical for fibre reinforced composites than that for polymeric samples. Nevertheless, the notch tip radius should be sharp i.e. less than about 50μm. The specimen dimension *B* will always be the moulding thickness, consequently, the dimension W will always be an in-plane dimension for the moulding.

In conducting the fracture mechanics test on each moulded sample it is necessary to test both "L-T" and "T-L" specimens.

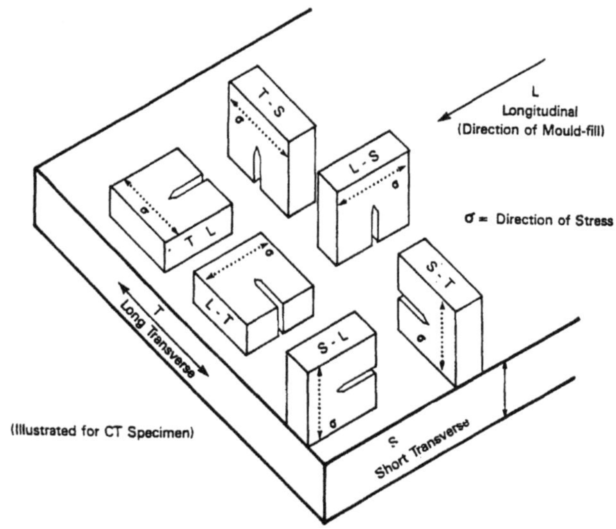

Figure 1. Specimen configuration for discontinuous composite sheet: only specimens "T-L" and "L-T" are relevant

Special Case for a Disc Moulding:-

NB. For an Anisotropic Disc Moulding the "L" direction will be a radial direction from the gate and the "T" direction will be the normal to the "L".

3.3 The fracture measurements

The fracture mechanics tests conducted on the SENB or CT specimens will provide a means of calculating K_c and G_c for the composite using the procedures described in the main protocol (see "K_c and G_c at Slow Speeds for Polymers"). As discussed earlier, it is important to ensure that the appropriate co-ordinates on the force-displacement signal are used for defining the initiation of the crack growth. There has been considerable research of this topic through instrumentation of the fracture specimens. This includes some useful observations from an acoustic emission study by one of the ESIS participating laboratories as part of our activities. The detail of such studies is beyond the scope of this paper, but Figure 2 covers the three contingencies. Either the peak force, or a 5% offset (see the main protocol) or a pre-peak force (termed "pop-in") value should be used as illustrated in Figure 2. One of these contingencies then provides a route to the calculation of K_c and G_c for the composite. The selection of the appropriate co-ordinates of crack initiation should proceed along the following steps:-

(i) Does "pop-in" occur (see lower curve in Figure 2). This can be recognised by a peak force followed by a drop in force (sometimes prior to a further and even significant rise in force). If the stiffness (slope of the force-displacement curve) reduces after the drop in the force, then it is likely that the crack has initiated. In this case, the co-ordinates of "pop-in" should be used for defining crack initiation.

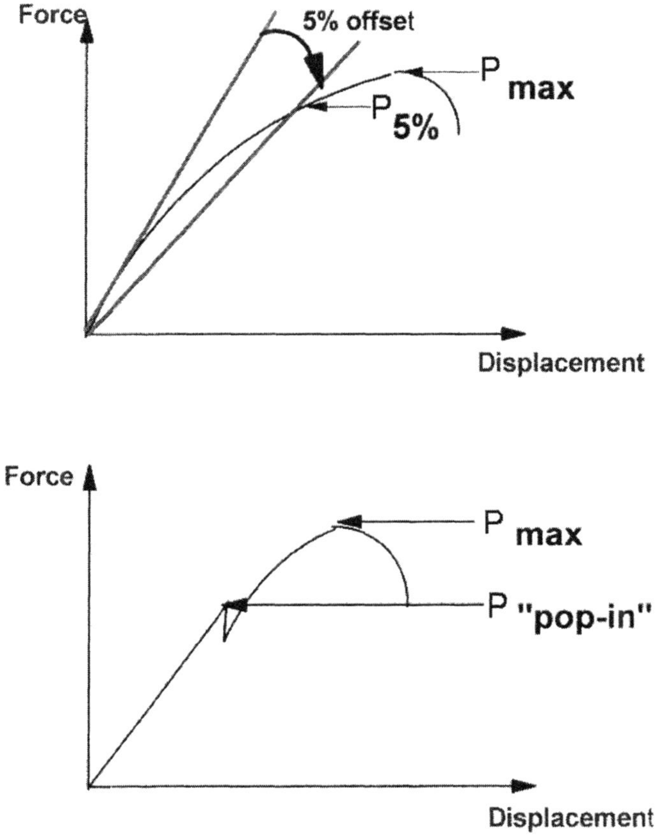

Force-displacement curves for crack initiation

Figure 2. Crack detection for fibre reinforced composites

(ii) If "pop-in" does not occur then there is a choice between using the peak force or the 5% offset co-ordinates for defining crack initiation (as shown in the top curve of Figure 2). In this case the recommendations of the protocol discussed in the paper entitled "K_c and G_c at Slow Speeds for Polymers" are used. To stay within the LEFM condition it was further specified that:

$$\frac{P_{max}}{P_{5\%}} < 1.1$$

i.e. a 10% non-linearity is allowed. If $P_{max} / P_{5\%} > 1.1$ then the test is invalid. If $P_{max} / P_{5\%} < 1.1$ then $P_{5\%}$ is used in calculation of K_Q unless P_{max} falls within the two lines, in which case P_{max} is used for the calculation of K_Q.

The test report must make clear which crack initiation value is used for the calculations and

this must be included in the table of results.
The next stage is to qualify the type of fracture in terms of the crack growth. If the composite is a homogeneous material then the crack would grow at right angles to the direction of applied stress i.e. as a colinear crack. However, for a discontinuous fibre reinforced composite the best that can be expected is an approximation to a colinear crack. This may occur when the fibres are all aligned in the same direction and the crack is able to initiate and propagate between the fibres. Unfortunately for injection moulded samples this idealisation will seldom, if ever, occur.
Therefore, it is necessary to examine the fractured surface and to measure the extent of non colinearity. In the first instance this can be done qualitatively. By observing the fractured surface side-on to the crack growth and then by examination of the plane of the crack. The side-on view will provide information of colinearity at the edge of the specimen i.e. in the "skin" layer. The observation of the plane of the crack will generate information on the relative overall flatness (or not) of the crack and hence will help to describe the influence of both "skin" and "core" layers on the fracture morphology. By microscopic examination of the plane of the crack it should then be possible to obtain a measure of the skin *(s)* and core *(c)* layer thicknesses, at least as described by the crack front. This is detailed in the section on "reporting".

3.4 Measurement of yield strength

The measurement of yield strength is a pivotal property for many fracture studies. It is necessary for the determination of a number of fracture functions such as plastic zone size and other expressions of size criteria. However, these size functions relate to homogenous materials and if we wish to apply them to heterogeneous and anisotropic materials then we have to do so with assumption.
The usual method for measuring yield strength is by tensile deformation. Indeed, a tensile yield strength is required for the calculations. However, a single failure process of yielding for discontinuous fibre reinforced composites is most improbable in tension because failure will be accompanied by other processes such as fibre pull out, fibre fracture, debonding and perhaps other mechanisms. A popular alternative to tension has been to conduct a compressive test where the accompanying mechanisms are usually eliminated [3].
If a square prismatic specimen of dimension a and length h is deformed in compression then there is a balance between friction and buckling such that the applied stress (σ_A) and the true stress (σ_T) are related:-

$$\sigma_A = \sigma_T (1 + \frac{\mu a}{4h}) \qquad (2)$$

Where μ is the coefficient of friction between specimen and anvil on the test machine.
If the ratio a/h is sufficiently small to avoid specimen buckling but sufficiently large to render the frictional term small, then applied stress can equate to true stress at yielding. (Otherwise a range of a/h values can be used to eliminate the frictional term). Moreover, if the stress is aligned in either the "T" or "L" direction, then an appropriate value of yield strength, in compression, can be determined for the composite. This value can be converted to a tensile yield strength by dividing by 1.3 [4]. In conducting these measurements, the direction of compressive stress will be either L or T.
This "tensile" yield strength may then be used in the determination of size criteria in order to qualify the fracture mechanics data. It should be emphasised that a direct tensile failure

strength is not acceptable as a yield strength.

3.5 Reporting and Presentation of results.

Reporting of the experimental results should be in three parts:-
(i) Test data as required by Tables 1 and 2 should be provided.

Laboratory	
Name	
Date	
Material	
Specimen configuration "T-L" or "L-T"	
Specimen type CT or SENB	
Test temperature (^{o} C)	
Test speed (mm/min)	
Compressive yield strength (MPa)	
Sample thickness (*B*) (mm)	
"Skin" thickness (s) mm	
"Core" thickness (c) mm	
Smooth fraction of fracture (α)	

Table 1 Test details K_c and G_c for discontinuous fibre reinforced composites

Specimen number		1	2	3	4	5
Specimen dimension	W (mm)					
Specimen dimension	*B* (mm)					
Notch tip radius	υ (μm)					
Force at crack initiation	P (N)					
Was this "peak", 5% offset or "pop-in"?						
Value of $P_{max}/P_{5\%}$ (if appropriate)						
K_c	(MPam$^{1/2}$)					
Mean K_c	(MPam$^{1/2}$)					
Size criterion? Is $B > B_{min}$						
Energy at crack initiation (J) Uncorrected						
Energy at crack initiation (J) Corrected						
G_c	(J/m^2)					
Mean G_c	(J/m^2)					

Table 2 Test results for K_c and G_c for discontinuous fibre reinforced composites

In table 2 the details of the size criterion, namely whether B is larger than B_{min}, should be included. In the main protocol this is interpreted as a validity criterion for the fracture data. However, in this protocol it is not a validity criterion, but can instead be considered as a quality criterion (which is a non-rigorous description). Nevertheless, all data can be accommodated and certainly used in the definition of the K_C versus smooth fraction of the fracture surface to be described later.

(ii) Examples of the force-displacement plots for each specimen type for each material should be included. In particular, there is a need to illustrate which of the three features described in Figure 2 are occurring.

(iii) The fracture surface should be described in terms of its "flatness" i.e whether the crack appears as co-linear. In addition, measurement of the skin and core layer thicknesses of the fractured surface should also be recorded. A plot of the measured K_c versus the smooth fraction of the fracture surface should be constructed as shown in Figure 6 and discussed in the next section.

The fracture morphology will be influenced by the type of specimen (L-T or T-L) and the value of *B*, the moulding thickness. This is illustrated in Figures 3 and 4 for the L-T and L-T specimens respectively. These illustrations make the assumption of a simple orientation for the fibres.

It is assumed that the fibres are parallel to the mould fill direction (L) in the surface, or skin, region and that in centre, or core, region the fibres are perpendicular to the direction of mould fill. This is approximately in line with observation, but an oversimplification of the detail.

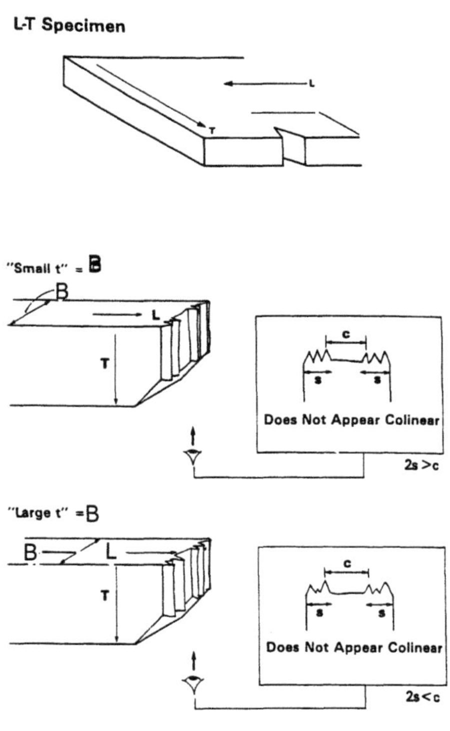

Figure 3 Fracture considerations for an L-T specimen.

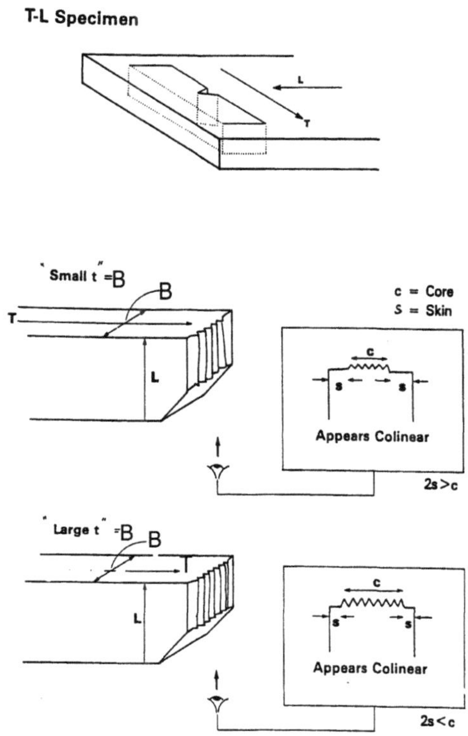

Figure 4 Fracture considerations for a T-L specimen

Observation then shows that the fracture surface is approximately smooth (colinear i.e. with an initial crack angle equal to 90°) when the crack is growing in a region between aligned fibres. However , if the crack is growing at right angles to the fibre alignment, then the fracture surface is quite rough and jagged (i.e. the initial crack angle is not equal to 90°). We try to schematically illustrate this effect in Figures 3 and 4. The size of the skin (and core) dimensions then defines the amount of smooth fracture that may occur in the four different specimens. If the size of the skin layer is designated 's' and the size of the core layer 'c', then the amount of smooth fracture for the T-L specimens will be equal to *2s/B* and for the L-T specimens it will be *c/B*. We will call these quantities the smooth fraction of the fracture surface, and denote these as α_{TL} for the T-L specimen and α_{LT} for the L-T specimen.
These measurements should be recorded in Table 1.
The final stage in reporting results should involve a plot of K_C versus smooth fraction of the fracture surface (α). Comparison of toughness can best be made in this form where highest toughness at a common value for "α" signifies best performance.

4. DISCUSSION OF EXPERIMENTAL RESULTS

The results presented here were obtained for a 50% w/w short glass fibre reinforced polyamide composite (Solvay IXEF)). Twelve laboratories (listed in an Appendix) conducted measurements on the 5 mm mouldings and some of them also made measurements on the 2 mm mouldings. The results are summarised in terms of a mean value for K_c and G_c together with their standard deviations. These are shown in Table 3 in combination with the colinearity of the crack by a "side-on" view of the fracture at initiation, and an observation as to whether the fracture surface is flat.

Specimen type	Was the crack colinear?	Flat fracture surface?	K_C (MPam $^{1/2}$)	G_C (kJ/m^2)
2 mm L-T	No	No	8.47 (1.71)	6.14 (2.73)
2 mm T-L	Yes	No	4.92 (1.36)	2.88 (1.91)
5 mm L-T	No	No	5.7 0 (0.96)	2.95 (1.08)
5 mm T-L	Yes	No	6.48 (1.00)	3.99 (1.42)

Table 3 Summary of fracture data.

Yield strength measurements were also made on these materials in uniaxial compression. For the L-T and T-L specimens, the first letter denotes the direction of compressive stress for the yield measurements. The tensile yield strength can then be obtained by dividing by 1.3. The quality of the fracture mechanics data as far as a specimen size criterion (equation 1) is concerned, may then be obtained. It is necessary for the ratio of B_{min} to B to be less than 1 for data that comply with a valid plane strain condition for homogeneous materials. These results and calculations are shown in Table 4.

Specimen type	Measured compressive yield strength (MPa)	Calculated tensile yield strength (MPa)	B_{min} (mm)	B_{min}/B (should be <1)
2 mm L-T	283	218	3.7	1.9
2 mm T-L	170	131	3.6	1.8
5 mm L-T	303	233	1.5	0.3
5 mm T-L	269	207	2.5	0.5

Table 4 Yield strength and size validity.

It is apparent from the data in Table 4 that the fracture results on the 5 mm thick mouldings would appear to meet this criterion, whilst those on the 2 mm do not, since the ratio B_{min} to B is greater than 1. However, it is implicitly assumed in these considerations that yielding of the resin can occur for fibre reinforced composites. Although it is known that failure can include yielding it is also clear that other mechanisms can also be involved. Nevertheless, some compressive yield results published by Leach and Moore [5] show that the relationship between yield strength and glass content for a range of glass fibre reinforced polyamide 6,6 composites can predict a resin yield strength that agrees with that measured value for the resin.

This implies that failure in compression in the composite is dominated by a yield process for the resin. Therefore, it can be helpful to apply the size validity checks for these inhomogeneous materials in order to comment on the quality of the fracture data, but not as a rigorous procedure.
The observations of the fracture morphology summarised in Table 3 enables a general picture to be constructed of the fracture process. This is done by combining the side on view with the view of the overall fracture surface i.e. the nature of colinearity. None of the fracture surfaces are smooth and flat, so none of the fractures are fully colinear. This should not influence the calculation of the stress field intensity factor, but will affect the strain energy release rate. Therefore, if modulus *(E)* was to be calculated from the fracture data using equation 3 (which in any case can only apply to homogenous materials), it is not likely to be correct, because although K_C might be valid G_C will be incorrect.

$$E = (1 - \nu^2)(\frac{K_C^{\ 2}}{G_C}) \qquad (3)$$

Where ν is a lateral contraction ratio. The fracture morphology is influenced by the type of specimen (L-T or T-L) and the value of *B*, the moulding thickness. This is illustrated in Figures 3 and 4 for the L-T and T-L specimens respectively and discussed in the section above. Optical microscopy of the fracture surfaces suggests that for the 2 mm moulding (actual size was 2.17 mm), s = 0.74 mm, whilst for the 5 mm moulding (actual size was 5.16 mm), s = 1.05 mm. Figure 5 shows a typical optical micrograph for one of the specimens and illustrates the "smooth" and "rough" morphology created by the fracture.

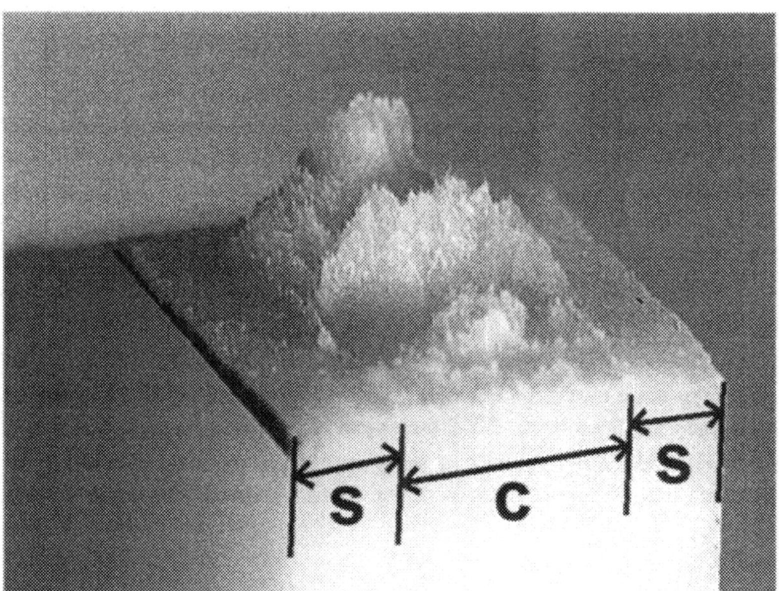

Figure 5 . An Optical Micrograph of the Fracture Surface of a 5 mm T-L Specimen

It would be reasonable to expect a relationship between K_c and the smooth fraction of the fracture surface. In fact, Friedrich [6] has analysed these fracture mechanisms for discontinuous fibre reinforced composites where in addition to the measurement of the K_c values and the values of ***c*** and ***s*** he also introduces efficiency factors that describe more rigorously the fibre alignment. However, for the purpose of a test protocol it is possible to forego the efficiency factor term and adopt the simplifying assumptions made above. Nevertheless, as shown in Figure 6, the singular values of fracture toughness are better resolved and more readily interpreted when considered in the form of a function of fracture toughness versus smooth fraction of the fracture surface.

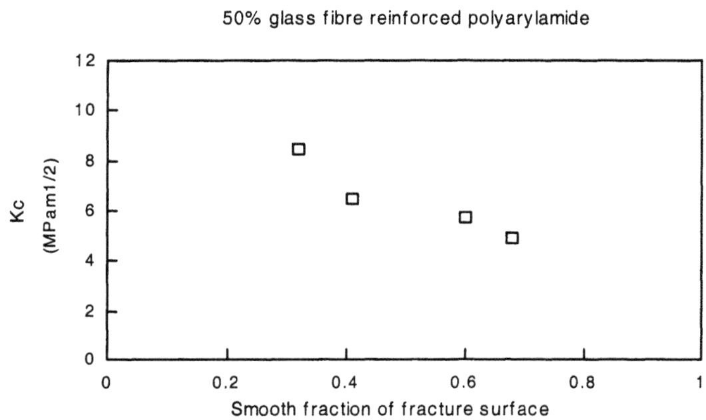

Figure 6 K_C versus Smooth Fraction of Fracture (α)

With reference to Figure 6 it can be expected that as the smooth fraction tends towards unity then the K_C value should tend towards the plane strain value for the resin; an anticipated value around 3.5 MPam $^{1/2}$ would seem reasonable. When the smooth fraction tends towards zero then the fracture process is dominated by fibre pull-out and fibre fracture. Consequently a large value for K_C would again seem to be reasonable.

It is therefore recommended that K_C should be measured with knowledge of the smooth fraction of the fracture surface.

The general use of the function K_C versus smooth fraction of the fracture surface helps comparisons of toughness for these discontinuous fibre reinforced composites. For example, if two different composite materials have been tested, then plots of K_C versus smooth fraction of the fracture surface can provide a simple means of toughness comparison by comparing the values of K_C at a common value of smooth fraction of the fracture surface. The larger K_C will infer higher toughness.

5. CONCLUDING COMMENTS

The conceptual idea of measuring K_C for discontinuous fibre reinforced composites is complex. In fact, it is likely that a rigorous fracture mechanics approach is not currently possible. However, with some simplifying assumptions, quite a useful methodology can be proposed. These assumptions involve a criterion for yielding in these composites similar to that for unreinforced polymers. In addition, if the fibre alignment is assumed to be simple, comprising aligned and perpendicular fibres, then the measured values of K_C can be usefully interpreted with a knowledge of the skin and moulding thicknesses. It is, of course helpful to conduct measurements on injection mouldings that are relatively thick in order to avoid some practical problems. This paper proposes a protocol for the measurement of the linear elastic fracture mechanics parameters, K_c and G_c. It has been shown that the stress field intensity factor, when plotted against the smooth fraction of the fracture surface, provides a powerful means of materials comparison and perhaps a helpful tool in engineering design.

6. REFERENCES

1 M Folkes "Short Fibre Reinforced Thermoplastics" Research Studies Press, J Wiley 1982.

2 A C Lowe, D R Moore, P M Rutter ESIS Pub 19 "Impact and dynamic fracture of polymers and composites" Ed by J G Williams & A Pavan, MEP Ltd (London) p383 (1995).

3 M Davies, D R Moore Comp Sci & Tech 40, 1991, p131.

4 S Gali, G Dolev, & O Ishai Int. J Adhesion & Adhesives, Jan 1981, 135.

5 D C Leach, D R Moore Composites 16,2, 1985.

6 K Friedrich, Comp Sci Technol 22, 1985, p43-74.

Appendix 1. ESIS TC4 Participants in the Work

ICI Plc, UK., (D R Moore, R S Hardy)
Cranfield University, UK. (I Partridge, C Durand)
Politecnico di Milano , Italy. (A Pavan)
Imperial College, London, UK. (B R K Blackman, H MacGillnray, J G Williams)
Institute of Polymer Research, Dresden, Germany (Martin, B Lauke)
Latvian Academy of Sciences, Latvia. (V Tamuzs)
Shell research BV, Amsterdam, The Netherlands. (A Cervenka)
Solvay S A, Belgium. (A Goldberg)
Politechnika Swietokrzyska, Kielce, Poland (L Golaski)
University of Twente , The Netherlands. (P E Reed)
Institute Technologico de Materiales, Lanera, Spain (F Javier Belzunce Varela)
University of Kaiserlautern, Germany (J Karger-Kocsis, K Friedrich)

DETERMINATION OF THE IMPACT FRACTURE TOUGHNESS K_{Id} OF PLASTICS AT HIGH RATES OF LOADING "> 1m/s"

W. BÖHME

1. INTRODUCTION

The procedure to determine the fracture toughness under quasistatic loading conditions is described in 'K_C and G_C at Slow Speeds for Polymers'. This procedure is based on fracture mechanics tests with precracked three point bend specimens, SENB, or compact tension specimens, CT, and the fracture toughness K_{IC} is evaluated from the measured load at fracture, and G_{IC} via the energy to fracture.

In principle, this procedure also applies to impact loaded, precracked three-point bend specimens at moderate loading-rates with impact velocities $v_o \leq 1$ m/s, especially if damping pads between the striker and specimen are used to reduce dynamic effects (see 'Determination of Fracture Toughness (G_{IC} and K_{IC}) at Moderately High Loading Rates'). In this case, so-called "instrumented" impact tests [1-4] have to be performed, where the striker is instrumented to measure the externally applied impact load as sketched by Fig. 1. The striker-instrumentation has to satisfy special requirements, e. g. the bandwidth should be not less than 100 kHz for metals according to [2-4] and 30 kHz for plastics according to ISO 179-2.

However, existing standards for impact testing devices and standardised impact tests are often related to impact velocities of several metres per second (e.g.: 2.9 m/s for plastics or 5 m/s for steels). When considering the crashworthiness of vehicles even higher speeds can be important and the resulting time-to-fracture, t_f, can be very short. For brittle materials and high impact velocities the onset of fracture is then usually observed in the very beginning of an impact test, where dynamic effects such as propagating elastic waves and, consequently, vibrations of the test sample, are dominant and loss of contact and bouncing can be observed (see e.g. [5-7]). These dynamic effects are indicated by oscillating load-signals with the first peak being termed the inertia-peak. Examples of such load-signals are given in the upper part of Fig. 2 in terms of K_I by applying static equations. Such externally measured load-signals can be misleading as explained in [8], since the actual crack-tip loading can be completely different, as demonstrated in the lower part of Fig. 2 by near crack tip strain gauge signals, again given in terms of K_I. Contrary to the evaluation of striker forces the evaluation of these local measurements results in meaningful, decreasing, toughness data with increasing loading rate (see Fig. 3).

Ireland [9], for example proposed limiting the range of applicability of quasi-static procedures to times-to-fracture, t_f, which are not less than the duration of three times the period of oscillation, τ, of the externally measured striker force $P_H(t)$. In practice, this restriction $t_f \geq 3\tau$ is similar to the above mentioned restriction $v_o \leq 1$ m/s. For shorter times-to-fracture, i.e. for brittle materials and/or high impact velocities, the consideration of the actual crack tip loading instead of the external load is recommended.

The crack tip loading history and the fracture initiation toughness at high loading rates can be measured directly, for example by near crack tip strain gauge instrumentation in combination with fast amplifiers (Fig. 1). Also by inertia-free optical methods such as the method of caustics in combination with high speed photography [5-8]. The involved effort in these direct measuring techniques is relatively high, and hence indirect methods with lower effort have been developed to predict the crack tip loading history of high rate tests.

One such scheme is to treat the specimen and machine compliance as a spring-mass model and thus correct the striker load to determine the load in the specimen [10]. Similar analyses have been proposed using various approximations and numerical schemes (see e.g. [11, 12]).

In order to further reduce the effort of determining the impact fracture toughness K_{Id} at commonly used impact velocities of several meters per second engineering approaches have been developed. The method of "Impact Response Curves (IRC)" was originally introduced by Kalthoff, Winkler, Böhme and Klemm (see e. g.: [13,14]). This procedure is based on a predetermination of the crack tip loading history, $K_I^{dyn}(t)$, e.g. by the optical method of caustics or by strain gauge instrumentation close to the crack tip. This curve has to be determined each time for new impact situations with different specimen sizes or different materials. In order to extend the range of applicability to various materials, specimen sizes and testing conditions the method of "Dynamic Key Curves (DKC)" has been developed by Böhme [6,15]. Based on a simple mass-spring model and basic measurements in model experiments, general rules have been developed to transfer these results to arbitrary materials and a wide range of testing conditions [6,15]. This procedure has been applied during a European round robin of ESIS TC4 with encouraging results sufficient for engineering purposes.

2. PRINCIPLE OF THE DKC-METHOD

It is the basic assumption of the DKC-method that the crack tip loading history $K_I^{dyn}(t)$ can be separated into a quasi-static part, $K_I^{qs}(t)$, and a dynamic correction function, $k^{dyn}(t)$ as sketched in Fig. 4 and described by:

$$K_I^{dyn}(t) = K_I^{qs}(t) * k^{dyn}(t). \qquad (1)$$

The first term, $K_I^{qs}(t)$, can be easily calculated by an analytically derived equation, which results from a simple mass-spring model [6,15]. The second term, $k^{dyn}(t)$, was determined once in model experiments by the evaluation of caustics which were obtained by utilising high-speed photography [6] resulting in a set of dynamic correction functions which in a normalised form are called "Dynamic Key-Curves (DKC)" [6,15].

This DKC-method describes the dynamic crack tip loading history, $K_I^{dyn}(t)$, for special types of SENB specimens and different materials, based only on a knowledge of the testing conditions. This procedure has been verified by other approaches to predict the dynamic crack tip loading which are based on more detailed mass-spring models (see e.g. [10-12]).

If the dynamic crack tip loading history $K_I^{dyn}(t)$ is known or predicted by one of these procedures, then during routine testing the measured time-to-fracture t_f determines the impact fracture toughness:

$$K_{Id} = K_I^{dyn}(t=t_f). \qquad (2)$$

according to [13-15]. The time-to-fracture, t_f, is the essential quantity to be determined during the tests. This time can be measured by different techniques, e.g., by an un-calibrated strain gauge attached close to the crack tip [13,16], or by conductive paint along the crack path [17,18], or eventually by contactless methods such as electric emission [19] for example.

At high impact velocities up to 8 m/s the DKC-method has been successfully applied to different materials such as steels [15], plastics [8] (see Fig. 3) and ceramics [17,18]. There might, however, be some limitations of the applicability for materials with strongly strain-rate dependent elastic moduli. The DKC-method should be considered as an engineering approach to determine impact fracture toughness K_{Id} even at short times-to-fracture where quasi-static procedures are no longer applicable.

3. GUIDELINE ON THE APPLICATION OF THE DKC-METHOD TO PLASTICS

For reasons of practical application and testing within an ESIS TC4 round robin a simplified guideline on the DKC-procedure and the corresponding evaluation was prepared in 1992 [20]. The results of this round robin exercise are presented in section 4. A revised version of the protocol considering the experience obtained during this round robin is given below. More details on the application of the DKC-procedure are given in [8,15].

Size of Specimens and Preparation

Based on the experience with dynamic effects of SENB specimens a special type of three point bend specimen was chosen, which was known to have minimised dynamic effects during impact. According to [6,15] reduced dynamic effects can be expected for specimens with the following relative dimensions:

relative initial crack length:	a/W	= 0.30 ± 0.02
relative specimen length:	L/W	= 5.50 ± 0.10
relative support span:	S/W	= 4.0 to 4.2

where W = specimen width.

The specimens and the cracks are prepared following the K_{IC} & G_{IC}-protocol ('K_C and G_C at Slow Speeds for Polymers'). Care should be taken on the rectangularity of the specimens to enable a perfect line-contact at both the impacting striker and the anvils.

Loading Devices

There is no restriction on the use of testing devices, except that the SENB-specimens must be loaded in three point bending. This means in fact a three-line contact between the specimen, the supports and the striker and no damping pads are allowed. Common pendulums, drop-weight towers and servo-hydraulic testing machines can be used to perform the impact tests. It is not essential, but very helpful, if the striker force is recorded during the tests. Since the quasi-static part of the DKC-approach is based on displacements the compliances of the loading system have to be taken into account, or the displacement has to be measured directly on the specimen itself. Both the specimen-compliance, C_s, and the machine-compliance, C_m, determine the loading history. The machine compliance is considered here in an integral manner, i.e. including the indentation of the specimen at the contact between the machine and the specimen, which for tests with plastics are usually the most compliant parts of the system apart from the specimen itself.

If the machine-compliance, C_m, is not negligible in comparison to the specimen-compliance, C_s, then instrumented and calibrated strikers can be used to determine the machine-compliance. A simple pre-test should be performed with an un-notched specimen of the material of interest. A low impact velocity v_o of about 0.2 - 0.5 m/s is applied and the striker force measured. The machine-compliance, C_m, can be determined from the slope of the mean load line, dP/dt_{MLL}, of the measured time-dependent, oscillating load-signal as sketched by the data-sheet as given in Fig. 5 and by applying the following equation:

$$C_m = \frac{V_o}{dP/dt_{MLL}} - C_{s,o} \tag{3}$$

where: $C_{s,o}$ = 20.1/(EB) = compliance of an un-notched specimen,
with E = elastic modulus of the specimen
(with rate-dependent materials such as plastics it is convenient to use values determined by vibration tests; an accuracy of 5% is acceptable)
B = thickness of the specimen.

Time-to-fracture measurements

The time-to-fracture, t_f, is the essential quantity to be determined during the tests. This characteristic time is defined by the difference between the moment of impact, t_o, and the time at fracture initiation, t_i:

$$t_f = t_i - t_o \tag{4}$$

In the DKC-method the moment of impact, t_o, is defined as that time were the load-transfer to the specimen starts neglecting initial settling effects. The moment of fracture initiation, t_i, is that time, when the crack starts to propagate. Especially for short times-to-fracture (< 100 μs), and depending on the applied measuring techniques, the observed times have to be corrected for example with regard to wave propagation effects:

t_o-determination

In principle, any method to detect the moment of impact as defined above is allowed. Two examples are given here:

i) An un-calibrated load cell (LC) may be used to detect the moment of impact, t_o. Therefore, at first the time $t_{o,LC}$ has to be evaluated from a measured load-signal by an extrapolation of the linear rising part of the inertia-peak to P = 0 as sketched by Fig. 6. Initial settling effects have to be neglected. Furthermore, delay-times of signals recorded by a load cell at a certain distance away from the point of impact have to be considered, too, and a corrected time has to be used finally during the evaluation:

$$t_o = t_{o,LC} - d_{LC}/c_o \qquad (5)$$

where d_{LC} = distance of the load cell from the nose of the striker, and c_o = wave propagation velocity of the striker material (for this approximate correction it is convenient to use $c_o = (E/\rho)^{1/2} = 5000$ m/s for steel)

ii) Conductive strips (CS) placed across the line of impact may be used to detect the moment of impact, t_o. The load transfer into the specimen starts usually somewhat delayed compared to the detected moment of contact, $t_{o,CS}$, depending on the thickness of these layers. Therefore, a corresponding corrected time has to be used during the evaluation:

$$t_o = t_{o,CS} + d_{CS}/v_o \qquad (6)$$

where d_{CS} = thickness of the conductive strips, and v_o = impact velocity.

t_i-determination

In principle, any method to detect the moment of crack initiation, t_i, is allowed. For large times-to-fracture >> 100μs the time of fracture initiation, t_i, can be approximately evaluated from a sudden drop of the load signal indicating the fracture event. However, at shorter times-to-fracture other measuring techniques are required. Two examples are given here:

i) The detection of crack initiation can be done by strain gauges attached near the crack tip. These signals usually indicate fracture initiation by a sudden drop of the signal (see Fig. 2, lower part). It is worth noting that for brittle materials with short times-to-fracture these signals are often smoothed (because the time to fracture and the time to accelerate and propagate the crack are comparable) and then preferably the first deviation from the nearly linearly rising part of the signal should be used as $t_{o,SG}$ (see Fig. 6). Depending on the distance of the strain gauge from the crack tip, d_{SG}, a corrected time-to-initiation should be used:

$$t_i = t_{i,SG} - d_{SG}/c_l \qquad (7)$$

where $c_l = \{E/(\rho(1-\nu^2))\}^{1/2}$ is the longitudinal wave propagation velocity of the specimen for plane stress: with ρ = mass density and ν = Poisson's ratio (e.g. c_l = 1766 m/s for the epoxy resin Araldite B [6]and 5390 m/s for steels).

ii) Crack propagation strips or conductive paint (CP) in front of the crack can be used to detect crack initiation times, $t_{i,CP}$. These measurements usually indicate crack initiation delayed for two reasons. One is crack tunneling below the strips due to the higher constraint in the mid-thickness of the specimen. Another is that it takes some time to open the crack and to break the layers. For example with thin silver paint and for specimens made from epoxy of thickness of 10 mm delay-times of approximately 5-10 μs with a significant amount of scatter were observed [17]. Such delay times have to be considered and a corrected time has to be used during the evaluation such as:

$$t_i \approx t_{i,CP} - 10\ \mu s \qquad (\pm 5\ \mu s!) \qquad (8)$$

This significant scatter limits the application of this procedure even as an approximate one, to times to fracture >> 50 μ s.

Other methods are allowed, but the results obtained should be once verified by comparison with results obtained by crack tip strain gauges.

Evaluation

Provided that the impact energy is large in comparison to the consumed fracture energy, then the impact fracture toughness can be determined by the measured time-to-fracture, t_f, and the following simplified equation [6,8,15]:

$$K_{Id} = \frac{E\,S\,f}{4\,W^{3/2}\,C_s^*(1 + C_m / C_s)} \left(v_o t_f\right) . k^{dyn}(t = t_f) \tag{9}$$

where a relative support distance S/W = 4.0 – 4.2 is used and:

$k^{dyn} = k^{dyn}(c_l t/W)$ = dynamic key curve (see next section),
E = see eq. (3),
f = f (a/W) is the well known static relationship for K_I-determination of three point bend specimens according to Srawley [21] and ASTM E 399:

$$f(\alpha) = \frac{6\alpha^{1/2}\left[1.99 - \alpha(1-\alpha)(2.15 - 3.93\alpha + 2.7\alpha^2)\right]}{(1+2\alpha)(1-\alpha)^{3/2}} \tag{9a}$$

with α = a/W,
C_s^* = $C_s^*(\alpha)$ = $EBC_s(\alpha)$ is the dimensionless specimen compliance after Bucci et al. [22], which is at short crack lengths somewhat larger than the function ϕ of the K_{IC} & G_{IC} - protocol:

$$C_s^*(\alpha) = 20.1 + 135\alpha^2 \begin{pmatrix} 1 - 2.11\alpha + 8.76\alpha^2 - 19.9\alpha^3 + \\ 41.4\alpha^4 - 67.7\alpha^5 + 92.1\alpha^6 - 76.7\alpha^7 + 35.6\alpha^8 \end{pmatrix}. \tag{9b}$$

C_s = C_s^*/EB = specimen compliance as calculated from C_s^* via eq. (9b),
C_m = machine compliance (see eq. (3)),
c_l = longitudinal wave propagation velocity for plane stress (see eq. 7).

Dynamic Key Curve $k^{dyn}(c_l t/W)$

For the special type of specimen chosen here with a/W=0.3, L/W=5.5 and S/W=4.0 the dynamic key curve k^{dyn} is shown in Fig. 7 taken from [6,15]. Three different time ranges have to be considered and the corresponding k^{dyn}-values are given by:

a) Initial time range $0 \le t \le 1.18\ W/c_l$:

$$k^{dyn} = 0 \qquad (10a)$$

Due to wave propagation effects no crack opening will occur in this time range, and only some crack closure caused by compressive waves can be observed [6,7]. Therefore, the time $1.18W/c_l$ is a lower limit for observable times-to-fracture for the chosen type of specimen.

In general, according to [6,7] this threshold is given by the time, when the first shear wave front approaches the crack tip: $t_{f,min}= (W\text{-}a)/c_t$. For example, for a specimen with W=10mm, a=3mm and Araldite B with c_l =1766 m/s and c_t =1022 m/s this threshold agrees approximately with the time $1{,}18W/c_l$ and results in a very short time of about 7 μ s. This is confirmed by measurements at IWM [17].

b) Fully dynamic time range $1.18\ W/c_l < t \le 9.2\ W/c_l$:

$$k^{dyn} = -\ 0.9096 + 0.8176(c_l t/W) - 0.1005(c_l t/W)^2 + 0.003765(c_l t/W)^3 \qquad (10b)$$

For the chosen type of specimen this is roughly the time range of the inertia peak. A dynamic evaluation with k^{dyn} as given by eq. (10b) has to be applied, if fracture is occurring in this time range.

c) Intermediate and quasi-static time range $t > 9.2\ W/c_l$:

$$k^{dyn} = 1 \qquad (10c)$$

can be used approximately in equation (9).

4. EXPERIENCE WITH THE DKC-METHOD

During a session of ESIS TC4 in Sardinia on the occasion of the ESIS-conference "Impact and Dynamic Fracture of Polymers and Composites" in September 1992 it was decided to investigate the applicability and accuracy of the DKC-method by a round robin exercise. This exercise focussed on the determination of the impact fracture toughness K_{Id} of plastics at high rates of loading at impact velocities "> 1m/s". The participants who contributed to this exercise are given in Table 1.

Materials

During the round robin exercise it was agreed to investigate two materials which cover a wide range of toughnesses. Furthermore, both materials should have different strain rate sensitivity. As an example of a tough and strain rate sensitive material a modified PVC was provided by EVC. The same material was investigated during a corresponding round robin at velocities of 1 m/s. As an example for a brittle and not very strain rate sensitive material the epoxy, Araldite B, was provided by IWM, Freiburg.

Size of Specimens and Preparation

According to the guideline on the application of the DKC-method the special type of three point bend specimens with reduced dynamic effects was chosen: a/W = 0.3, L/W = 5.5 and S/W = 4.0. Furthermore, it was decided to use specimens with a width W = 10 mm and a thickness B = 10 mm corresponding to standardised Charpy specimens. The specimens and the cracks of a length of 3 mm had to be prepared following the K_{IC} & G_{IC}-protocol (see 'K_C and G_C at Slow Speeds for Polymers'). The participants were told to take care on the rectangularity of the specimens to enable a perfect line-contact at both impacting tup and anvils. The PVC-specimens had to be precracked by sharpening an initial notch with a sliding razor blade and the epoxy-specimens by impact tapping.

Loading Devices

There was no restriction on the use of the testing devices, except that the SENB specimens must be loaded in three point bending, which means in fact a three-lines contact between specimen, supports and striker. Therefore, common pendulums, drop-weight towers and servo-hydraulic testing machines have been used to perform the impact tests.

According to the guideline the compliance of the machines was determined by the participants by pre-experiments at reduced velocity. The C_m values obtained are listed in Table 2. Apart from two results these values did not differ very much and are close to 0.2 m/MN (a variation of 20% would be of minor importance on the final evaluation). This result is in agreement with the expectation, that similar contact radii will result in similar C_m values.

Time-to-fracture measurements

The participants of the round robin were free to use appropriate methods to detect these times. Four simultaneously measured signals of an experiment at IWM are presented in Fig. 8. Some participants determined t_o from the increase of the force signal measured at the striker (see Fig 2.1a). Other participants determined t_o from signals of conductive strips on the specimens at the region of the striking tup (see Fig. 8c).

The time at initiation of fracture, t_i, was determined by several participants with strain gauges attached on the specimens near the crack tip (see Fig. 8b). Other participants used conductive paint across the crack path (see Fig. 8d).

Reference data

In order to have reference data for comparison, some experiments were performed at IWM with direct measurements with strain gauges attached close to the crack tip. A quasi-dynamic calibration-factor was determined with a few experiments at a reduced velocity of 0.5m/s by comparison with the externally measured striker force. This calibration-factor was then applied at higher rates of loading to evaluate the impact fracture toughness K_{Id} from the fracture initiation point of the strain gauge signals. The validity of this procedure has been verified several years ago by comparison with results of caustics obtained by high speed photography (see [16]).

Results of PVC

The results for impact tests with PVC and evaluations according to the guideline are given in Fig. 9. The filled black circles are the reference data. All other data were evaluated by using the IWM-program and the input-data of the individual laboratories in order to avoid some differences of the individual evaluations. The corresponding times-to-fracture vary between about 400 μ s at 1 m/s down to about 30 μ s at 8 m/s. There is a significant amount of scatter which might be due to different procedures in measuring the time-to-fracture. Most of the data are close to the reference data and show clearly a decreasing toughness with increasing loading rate.

Results of Araldite B

The results for impact tests with Araldite B and evaluations according to the guideline are given in Fig. 10. Again, the filled black circles are the reference data. All other data were evaluated by using the IWM-program and the input-data of the individual laboratories. The corresponding times-to-fracture vary between about 150 μ s at 1 m/s down to about 20 μ s at 8 m/s. One set of data shows a reasonable agreement with the reference data. The results of the other participants are on the average about 30% higher. Nevertheless, these results indicate, that the toughness of this epoxy is less dependent on the strain rate than PVC.

Discussion

Extended discussions during several sessions of the ESIS TC4 committee in Les Diablerets focussed on the following topics:

For the investigated high loading rates and impact velocities > 1 m/s the evaluated toughness values are more meaningful than misleading quasi-static evaluations of externally measured striker forces as can be seen in Fig. 3. The results are meaningful especially for PVC, but compared to more precise direct measurements most of the values obtained for the epoxy are systematically 30% too high. Two possible reasons were considered:

- One reason for the systematic deviations obtained of about +30% could be the pre-cracking procedure: lower bound toughness values will be observable only with natural sharp cracks. Especially for an epoxy it is sometimes very difficult to produce a natural crack, even by impact tapping. Therefore, during a next round robin all specimens should be precracked by only one participant.

- Another systematic deviation could be caused by different time-to-fracture measurements. This might be due to the fact, that time-corrections were not included in the first draft of the guideline. However, a precise determination of the time-to-fracture, t_f, is essential for the final result, especially at very short times-to-fracture < 100 μs as observed with the epoxy. Therefore, the guideline [20] was modified. More precise definitions to measure the time-to-fracture t_f are given in the new version and time-corrections are now included (see section 3 above).

The round robin exercise demonstrated in principle the application of the DKC-method on the determination of the impact fracture toughness K_{Id} of plastics at high rates of loading, where external force measurements would be misleading. In detail the following conclusions can be drawn:

- The equations to determine K_{Id} are meaningful.
- The determination of the machine compliance is sufficiently accurate.
- The DKC-method applies well for plastics such as PVC,
- at different loading rates, and
- at different levels of toughness.

An improved accuracy of this engineering approach, especially for materials like epoxy with lower toughness values, can be expected from an improved pre-cracking procedure.

The method of dynamic key curves has to be considered as an engineering approach. The accuracy of this DKC-approach is estimated by current experience to be about 10 %. This is often acceptable, especially at high impact velocities, where external force-measurements are completely misleading. More complicated measurements close to the events of interest would be necessary to achieve higher precision, but the effort will be greatly increased.

REFERENCES

[1] Instrumented Impact Testing, ASTM STP 563, American society for Testing and Materials, Philadelphia, 1969

[2] DVM-Merkblatt 001, Meßtechnische Anforderungen beim instrumentierten Kerbschlagbiegeversuch, DVM, Berlin, 1986

[3] Proposed standard method for the instrumented Charpy-V impact test on metallic materials, Draft 10, prepared by ESIS TC5 Technical Committee on Dynamic Testing at Intermediate Strain Rates, January 1994

[4] ISO 14 556, Steel - Charpy V Pendulum impact test - Instrumented test method, 2000

[5] Böhme, W., Kalthoff, J.F., Int. Journal of Fracture, Vol. 20, 1982, pp. R139-R143

[6] Böhme, W., Experimentelle Untersuchungen dynamischer Effekte beim Kerbschlagbiegeversuch, PhD thesis, Darmstadt, 1985, and: scientific report W1/85, Fraunhofer-Institut für Werkstoffmechanik, Freiburg, 1985

[7] Böhme, W., The Influence of Stress Waves on the Dynamic Crack Tip Loading in Three-Point Bend Impact Testing, in: Proc. Int. Conf. IMPACT '87, Bremen, Germany, 1987, Ed.: C.Y. Chiem et al., DGM, Oberursel, Vol. 1, 1988, pp. 305-311

[8] Böhme, W, Application of Dynamic Key Curves for the Determination of the Impact Fracture Toughness of Polymers at High Rates of Loading, in: *Impact and Dynamic Fracture of Polymers and Composites,* ESIS 19, Eds.: J. G. Williams and A. Pavan, MEP, London, pp. 59-71, 1995

[9] Ireland, D.R.: Critical Review of Instrumented Impact Testing, Proc. Int. Conf. on Dynamic Fracture Toughness, London, 1976

[10] Williams, J.G., Adams, G.C., Int. J. Fract., 33, 1987, pp. 209-222

[11] Peuser, T., in: Proceedings AFMMS, Int. Conf., Freiburg, Germany, Eds.: G.C. Sih et al., Martinus Nijhoff, 1983, pp. 455-465

[12] Rokach, I.V., J. of Theoretical and Applied Mechanics, 1, 32, 1994

[13] Kalthoff, J.F., Winkler, S., Böhme, W., Klemm, W.: Determination of the Dynamic Fracture Toughness K_{Id} in Impact Tests by Means of Response Curves, in Adv. in Fracture Research, Eds.: D. Francois et al., Pergamon Press, Oxford, New York, 1980, pp. 363-373

[14] Kalthoff, J.F., Winkler, S., Böhme, W., A Novel Procedure for Measuring the Impact Fracture Toughness K_{Id} with Precracked Charpy Specimens, Journal de Physique, Coll. C5, No. 8, Tome 46, 1985, pp. 179-186

[15] Böhme, W.: Dynamic Key-Curves for Brittle Fracture Impact Tests and Establishment of a Transition Time, Fracture Mechanics: 21. Symp., Annapolis, USA, ASTM STP 1074, 1990, pp. 144-156

[16] Böhme, W., Kalthoff, J.F., Der Einfluß der Probengröße auf dynamische Effekte bei der K_{Id}-Bestimmung im Kerbschlagbiegetest, IWM-Bericht W3/83, Freiburg, 1983

[17] Gerster, T., Messung der Rißzähigkeit von spröden Werkstoffen mit Schlagbiegeversuchen, Diplomarbeit, Fraunhofer-Institut für Werkstoffmechanik (IWM), Freiburg, 1996

[18] Böhme, W., Bestimmung des dynamischen Bruchverhaltens nichtmetallischer Werkstoffe mit instrumentierten Pendelschlagwerken und unterschiedlichen Meßverfahren, 24. Vortragsveranstaltung des DVM-Arbeitskreises Bruchvorgänge, 18./19. Feb. 1992, Aachen, Germany, DVM, Berlin, 1992, pp. 403-414

[19] Winkler, S.: Brucherkennung mit elektrischer Emission, Fh-IWM Report T 10/90, Freiburg, Germany, 1990

[20] Böhme, W., Short guideline on the application of the method of "Dynamic Key Curves" to the determination of the impact fracture toughness K_{Id}, 1. Draft, Freiburg, Feb. 1992

[21] Srawley, J.E.: Int. Journal of Fracture, Vol. 12, 1976, pp. 475-476

[22] Bucci, R.J., Paris, P.C., Landes, J.D., Rice, J.R., in ASTM STP 514, American Society of Testing and Materials, 1972, pp. 40-69

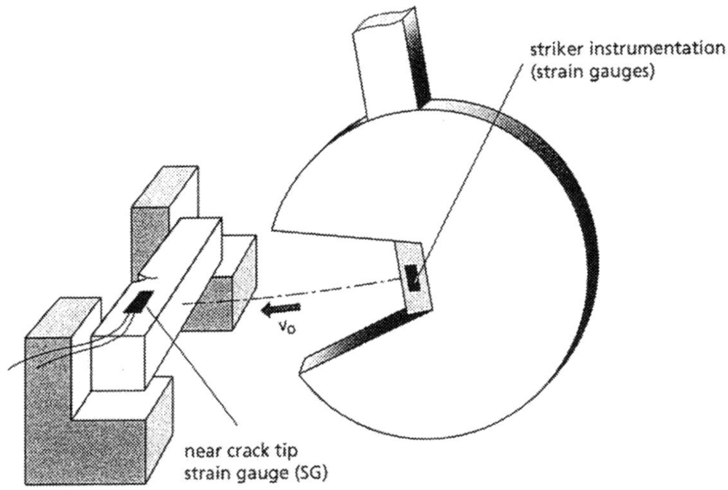

Fig. 1: Instrumented impact test

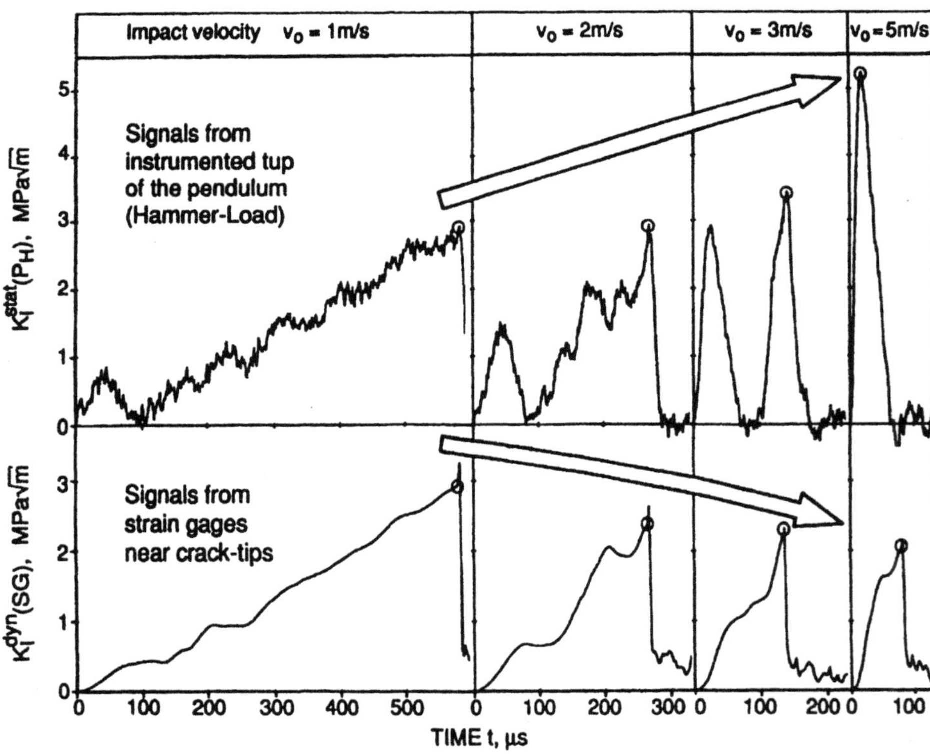

Fig. 2: Signals of an instrumented striker (upper part) and near crack tip instrumentation (lower part) for PVC at different impact velocities and given in terms of K_I

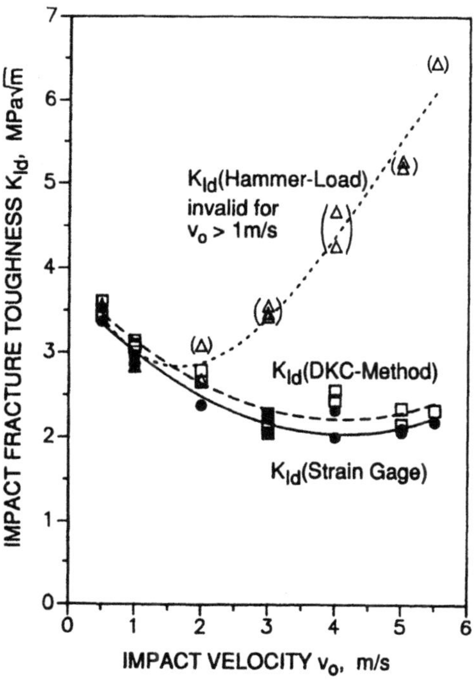

Fig. 3:
Impact fracture toughness data for PVC obtained by three different measuring and evaluation procedures according to [8]:

- hammer-load
- near crack tip strain gages
- Dynamic Key Curve (DKC)

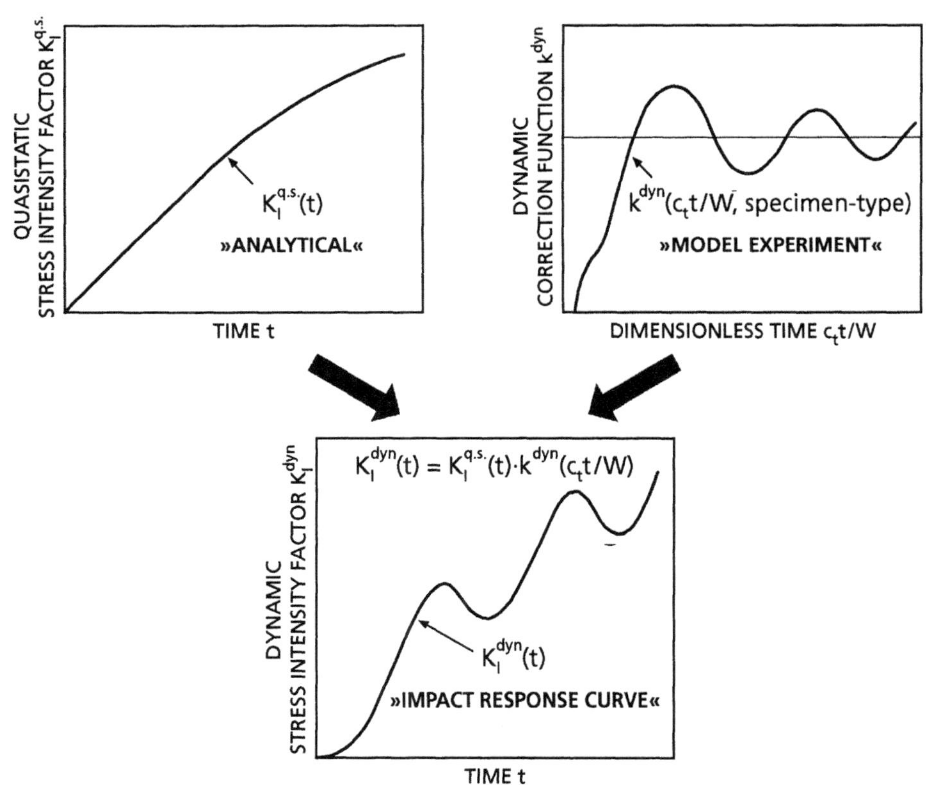

Fig. 4: Principle of the DKC-method (schematic)

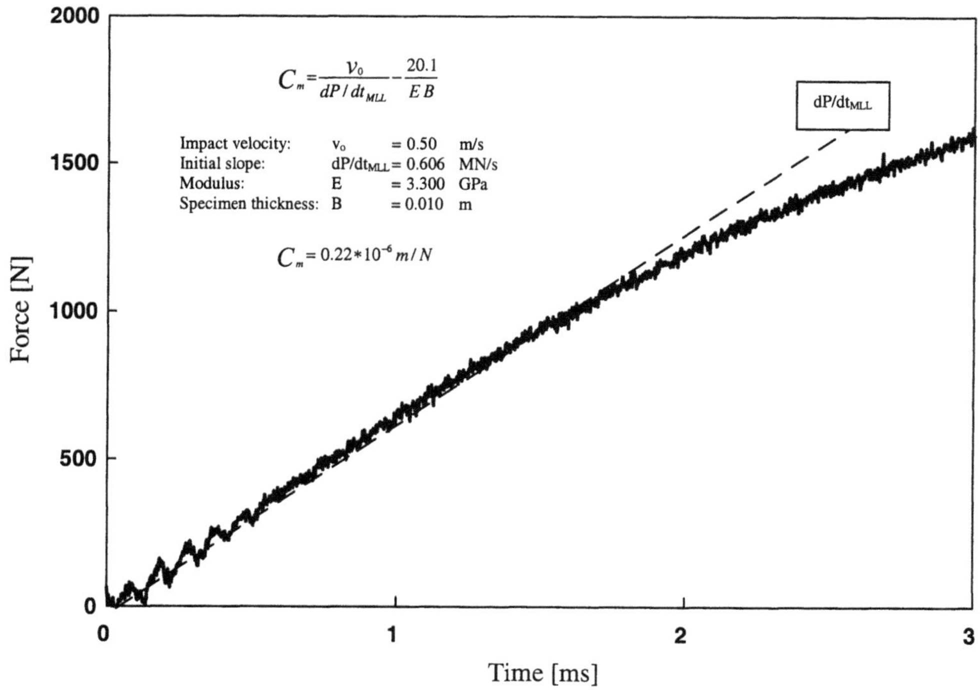

Fig. 5: Example of the determination of the machine-compliance C_m by a test at reduced velocity with an unnotched specimen (PVC)

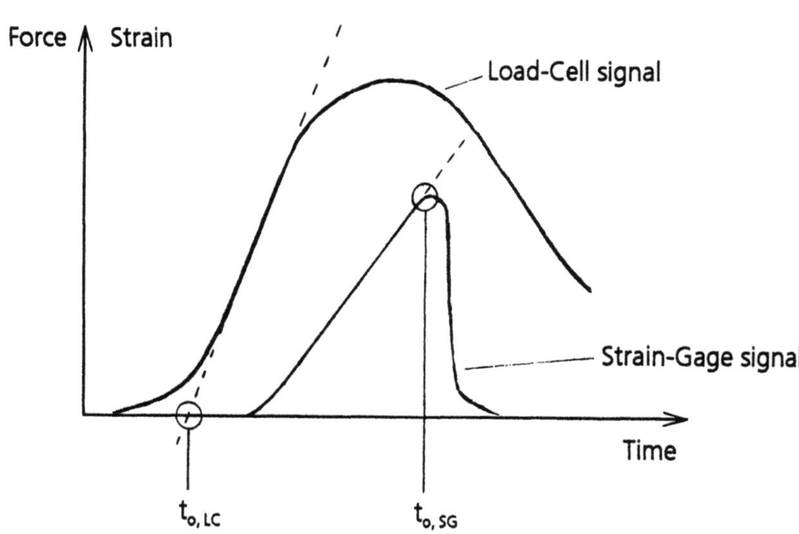

Fig. 6: Determination of the time-to-fracture (schematic)

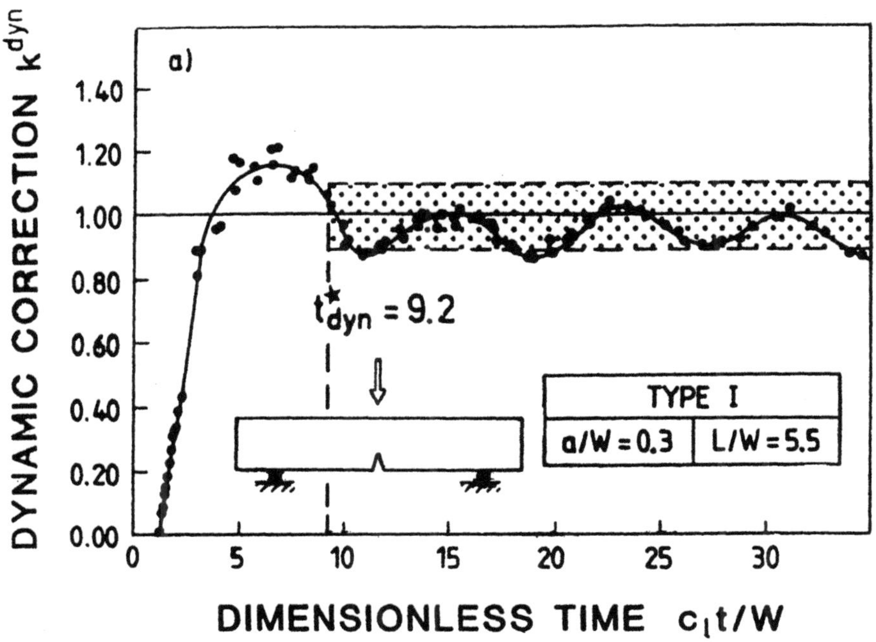

Fig. 7: Dynamic key curve for a bend specimen with a/W=0.3, S/W=4 and L/W=5.5

Participants	Organisation
Ph. Bèguelin	EPFL, Lausanne (CH)
W. Böhme, M.Hug, T. Gerster	IWM, Freiburg (D)
Ir. A. Cervenka	SHELL, Arnhem (NL)
Ir. J.C. Dekker	University, Delft (NL)
Z. Major	University, Leoben (A)
D.R. Moore	ICI, Wilton (UK)
P.E. Reed	University, Twente (NL)
J.G. Williams, M. Chong	Imperial College, London (UK)

Table 1: Participants in the ESIS TC 4 round robin exercise ">1 m/s"

Organisation	Machine-Compliance C_m (m/MN): PVC	Machine-Compliance C_m (m/MN): Epoxy
EPFL, Lausanne (CH)	0,33	0,22
FhG-IWM, Freiburg (D)	0,22	0,20
Univ. Delft (NL)	---	0,12
IWPK, Leoben (A)	0,24	0,24
ICI, Wilton (UK)	0,24	0,20
Univ. Twente (NL)	0,23	---
Imp. Coll., London (UK)	0,19	0,19

Table 2: Obtained values for the machine-compliance C_m

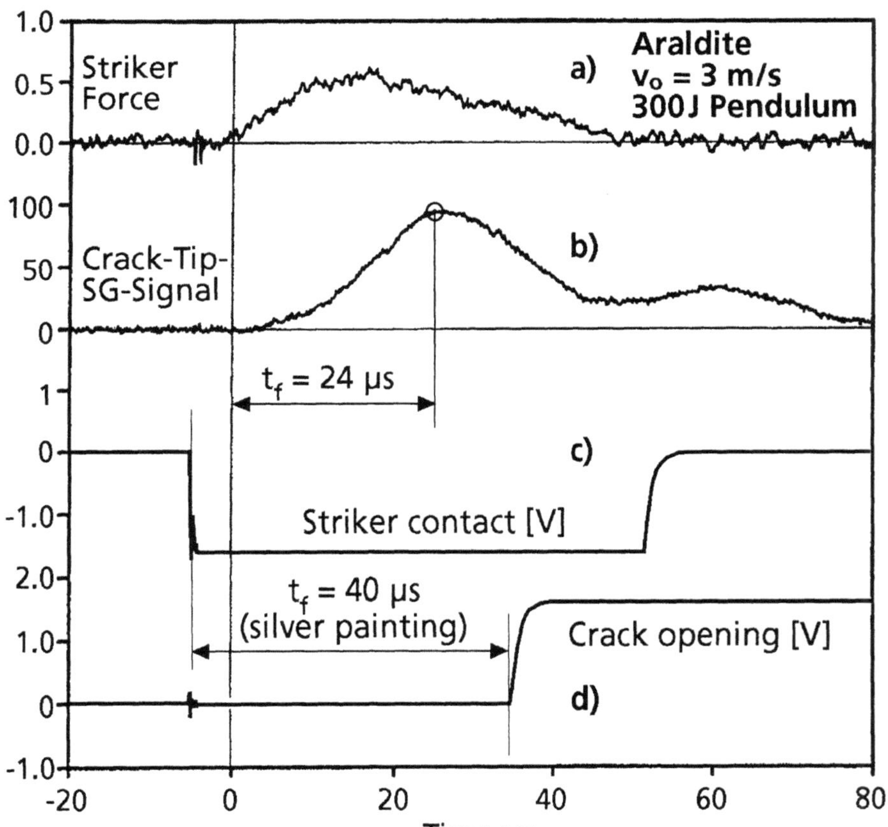

Fig. 8a)-d): Example of measured signals during an impact test

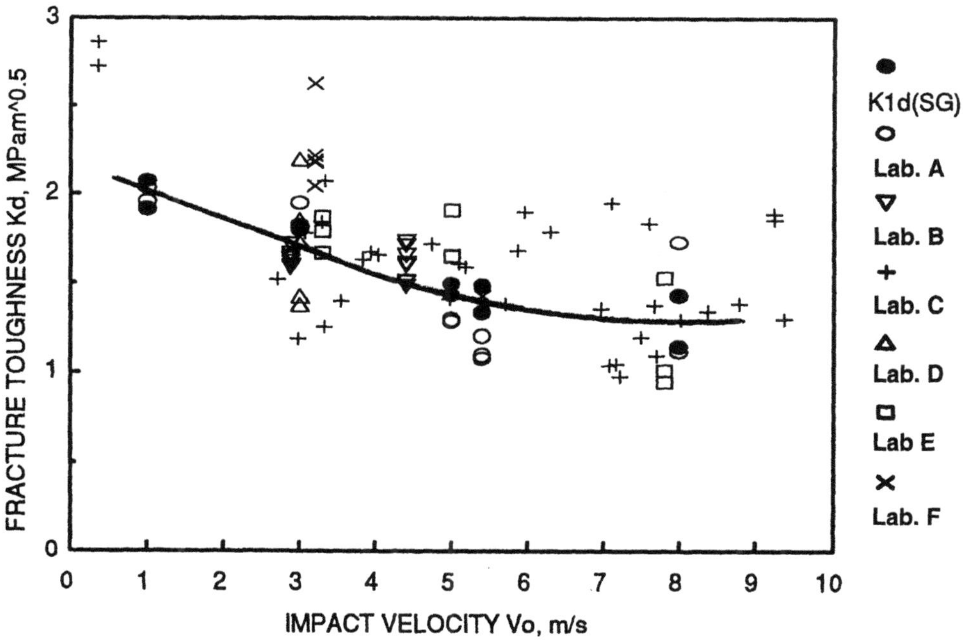

Fig. 9: Impact fracture toughness data of PVC

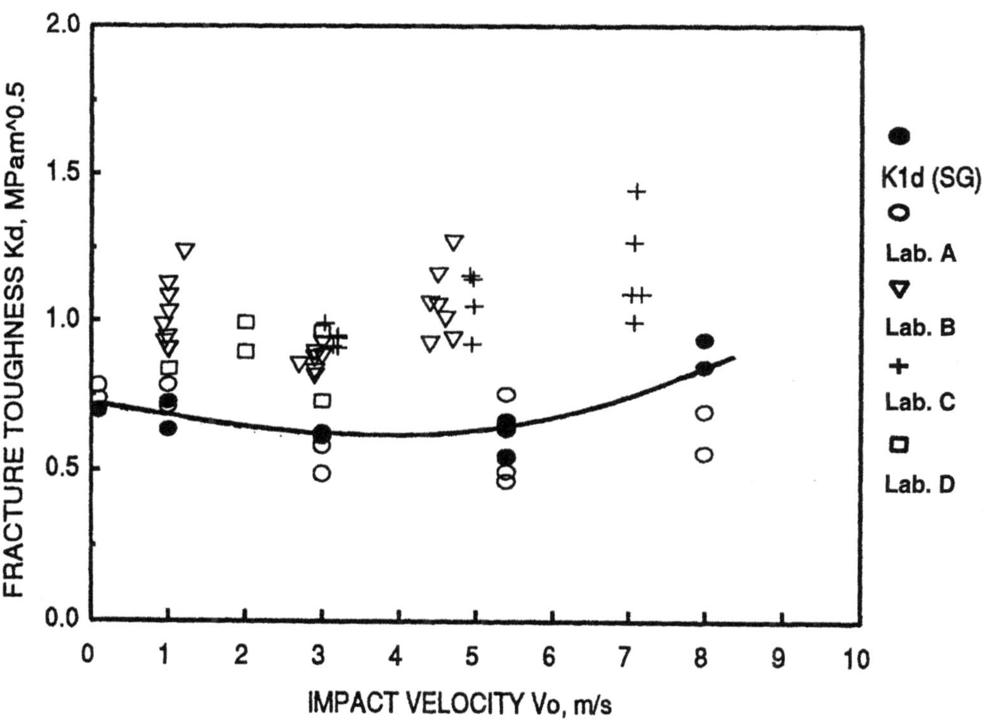

Fig. 10: Impact fracture toughness data of Araldite B

FATIGUE CRACK GROWTH OF POLYMERS

L. CASTELLANI and M.RINK

1. INTRODUCTION

Fracture resistance under fatigue loading has been widely studied on metallic materials and it is described either using stress versus number of cycles to failure curves (S-N curves) or analyzing fatigue crack propagation (FCP) following the fracture mechanics approach using crack speed versus applied stress intensity factor range curves (da/dN-ΔK curves). For metallic materials these approaches are adopted both for the characterisation of the material's behaviour and in generally accepted design procedures. S-N curves can be obtained on notched or unnotched specimens following a number of standard practices (see for example ASTM E466 and E468). For da/dN-ΔK curves, specimens containing a sharp notch (or crack) are adopted, and standards such as ASTM E647 may be used.

In the case of polymeric materials, the problem of fatigue fracture has received attention more recently than for metals. Only the S-N approach has been standardised (ASTM D671). Fracture mechanics has been, in the last decades, often applied to fatigue fracture of polymers[1,2], generally following the procedures for metals, but no standard has yet been proposed. Both these approaches have been adopted to generate data on polymeric materials but in any case, this data is mainly used to find structure to properties relationships or to compare different materials under the same testing conditions, rather than directly for design purposes. Fatigue life of plastic components is generally forecasted on the basis of tests performed directly on the components.

Based on these considerations, a new activity on the definition of a test protocol for fatigue crack propagation testing of polymers was initiated by ESIS TC4 in 1997. Aims of the project are:

- a critical analysis of the current test procedures, and the issuing of test guidelines which can accomodate as many as possible of the existing experimental techniques, but still ensure the correct evaluation, within the LEFM framework, of the fatigue fracture resistance.
- an assessment of the reproducibility of the results obtained in accordance with the given guidelines, and, therefore, of the validity of the measured data when used for material selection and for design.

Version 2 of the protocol, issued on February 2000, is enclosed as Appendix B to this paper. A summary and discussion of the present status of inter-laboratory fatigue crack propagation testing within ESIS TC4 are given in section 3.

2. ISSUES TO BE CONSIDERED WHEN APPLYING THE PROTOCOL

The first version of this protocol was largely based on the ASTM E647. On account of the stress-strain characteristics of polymers, generally exhibiting pronounced viscoelasticity and larger departures from linearity with respect to metals, restrictions were made on the crack length measurement methods: only direct video-recording was considered, excluding indirect techniques (crack length determination from specimen compliance and the use of crack gauges bonded to the specimen surface). Moreover, some modifications were introduced in the

allowed specimen geometries and notching procedures, based on the existing literature and on general experience in fracture testing of plastics.
After extensive discussion and a first inter-laboratory test run in the ESIS TC4 committee, the present, second version of the protocol was issued. Most of the above mentioned restrictions in crack length monitoring were released, recognizing the advantages of specimen compliance and crack gauge techniques in terms of test automation and measurement resolution.
TC4 activity on this topic is still in progress, aiming to test the proposed protocol on an extended range of polymeric materials and to refine the procedure in order to achieve a better reproducibility of the results.
A number of questions are still open, and will require further investigation, with reference to some distintive features of the mechanical behaviour of polymers which are likely to be not properly considered when using test procedures originally conceived for metallic materials. The main issues to be considered are:

Hysteretic heating

Energy dissipation related to viscoelasticity gives rise to heat generation which, if heat cannot be dissipated, results in a temperature increase. This may alter the material's response to fatigue crack propagation and ultimately result in the so called thermal failure in which the material fails prevalently through viscous flow. In notched specimens, used for fatigue crack propagation tests, hysteretic heating is generally confined to the vicinity of the crack tip but under some conditions it may become more generalized. In any case attention must be paid to this issue which may be reduced or suppressed by limiting the applied stress, using low frequencies and increasing the surface to volume ratio of the specimen.

Creep

The viscoelastic response to load application can produce a significant crack growth rate per cycle. This crack extension, related to the time the specimen is under load in each cycle, can be in some cases comparable or even larger than the true fatigue crack growth due to the damage induced by each loading cycle. This effect depends on the material (and therefore the relation between test temperature and the thermal transitions of the polymer), and also on the testing variables which determine the time under load, e.g. the ratio R between minimum and maximum load, frequency, wave form [2, 6].

Size effects

The basis of da/dN vs. ΔK curves is the fact that linear elastic fracture mechanics can be applied and that the stress intensity factor range controls the crack propagation rate. Although the size requirements normally applied to metallic materials (ASTM E647) are, as a first approximation, proposed in the test protocol (see Appendix B), they may not be adequate for polymers and it should be verified under which conditions ΔK is actually controlling the fracture phenomenon. Determination of the yield stress value to be used in specimen size validation, and particularly the strain rate at which it is to be measured, are also open items for discussion.

Crack initiation from a notch

Fatigue testing on metallic materials is made using cracked specimens and the initial crack is introduced by fatigue under particular conditions. This type of procedure is not fit for polymeric materials because too much damage would take place at the crack tip. Therefore, as for static fracture toughness measurements, some notching procedures are proposed in the protocol. From this starting notch crack initiation will take place and then the crack will propagate throughout the specimen[7]. Characterization of material's resistance to crack initiation under fatigue loading is not presently considered in the protocol, and can be a topic for future work3. inter-laboratory testing

3. INTER LABORATORY TESTING

A round robin test was carried out within ESIS Technical Committee 4, during years 1998 and 1999, on fatigue crack propagation testing in accordance with version 1 of the test protocol. Test material was a commercial extrusion-grade Acrylonitrile-Butadiene-Styrene copolymer (ABS), with the following characteristics:

- Young modulus about 1800 MPa (from ISO 527 tensile tests)
- Tensile yield stress about 35 MPa (from ISO 527 tensile tests)
- Acrylonitrile content in the Styrene-Acrylonitrile (SAN) matrix: 25 wt. %
- SAN matrix wt. average molecular weigth Mw = 140000.

Rectangular samples, about 300 x 200 mm, cut from a 6 mm thick extruded sheet were supplied to each of the participants.
Round robin test conditions were set as follows:

- Frequency: 1 Hz
- Waveform: sine
- R-ratio: 0.1

Specimen type(s) and dimensions, notching, load level and the technique used to measure the crack length were to be chosen by the participants within the limits and prescriptions given in the protocol.
Specimen preparation (machining of the specimens from the supplied sheets, notching, conditioning) was to be carried out by the participants.

Six laboratories participated from four European countries (France, Germany, U.K and Italy). An alphabetical list is given in Appendix A. In the following, participants are referred to as Lab 1, Lab2,...Lab 6, with a different ordering with respect to that used in Appendix A.
Table 1 shows the specimen geometry and test conditions adopted by each participant.
Lab 1 and Lab 4 made use of the specimen compliance method for measuring the crack length, together with visual techniques (travelling microscope and video recording respectively) for calibration and comparison. Crack gauge and video recording techniques were compared by Lab 6. Consistent results were found between the different methods within each laboratory.
Notching was performed by the methods listed in the protocol, with the exception of Lab 1, where fatigue pre-cracking was used to sharpen the initial notch.

A summary of results is here reported. Data have been fitted to the Paris-Erdogan equation[8]:

$$\frac{da}{dN} = A \cdot \Delta K^m \qquad 1)$$

ΔK range and the average values of coefficients A and m in eq. 1 are listed in Table 2. A plot of the results is also shown in Figure 1.

Participant	Lab 1	Lab 2	Lab 3	Lab 4		Lab 5	Lab 6
Specimen type	CT	CT	CT	CT	SENT	SENT	CT
Nr. of specimens	3	4	6	3	2	5	2
Specimen size, mm	W=32	W=52	W=56	W=50	W=40 L=120	W=35 L=100	W=28
Force range ΔP, N	83 to 96	398 to 406	180, 225, 270	360, 450, 540	1600, 2000	1863 to 1894	178
Crack length measurement	Compliance + travelling microscope	crack gauge	travelling microscope	video + compliance		video	crack gauge + video
Notching method	fatigue pre-cracking	razor tapping	razor tapping	razor tapping		razor sliding	razor tapping

Table 1 – Test conditions within each laboratory

	ΔK range MPa $m^{1/2}$	A(x10^4)	m
Lab 1	0.6 – 2.9	5.1 ± 1.6	3.86 ± 0.06
Lab 2	1.5 – 8.9	3.0 ± 1.2	4.12 ± 0.39
Lab 3	0.9 – 2.6	7.1 ± 1.3	4.16 ± 0.61
Lab 4			
CT	1.2 – 5.0	1.8 ± 0.6	4.70 ± 0.30
SENT	1.1 – 3.9	5.8 ± 1.6	4.72 ± 0.12
Lab 5	1.1 – 4.4	5.7 ± 3.5	4.86 ± 0.86
Lab 6	1.2 – 1.9	6.3 ± 1.2	4.81 ± 2.11

Table 2 – Round robin test results: range of applied ΔK and average values of Paris law parameters A and m

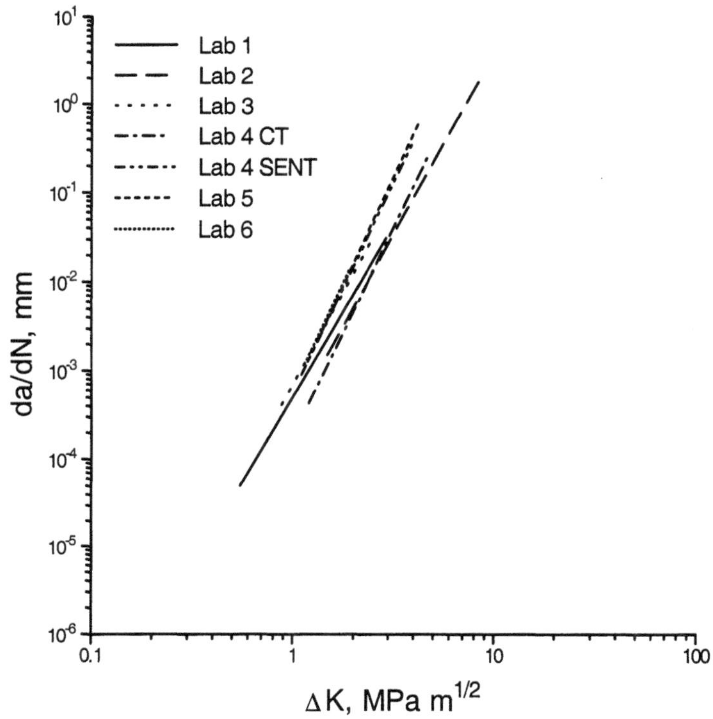

Figure 1 – Round Robin test results: double logaritmic plot of data listed in Table 2.

The following considerations can be drawn from the results:

- Variability of results from different laboratories is not significantly higher than the scattering of data obtained within each single laboratory. The proposed test protocol appears therefore to include all the essential details which are needed to prevent the occurrence of large systematic inter-laboratory deviations. On the other hand, the existing data scattering shows that determination of crack propagation rate at a given ΔK is presently affected by a considerable uncertainity, which has to be reduced by a critical analysis and refinement of the test procedure.

- Results for parameter A in the Paris equation best fits to the data exhibit a greater variability than those for parameter m. Lack of reproducibility appears therefore to be mainly due to errors in evaluating the magnitude of ΔK and da/dN. A careful re-examination of the experimental uncertainities involved in these evaluations and of possible additional requirements to be included in a new version of the protocol will be considered.
- Large differences are found in Lab 4 between results obtained with CT and SENT specimen geometries. Gripping of SENT specimens is a possible source of experimental errors because of the no-rotation requirement imposed by the protocol: even small rotations can strongly impair the correctness of the K-calibration equation given for this geometry.

4. CONCLUSIONS

A test protocol for determination of fatigue crack propagation (FCP) resistance of polymeric materials has been developed and subjected to a first verification through an inter-laboratory testing programme carried out in ESIS TC4.
Most of the commonly adopted experimental techniques for FCP testing of plastics have been included in the present version of the protocol. Different crack length measurement methods, in particular, were tested and found to provide acceptably consistent results.
Variability of the experimental results measured by different laboratories was found to be not significantily greater than the scattering of data obtained within each single laboratory.
A number of open issues, mainly involving the distinctive features of the mechanical behaviour of polymers as compared to metals, have been identified and are to be considered in order to improve the reproducibility of data obtained with the proposed protocol. During next stages of ESIS activity on fatigue fracture of polymers, particular attention will be given to:

- size criteria used to ensure the correctness of the LEFM approach used in this test procedure.
- a possible revision of the precision requirements concerning the measurement of the basic test variables (force, frequency, crack length).
- the static (creep) contribution to crack growth, strictly related to the "time under load" variable and therefore to waveform, maximum force, R-ratio and frequency.
- evaluation of the hysteretic heating occurring during the test.
- specimen clamping and stiffness of the calmping system and of the testing machine, particularly with reference to the conditions for validity of the stress intensity calibration equations.

5. ACKNOWLEDGEMENTS

The authors wish to thank Dr. B. Blackman, Dr. R. Moore, Prof. A. Pavan Dr. F. Ramsteiner, Prof. H. Sautereau and Prof. J.G. Williams for contributing to inter-laboratory testing. All the members of ESIS TC4 are gratefully acknowledged for the helpful and stimulating discussions during the development of the protocol.

6. REFERENCES

1- R.W. Herzberg and J.A. Manson, "Fatigue of Engineering Plastics", Academic Press, 1980.
2- J.G Williams, "Fracture Mechanics of Polymers", Ellis Horwood Ltd, Chichester, 1984
3- ASTM E 647-95 "Standard Test Methods for Measurement of Fatigue Crack Growth Rates"
4- D.R. Rooke and D.J. Cartwright, "Compendium of Stress Intensity Factors", Hillingdon Press, Uxbridge, 1976
5- A. Saxena and S.J. Hudak, Jr., Int. J. of Fracture, vol. 14, n. 5, pp.453-468, 1978
6- M.G. Wyzgoski, G.E. Novak, D.L. Simon, J. Mater. Sci. vol. 25, pp. 4501-10, 1990
7- M. Rink, F. Briatico-Vangosa, L.Castellani, 2nd ESIS TC4 Conf. , 13-15 Sept. 1999, Les Diablerets, Switzerland.
8- P.C. Paris, F. Erdogan, J. Basic Eng., vol. 85, pp 528-34, 1960

APPENDIX A – List of participants to inter-laboratory testing

- BASF – Abt. Polymerphysik, Festkorperphysik – Ludwigshafen, Germany - Dr. F. Ramsteiner.
- Enichem – Research Centre – Mantova, Italy – Dr. L. Castellani.
- ICI Plc – Wilton, U.K. – Dr. D.R. Moore.
- Imperial College – Dept. of Mechanical Engineering – London, U.K. – Prof. J.G. Williams, Dr. B. Blackman.
- INSA Lyon – Lab. des Materiaux Macromoleculaires – Villeurbanne, France – Prof. H. Sautereau.
- Politecnico di Milano– Milano, Italy – Prof. M. Rink, Prof. A. Pavan.

APPENDIX B – Test Protocol

ESIS TC4 - Technical Committee on Polymers and Composites

A PROTOCOL FOR TENSION-TENSION FATIGUE CRACK PROPAGATION TESTING OF PLASTICS

Version 2 – February 2000

Contents:

1 SCOPE

This test protocol specifies a method for measuring the propagation of a crack in a notched specimen subjected to a cyclic tensile load varying between constant minimum and maximum positive values. Test results include crack length as a function of the number of load cycles and the crack length increase rate as a function of stress intensity factor and energy release rate at the crack tip. The possible occurrence of discontinuities in crack propagation is detected and reported.
The method is suitable for use with the following range of materials:
rigid and semi-rigid thermoplastic moulding and extrusion materials, including filled and reinforced compounds in addition to unfilled types; rigid and semi-rigid thermoplastic sheets.
rigid and semi-rigid thermosetting materials, including filled and reinforced compounds; rigid and semi-rigid thermosetting sheets.

2 TERMS AND DEFINITIONS

For the purposes of this test protocol, the following terms and definitions apply.

Cycle

The smallest segment of the force- or stress-time function which is repeated periodically. The terms fatigue cycle, load cycle and stress cycle are also commonly used.

Number of elapsed cycles N

The number of load cycles since the beginning of the test.

Waveform

The shape of the force-time function within a single cycle. Sine function is commonly used, but any different waveform, e.g. triangular, square, etc. , may be employed

Maximum force P_{max}

The force having the largest value in the cycle. It is expressed in newtons, N. Only positive, i.e. tensile, forces are used in this test method.

Minimum force P_{min}

The force having the minimum value in the cycle. It is expressed in newtons, N.

Range of force ΔP

The difference between the maximum and the minimum forces in one cycle, expressed as:

$$\Delta P = P_{max} - P_{min}$$

Force ratio (also called stress ratio) R

The ratio of the minimum to the maximum forces in one cycle, that is:

$$R = P_{min} / P_{max}$$

Stress intensity factor K

Limiting value of the product of the stress $\sigma(r)$ perpendicular to the crack area at a distance r from the crack tip multiplied by the square root of $2\pi r$, if r becomes small:

$$K = \lim_{r \to 0} \sigma(r) \sqrt{2\pi r}$$

It is expressed in pascals root metre, Pa $m^{1/2}$.

Maximum stress intensity factor K_{max}

The highest value of the stress intensity factor in one cycle.

Minimum stress intensity factor K_{min}

The lowest value of the stress intensity factor in one cycle

Range of stress intensity factor ΔK

The difference between the maximum and minimum stress intensity factors in one cycle, expressed as:

$$\Delta K = K_{max} - K_{min}$$

Energy Release Rate G

The change in the external work δU_{ext} and strain energy δU_S of a deformed body due to enlargement of the cracked area δA

$$G = \frac{\delta U_{ext}}{\delta A} - \frac{\delta U_S}{\delta A}$$

it is expressed in joules per square meter, J/m^2.

Under linear elastic assumptions, the following relationship between stress intensity factor K and energy release rate G holds:

$$G = \frac{K^2}{E'}$$

where

$E' = E$ for plane stress, and $E' = \dfrac{E}{1-\nu^2}$ for plane strain conditions.

E and ν are the tensile modulus the Poisson ratio respectively.

Maximum energy release rate G_{max}

The highest value of the energy release rate in one cycle.

Minimum energy release rate G_{min}

The lowest value of the energy release rate in one cycle

Range of energy release rate ΔG

The difference between the maximum and minimum energy release rate in one cycle, expressed as:

$$\Delta G = G_{max} - G_{min}$$

Notch

A sharp indentation, generally made by a razor blade or a similar sharp tool, performed on the specimen before the test and intended to be the starting defect where the fatigue-induced crack initiates.

Initial crack length a_0

The length of the notch. For CT specimens, it is measured from the line connecting the load bearing points (i.e. the line through the centers of the holes) to the notch tip. For SENT specimens, it is measured from the specimen edge to the notch tip. It is expressed in metres, m.

Crack length a

The total crack length at any time during the test, given by the initial crack length a_0 plus the crack length increment due to fatigue loading. It is expressed in metres, m.

Fatigue crack growth rate da/dN

The rate of crack extension caused by fatigue loading and expressed in terms of average crack extension per cycle.

Stress intensity calibration

A mathematical expression, based on empirical or analytical results, that relates the stress intensity factor to force and crack length for a specific specimen planar geometry.

Gage length L_0

For the SENT specimen, the free distance between the upper and lower grips after the specimen is mounted in the testing machine. It is expressed in metres, m.

3 PRINCIPLE

A constant-amplitude tensile force wave imposed to the specimen, with a suitable choice of the test conditions (specimen shape and size, notching, maximum and minimum forces, force wave frequency, etc.), causes the propagation of a crack from the initial notch.
Crack length a during the test is monitored and recorded as a function of the number of elapsed loading cycles N.
Numerical differentiation of the experimental a(N) function provides the fatigue crack growth rate da/dN which is reported as a function of stress intensity factor and energy release rate at the crack tip.

4 SIGNIFICANCE AND USE

Fatigue crack propagation, particularly when expressed as the fatigue crack growth rate da/dN as a function of crack-tip stress intensity factor range ΔK or energy release rate range ΔG, characterizes a material's resistance to stable crack extension under cyclic loading. Background information on the fatigue behaviour of plastics and on the fracture mechanics approach to fatigue for these materials is given in Refs. [1] and [2].

Expressing da/dN as a function of ΔK or ΔG provides results that are independent of planar geometry, thus enabling exchange and comparison of data obtained from a variety of specimen configurations and loading conditions. Moreover, this feature enables da/dN versus ΔK or ΔG data to be utilized in the design and evaluation of engineering structures. The concept of similitude is assumed, which implies that cracks of differing lengths subjected to the same nominal ΔK or ΔG will advance by equal increments of crack extension per cycle.

Fatigue crack propagation data are not geometry independent in the strict sense since thickness effects generally occur. The potential effects of specimen thickness have to be considered when generating data for research or design.

Anisotropy of the molecular orientation or of the orientation of second phase domains, and the presence of residual stresses, can have an influence on fatigue crack propagation behaviour. The effect can be significant when test specimens are removed from semi-finished (e.g. extruded sheets) or finished products. Irregular crack propagation, namely excessive crack front curvature or out-of-plane crack growth, generally indicates that anisotropy or residual stresses are affecting the test results.

This test method can serve the following purposes:

To establish the influence of fatigue crack propagation on the life of components subjected to cyclic loading, provided data are generated under representative conditions and combined with appropriate fracture toughness data and stress analysis information.

To establish material selection criteria and inspection requirements for damage tolerant applications.

To establish, in quantitative terms, the individual and combined effects of material's structure, processing conditions and loading variables on fatigue crack propagation.

5 TEST SPECIMENS

5.1 Shape and size

5.1.1 Specimen geometry.

Two test specimens can be used: single edge notched tension (SENT) and compact tension (CT). Figures 1 and 2 describe their geometrical characteristics.

5.1.2 Thickness and Width

When thickness B is too small compared to the specimen width W it is difficult to avoid lateral deflections or out-of-plane bending of the specimen; with very thick specimens, conversely, through-thickness crack curvature corrections are often necessary and difficulties may be encountered in meeting the through-thickness straightness requirement of 7.1.

On the basis of these considerations the following limits for B and W are recommended:

a- For CT specimens $W/10 \leq B \leq W/2$.

b- For SENT specimens $W/20 \leq B \leq W/4$.

It has to be noticed that the test results are in general thickness dependent: specimens obtained from the same material but having different thickness are likely to give different responses (see section 4).
It is usually convenient to make the thickness B of the test specimens equal to the thickness of the sheet sample from which the specimen is cut.

5.1.3 Size requirements

In order for the results to be valid according to this test method it is required that the material behaviour be predominantly linear elastic at all values of applied force and crack length. Deviations may arise from either viscoelastic behaviour of the material or large scale plasticity ahead of the crack tip. The former may result in significant non-linearity of the mechanical behaviour, possibly enhanced by a progressive rise of the specimen temperature during the test. The testing procedure outlined in this protocol is therefore recommended only for materials exhibiting very limited viscoelasticity under the loading frequency adopted and for the expected test duration. Large scale plasticity of the ligament can be avoided by ensuring that the plastic zone around the crack tip be small compared with the uncracked ligament size (W-a). On the basis of previous experience on metallic materials [3], it is required that the following size limits be satisfied in order for the test results to be valid according to this method:

$$(W - a) \geq (4/\pi)(K_{max}/\sigma_y)^2 \tag{1}$$

where (W-a) is the uncracked specimen ligament and σ_y is the tensile yield stress.

The same size limits are expressed in graphical form in fig. 3, where the dimensionless quantities $K_{max}/(\sigma_y\sqrt{W})$ and a/W are plotted versus each other: all combinations of specimen size, crack length, material yield stress and stress intensty factor which fall below the curve in fig. 3 satisfy the specimen size requirements of this test method.

5.2 Notching

A sharp notch or, when feasible, a natural crack, intended to be the defect from which the fatigue-induced crack initiates and propagates, is introduced into the specimen in the locations depicted in figures 1 and 2, either in a single step or by sharpening the tip of a blunt slot or notch previously obtained by machining.
It is required that the initial crack length a_0 in the CT specimen be at least 0.2 W in length so that the K-calibration is not influenced by small variations in the location and dimensions of the loading-pin holes. Notch length in CT specimens shall be chosen accordingly.
The notch in both the CT and SENT specimens shall be centered with respect to the specimen centerline to within $\pm$ 0.01 W.
When sharpening a previous blunt notch, the length of the sharp notch shall be larger than four times the machined blunt notch tip radius. Methods a, b, c and d can be used to create a natural crack or a sharp notch (see also the ESIS-TC4 protocol for determining K_C and G_C):

a- Machine a sharp notch into the test specimen and then generate a natural crack by tapping on a new razor blade placed in the notch (it is essential to practice this since, in brittle test specimens, a natural crack can be generated by this process but some skill is required in avoiding too long a crack or local damage).

b- For some brittle test specimens, if difficult control or repeatibility of the crack performed with method a) is experienced, it can be in some case advantageous to generate a sharp notch by pressing a new razor blade at a temperature close to, but lower than, the glass transition temperature of the sample. Any specimen deformation or damage during the application of this notching procedure must be avoided by proper handling of the specimen and correct choice of the temperature. Use a new razor blade for each test specimen.

c- If a natural crack cannot be generated, as in tough test specimens, then the notch shall be sharpened by sliding a razor blade across the notch. Use a new razor blade for each test specimen.

d- With tough materials, cooling the test specimens and then performing razor tapping is sometimes successful.

It may be useful to check the effectiveness of the notching procedure by performing ramp tests at constant displacement or loading rate on specimens notched with different methods. The best notching should give the lower K value at crack initiation.

5.3 Side grooving

Specimens may need side grooves to avoid deviations of the crack path from the plane of symmetry (see 7.4), and to promote straighter crack fronts during testing. Side grooves may also, in some cases, improve the visibility of the crack tip when using visual methods for crack length measurement.
The side grooves must be equal in depth and have an included angle of $0.45 \pm 5°$ with a root radius of 0.25 ± 0.05 mm.
The total thickness reduction due to side grooving must not exceed 0.2 B.
When using side grooves, the specimen thickness B shall be measured as the distance between the roots of the side grooves.

6 TESTING EQUIPMENT

6.1 Testing machine

The machine shall be able to impose a prescribed force to the specimen (i.e. to operate in "load control" mode), and to vary the force with time according to a controlled waveform. The force distribution has to be symmetrical to the specimen notch. Hydraulically driven testing machines with electronic control are generally suited to this purpose. Mechanically driven machines can be used taking into account their lower versatility in terms of types of waveform and frequency range.

6.1.1 Waveform.

The most commonly employed force waveform is a sine wave, but other types, e.g. triangular or square waves, may be used when simulating service conditions or investigating the effects of waveform itself. Two important test variables, namely maximum force, P_{max}, and force ratio, R, characterize the force waveform and significantly affect test results. Force as a function of time shall be controlled with an accuracy of ± 1%, and the maximum and minimum force values shall be constant, during the entire test, within 1%.

6.1.2 Frequency.

The frequency of the force wave is a test variable that may be adjusted according to different criteria, such as the simulation of service conditions or the investigation of the effects of the frequency on test results. High frequency values (> 5 Hz) are likely to induce hysteretic heating: this must be taken into account when evaluating the test results. The frequency of the force wave must be determined, before the test, with an accuracy of 1%.

6.1.3 Cycle counter

The testing machine shall be equipped with a cycle counter displaying at any time during the test the number of elapsed loading cycles.

6.2 Grips

Conventional grips for tensile testing are suitable for use with the SENT test specimen, provided they can accomodate the fatigue specimens which are usually larger than the standard tensile test specimens.
The compact tension specimens (CT) are loaded by two pins in the holes. Pin diameter shall be 0.230W ± 0.005W, where W is the specimen width (see fig. 2). Pins must be free to rotate in the specimen holes during the test.
Careful alignment of the gripping fixtures and of the whole loading train must be ensured to avoid out-of-plane displacements of the specimen.

6.3 Crack length measurement

Fatigue crack length measurement shall be made as a function of elapsed cycles with a resolution of at least 0.1 mm or 0.002W, whichever is greater.
Crack length data readings shall be taken at fixed crack length increments Δa. Minimum increment Δa_{min} must be greater than 0.5 mm or five times the crack length measurement resolution, whichever is greater. At least 20 Δa measurements shall be made between the initial crack length a_0 and the final crack length at the end of the test a_f so that the maximum increment value will be $\Delta a_{max} \leq (a_f - a_0)/20$. If the above requirements cannot be satisfied ($\Delta a_{max} < \Delta a_{min}$) the specimen dimensions are not suitable for subsequent testing and larger specimens have to be employed.
In correspondence to each Δa the number N of cycles elapsed since the beginning of test shall be recorded.
Fatigue crack length measurements shall be made without interrupting the test.
Crack length shall be measured during the test by means of at least one of the following techniques.

6.3.1 Travelling microscope

A low power (approximately 15 to 30x) travelling microscope can be used for fatigue crack length measurement. Crack length and corresponding elapsed cycles number readings shall be recorded in accordance with 6.3.
It is recommended that, prior to testing, reference marks be applied to the test specimen surface at precisely determined locations along the direction of cracking. Using reference marks eliminates potential errors due to accidental movement of the travelling microscope.
If the specimen surface is marked, along the expected crack path, with precision grids or scales complying with the resolution requirements given in 6.3, crack length can then be determined directly with any magnifying device having suitable resolving power.
Marks applied on the specimen surface must not affect crack initiation or propagation.

6.3.2 Video recording

Crack length during the test can be automatically monitored by means of a video camera, equipped with a low magnification lens (approximately 15 to 30x) and connected to a video recording device.
The video recording device shall be synchronized with the cycle counter of the testing machine (see 6.1.3) in order that the number of elapsed loading cycles corresponding to each video-recorded image can be determined.
When using the video recording technique, accurate calibration of the length readings on the recorded images shall be performed, before the test, in order to ensure that resolution requirements of 6.3 are fulfilled.
Alternatively, the specimen surface, along the expected crack path, shall be marked with grids or scales allowing the direct reading, on the video recorded images, of the crack length with the required resolution (see 6.3) .
Marks applied on the specimen surface must not affect crack initiation or propagation.

6.3.3 Specimen compliance

When using the CT specimen, crack length during fatigue crack propagation testing can be measured by monitoring the specimen compliance.
Specimen compliance is determined as the slope of the linear relationship between the crack mouth opening displacement, V, and the applied force, P, within a loading cycle. This can be simply accomplished by monitoring the peak displacement, on account of the fact that peak force is constant during the test. Such a procedure, however, may lead to an incorrect estimate of the specimen compliance due to non-linearities in the force-displacement curve. A more accurate evaluation is obtained when recording force and displacement signals within a single loading cycle with sufficient detail to recognize possible non-linear portions of the curve and to exclude them from a linear fit. If using this procedure, it is recommended that either the loading or the unloading portion of the fatigue cycle are consistently used for calculations throughout the test.
After maesuring V and P, the normalized compliance C_N is obtained by the following expression:

$$C_N = \frac{BEV}{P} \qquad 2)$$

where B is the specimen thickness and E is the tensile modulus. The elastic modulus measured on plastic specimens can be affected by processing-induced orientation; it is therefore recommended that tensile test specimens for modulus determination are as similar as possible, with respect to processing conditions and orientation, to the fatigue test specimens. Usually fatigue specimens are machined out of sheets or flat moulded parts: tensile specimens can then be machined from the same piece, taking care that the orientation is the same (length of the tensile specimen has to be parallel to the line joining the two holes in the CT specimen).
Four different locations are considered for measuring the crack mouth opening displacement on the CT specimen. Their position are defined in figure 4.
Selection of displacement measurement gauges, attachment points and methods of attachment are dependent on the test conditions and on the material to be tested. Gauges must be linear over the displacement range to be measured, and must have sufficient resolution and frequency response. Attachment points must be accurately and repetitively placed on the specimen, and must not be susceptible of wearing during the fatigue cycling.
Polynomial expressions describing the normalized crack length a/W as a function of the normalized compliance of the CT specimen, measured at the above defined locations, have been estabilished for metallic materials [3,5] and have been proved to be valid for polymeric materials as well. They are given by:

$$a / W = C_0 + C_1 U_x + C_2 (U_x)^2 + C_3 (U_x)^3 + C_4 (U_x)^4 + C_5 (U_x)^5 \qquad 3)$$

where

$$U_x = \frac{1}{\left(\frac{BEV_x}{P}\right)^{1/2} + 1} \qquad 4)$$

and the coefficients C_0, C_1, ..., C_5 assume the following values in correspondence to the four measurement locations:

Measurement location	C_0	C_1	C_2	C_3	C_4	C_5
V_{X1}	1.0012	-4.9165	23.057	-323.91	1798.3	-3513.2
V_0	1.0010	-4.6695	18.460	-236.82	1214.9	-2143.6
V_1	1.0008	-4.4473	15.400	-180.55	870.92	-1411.3
V_{LL}	1.0002	-4.0632	11.242	-106.04	464.33	-650.68

The number of compliance measurements perfomed during the test and their spacing must ensure that the crack length measurement requirements given in 7.4 are satisfied. The compliance method for crack length measurement lends itself to automatic data acquisition and a large number of readings is commonly obtained.
At least two visual crack length readings shall be taken, at crack tip positions at least 0.2W distant from each other, during the test. The visual readings must be adjusted for curvature to obtain the physical crack lengths using the procedure outlined in 8.2. Any difference between

the physical and compliance crack lengths must be used to adjust all the compliance crack lengths. This is accomplished by calculating an effective modulus of elasticity, E*, and using this in the compliance equation (eq. 4) to adjust all crack length calculations. If the effective modulus E* differs from the tensile modulus E by more than 20%, then the test equipment is improperly set-up and data generated from such records are invalid by this method.
SENT specimens are at present not recommended for use with the specimen compliance method.

6.3.4 Crack gauges

Crack gauges for crack growth measurement are commecially available and commonly used in fatigue testing. They generally consist of electrically conductive thin foils, bonded to the specimen surface over the expected crack path, which are progressively cut into two parts as the crack propagates. The electrical resistivity measured across the crack path changes form a minimum value corresponding to the uncracked foil to increasingly greater values as the crack grows. The electrical resistance can thus be used as an indirect measure of the crack length.
The adhesive used to bond the crack gauge to the specimen surface has to ensure that the crack length on the crack gauge is exactly equal to the crack length on the specimen surface. The adhesive must not affect the fatigue response of the specimen itself.
Calibration of the crack gauges shall be perfomed by means of at least two visual crack length readings taken at crack tip positions at least 0.2W distant from each other.

6.4 Test atmosphere

As the test duration may be large, particular attention must be given to the constancy of the various parameters characterizing the test atmosphere (temperature, humidity, etc.)

7 TESTING PROCEDURE

7.1 Specimen dimensions

Before the test, the specimen thickness B and width W and the initial crack length a_0 shall be measured with an accuracy of 0.05 mm. Specimen dimensions shall be within the tolerances given in figs. 1 and 2. The initial crack length a_0 shall be measured on the front and back surfaces of the specimen: if the two readings differ by more than 0.25 B, the notching operation is not suitable and subsequent testing would be invalid under this test method. If the notch departs more than the allowable limit from the plane of symmetry (see 7.4) the specimen is not suitable for subsequent testing.

7.2 Specimen mounting

Loading pins shall be inserted into the CT specimen holes taking care that load line is parallel to the specimen edge (b1 in fig. 2) and that the pins are free to rotate in the holes.
SENT specimens shall be fixed so that the distances between the notch plane and the upper and lower grips shall be equal within ± 0.02 W; the gage length L_0 (that is the free distance between the grips) shall be greater than 2W.

7.3 Loading

Loading of the test specimen has to be performed in a relatively short time (that is, short with respect to the duration of the test) to avoid creep effects before cyclic loading. A loading time shorter than 1 minute is usually feasible and adequate. During this stage the applied force must be kept lower than the maximum force used during the test, to avoid retardation effects on crack propagation.

7.4 Out-of plane crack propagation

If at any point in the test the crack deviates more than ± 20° from the plane of symmetry over a distance of 0.1 W or greater, the data are invalid according to this test method.

7.5 Discontinuous crack propagation

When irregularities in the crack propagation are observed, crack length readings will be taken so as to describe the irregularities as accurately as possible. Polymeric materials subjected to fatigue frequently exhibit discontinuous crack propagation: the crack is observed to occasionally stop and then resume the propagation, sometimes with a sudden acceleration, after several cycles. In that case data readings will be taken as close as possible to crack arrest and re-start, in order that the discontinuity will be clearly apparent in a crack length (a) vs. elapsed cycles (N) plot.

7.6 Number of tests

It is a good practice to conduct replicate tests. Multiple tests can be planned such that regions of overlapping da/dN versus ΔK or ΔG are obtained.

8 CALCULATION AND INTERPRETATION OF RESULTS

8.1 Crack length vs. number of cycles

Recorded crack length increments added to the initial crack length a_0 will provide the crack length values which will be plotted against the corresponding values of cycle number N. In the case of discontinuous crack propagation, crack length readings have to be taken according to 7.5.

8.2 Crack curvature correction

Through-thickness curvature of the crack front may occur during crack propagation. Crack measurements carried out according to methods described in 6.3.1, 6.3.2 and 6.3.4 are taken on the specimen surface, and a correction may be needed to account for crack curvature. When using the specimen compliance method for crack length measurement (6.3.3), correction for crack curvature is incorporated in the calibration of the technique: visual readings used for calibration, however, are taken on the specimen surface and may need to be corrected for crack curvature.

After completion of testing, examine the fracture surfaces, preferably at two locations to determine the extent of through-thickness crack curvature. If a crack contour is visible, calculate the average through-thickness crack length as the average of the measurements

obtained at the surfaces and at the center of the specimen. Then calculate the difference, δ, between the average through-thickness crack length and the corresponding crack length measured during the test. Crack curvature correction is performed by adding δ to the crack length values measured during the test.
If the crack curvature correction results in a greater than 5% difference in calculated stress-intensity factor at any crack length, then employ this correction when analyzing the recorded test data. When the magnitude of the crack curvature correction either increases or decreases with crack length, use a linear interpolation to correct intermediate data points.

8.3 Crack growth rate da/dN

The rate of fatigue crack growth is to be determined from the crack length versus elapsed cycle data (see 9.2) by numerical differentiation. A simple secant procedure, based on the calculation of the slope of the straight line connecting two adjacent data points, is generally adequate. According to this procedure the crack growth rate at any average crack length $\bar{a} = (a_i + a_{i+1})/2$ is given by:

$$(da/dN)_{\bar{a}} = \frac{(a_{i+1} - a_i)}{(N_{i+1} - N_i)} \qquad 5)$$

$\bar{a}$,which is the average crack length value within the $(a_{i+1} - a_i)$ increment, is used to calculate ΔK by means of the equations 6) or 7) (see 8.4).
If discontinuous crack propagation is observed, the crack growth rate shall only be calculated within the continuous regions of the a(N) curve.

8.4 Stress intensity factor range ΔK

Use the average crack length values $\bar{a}$ of 8.3 to calculate the corresponding stress intensity factor range values according to the following stress intensity calibration expressions:
For the CT specimen ΔK is given by:

$$\Delta K = \frac{\Delta P}{B\sqrt{W}} \frac{(2+\alpha)}{(1-\alpha)^{3/2}} (0.886 + 4.64\,\alpha - 13.32\,\alpha^2 + 14.72\,\alpha^3 - 5.60\,\alpha^4) \qquad 6)$$

where $\alpha = a/W$; the expression is valid for $a/W \geq 0.2$.

For the SENT specimen ΔK will be calculated by [4]:

$$\Delta K = \frac{\Delta P}{B\sqrt{W}} \frac{5\sqrt{\pi\alpha}}{(20 - 13\,\alpha - 7\,\alpha^2)^{1/2}} \qquad 7)$$

where $\alpha = a/W$.

8.5 Energy release rate range ΔG

Energy release rate range ΔG is calculated from the stress intensity factor range ΔK by means of the following equation:

$$\Delta G = \frac{(\Delta K)^2}{E}\frac{1+R}{1-R} \qquad 8)$$

where E is the tensile modulus.

On account of the experimental uncertainities involved in the determination of ΔK, R and E, the difference between plane stress and plane strain expressions for G (see section 2) is neglected for the present calculation.

9 TEST REPORT

The test report shall include the following information:

1. Specimen type and dimensions, including thickness B and width W. For SENT specimens, the gage length L_0 .
2. The yield stress value used to determine specimen size according to 5.1.3 (eq. 1).
3. The method used to create the notch, and the value of the initial crack length a_0.
4. Description of the testing machine and of the grips and fixtures used.
5. Description of the method used to measure the crack length, including the measurement precision.
6. Test loading variables, including ΔP, R, frequency of the force cycle and waveform.
7. Maximum and minimum temperature and humidity during the test.
8. The occurrence of crack curvature, the procedure used to correct it and the magnitude of the correction.
9. The occurence of discontinuous crack propagation.
10. A plot of a versus N.
11. A plot of log (da/dN) versus log (ΔK).
12. A plot of log (da/dN) versus log (ΔG).
13. A table of the test results, including a, N, ΔK, ΔG and da/dN.

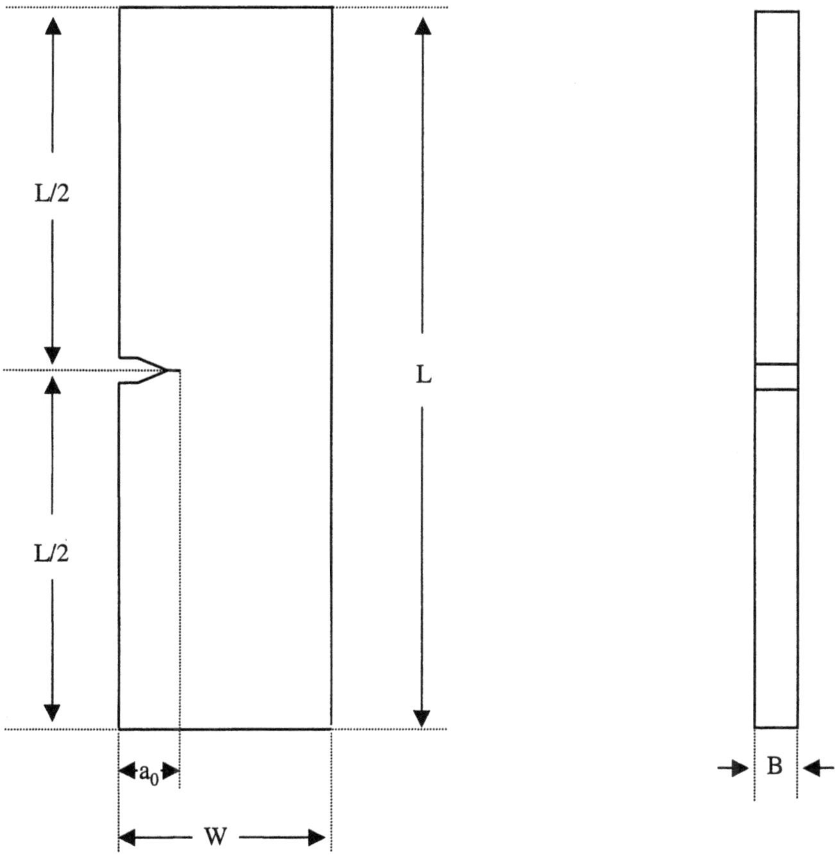

W	Width	W/20 ≤ d ≤ W/4
L	Length	L > 2.5 W
B	Thickness	
a_0	Initial crack length	

The machined notch shall be centered with respect to the specimen centerline to within ± 0.01 W

Figure 1 — Single Edge Notched Tension (SENT) specimen for fatigue crack propagation testing.

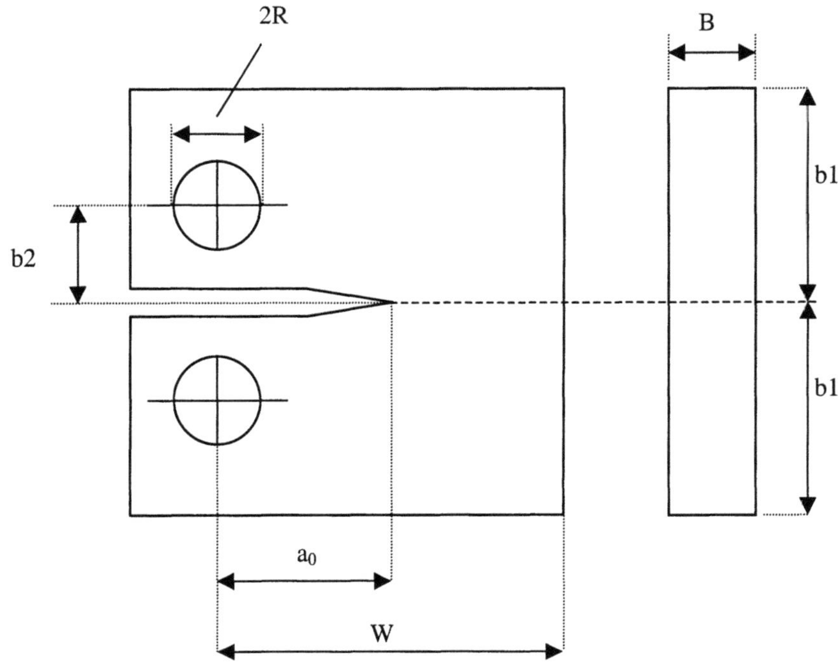

W	Width	$W/10 \leq B \leq W/2$
L	Overall length	$1.25\ W \pm 0.01\ W$
b1	Half breadth	$0.6\ W \pm 0.005W$
b2	Distance between the centres of the two holes and the crack plane	$0.275\ W \pm 0.002\ W$
R	Radius	$0.125\ W \pm 0.005\ W$
B	Thickness	
a_0	Initial crack length	$a_0 > 0.2W$

The machined notch shall be centered with respect to the specimen centerline to within ± 0.01 W

Figure 2 — Compact Tension (CT) specimen for fatigue crack propagation testing.

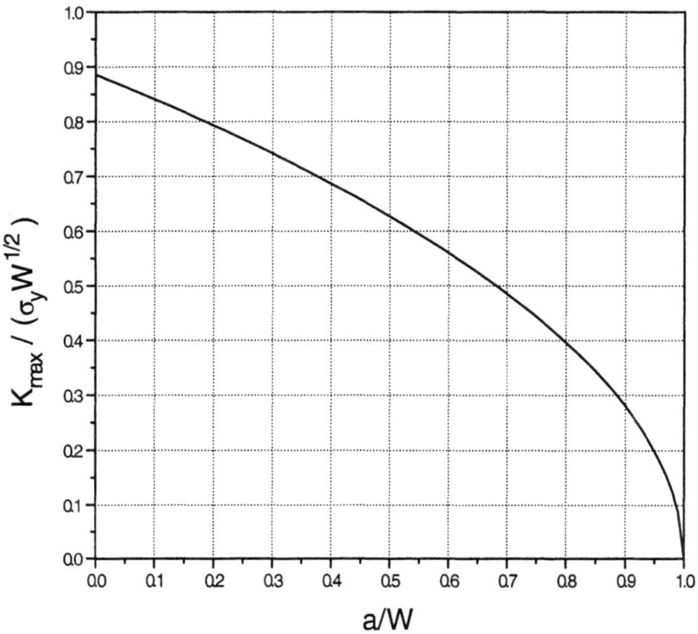

Figure 3 — **Normalized size requirements for fatigue crack propagation specimens. Values which fall below the curve satisfy the specimen size requirements of this method.**

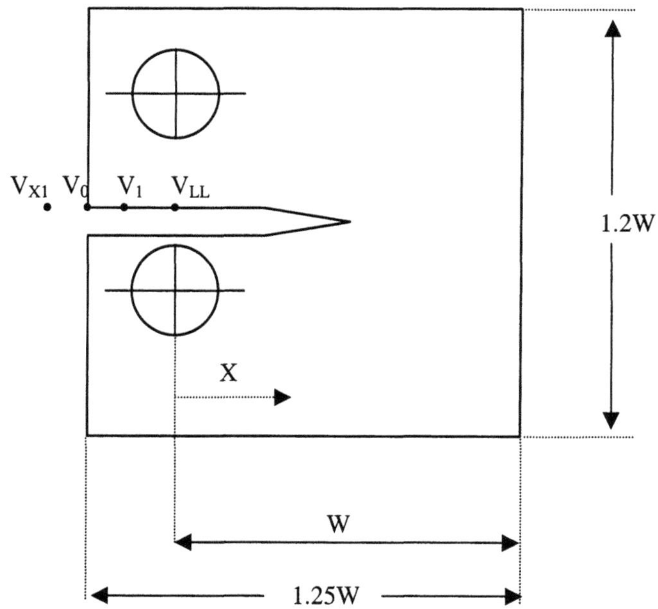

V_{LL}	Load line	X / W = 0
V_1	Location "1"	X / W = - 0.1576
V_0	Front face of the specimen	X / W = - 0.25
V_{X1}	Location "X1"	X / W = - 0.345

Figure 4 — Locations for measurement of crack mouth opening displacement in CT specimens.

CHAPTER 2

Elastic-Plastic Fracture Mechanics

INTRODUCTION TO ELASTIC-PLASTIC FRACTURE MECHANICS

J.G. WILLIAMS

As discussed previously many fracture tests and practical situations of interest for polymers and composites can be analysed using LEFM. Some cases, however, involve more extensive plastic deformation and it is necessary to extend the elastic analysis to incorporate additional dissipated energy. All the materials to be discussed here are the tougher polymers and particularly polyethylenes and rubber modified systems. These have been deliberately designed to have rather low yield stresses, so that any fracture is accompanied by extensive plastic deformation giving high energy dissipation and thus high toughness.

The most common parameter used is J_c which is an energy per unit area as is G_c. Indeed the definition of J_c is actually the energy release rate for a non-linear elastic solid. It is used, however, for materials where the load—displacement relationship is non-linear, but usually this is brought about by plastic, and not elastic, deformation. Generally the specimens used are fully plastic and stable crack growth is generated. J_c is calculated from the energy dissipated using a form equivalent to the second of equation (7) and equation (14).

$$J_c = \eta \frac{U_t}{B(W-a)} \tag{1}$$

where U_t is now the total energy and η is a calibration factor. The J_c tests described here used deep notched $(a/W \approx 0.5)$ three point bend specimens for which $\eta = 2$. A set of nominally identical specimens is loaded to give different amounts of stable crack growth, Δa. The energy is measured and J_c determined to give an 'R' curve in the form of J_c versus Δa.

The deep notched bend configuration was originally used for metals because it gave a high constraint and the initiation value of $J_c(\Delta a = 0)$ was believed to be that which would be obtained for G_c from a valid, LEFM, specimen. This was a very important notion when characterising very tough materials for which the LEFM size requirements were large and difficult to meet practically. The validity criteria for thickness for J_c is,

$$B > 25 \frac{J_c}{\sigma_y} \tag{2}$$

which is about a factor of five less than equation (3). It was also noted in these highly ductile crack situations that crack tip blunting occurred and that a correction for this was necessary. This blunting was assumed to be semi-circular, so that the crack growth due to blunting was,

$$\Delta a_b = \frac{\delta}{2} = \frac{J_c}{2\sigma_y}$$

ie $$J_c = 2\sigma_y \ \Delta a_b \tag{3}$$

is the blunting line. The true initiation value was thus determined when this line intercepted the crack growth curve. The crack growth curve was usually assumed to be linear, but of lower slope than the blunting line, so two lines could be drawn and the intercept gave the initiation value. All these processes are shown diagrammatically in fig 1.

This scheme seems to work well for moderately tough materials where there is a clear difference in the two slopes. In the limit for very brittle fractures the R curve has a zero slope leading to unstable initiation which is easy to define. For very tough polymers, such as pipe grade polyethylenes, the R curve is very steep and it is difficult to distinguish crack growth from blunting and a definition of initiation becomes problematic. The solution adopted for both polymers and metals is to abandon the notion of blunting and a true initiation value and to characterise the material with a power law R curve;

$$J_c = A(\Delta a)^N \tag{4}$$

and an "initiation" value when $\Delta a_o = 0.2mm$. This is an arbitrary value which is small and close to the lower limit of detectable values of Δa and is thus a sensible practical definition of initiation, since it is more easily defined than the intercept of two sloping lines. The protocol given here uses this scheme and gives an R curve plus the J_c at 0.2mm as characterising parameters. It is now generally accepted that such R curves are size dependent and do not represent fundamental material properties. They are, however, useful for comparing materials when the size constraints in the protocol are used. The usual concerns over the initiation value pertain, but it represents a useful, and fundamentally important, value.

Equation (4) may be written in a form for the true crack growth,

$$J_c = A(\Delta a_o + \Delta a)^N \tag{5}$$

and the initiation value is $J_o = A\Delta a_o{}^N$, ie when $\Delta a = 0$. Thus for small values of N and/or $\Delta a << \Delta a_o$ this has a linear form,

$$J_c = J_o + N\,J_o.\frac{\Delta a}{\Delta a_o} \tag{6}$$

while for $\Delta a >> \Delta a_o$,

$$J_c = J_o\left(\frac{\Delta a}{\Delta a_o}\right)^N \tag{7}$$

Both of these expressions are useful in interpreting the protocol using the concept of essential work. This scheme has proved useful in characterising thin films and uses double edge notch sheets of varying ligament length ℓ, loaded in tension. The ligament becomes fully plastic and then fails completely and the total energy to fracture, U_t, is determined as a function of ℓ. It is then assumed that the fracture process is in two parts. There is an inner process zone where the essential work of fracture w_e permit areas is dissipated and that this is surrounded by a larger zone governed by ℓ in which the rest of energy, termed non-essential, is dissipated. Thus the total energy is given by,

$$U_t = w_e B\ell + \beta w_p.B\ell^2 \tag{8}$$

where w_p is the non-essential work per unit volume and β is a shape factor for the ligament zone. Thus we may write

$$\frac{U_t}{B\ell} = w_e + \left(\beta w_p\right).\ell \tag{9}$$

and a plot of $\left(U_t / B\ell\right)$ versus ℓ should be linear with an intercept of w_e. This is often quite a good representation of the data, as will be shown later. w_e, of course, has the same units as G_c and J_c and one would expect some connection though, the stress state is usually plane stress for films giving high values.

The analysis can be directly related to J_c via equation (6) which may be integrated to find U_t;

$$U_t = 2B\int_0^{\frac{\ell}{2}} J_c d\Delta a = B\ell J_o + \frac{B\ell}{4} N \frac{J_o}{\Delta a_o}$$

ie

$$\frac{U_t}{B\ell} = J_o + \left(\frac{N}{4}\frac{J_o}{\Delta a_o}\right)\ell \tag{10}$$

and $J_o \equiv w_e$. Such conditions only apply for low values of N and for larger values the form tends to,

$$\frac{U_t}{B\ell} = \frac{J_o}{N+1}\left(\frac{\ell}{2\Delta a}\right)^{N+1} \tag{11}$$

ie non-linear in ℓ with no intercept. This is really the same problem as defining initiation in tough systems and can cause problems in essential work in terms of defining linearity and an intercept. The method is very robust, however, for moderately tough materials and has been found to be useful, but care must be taken when the toughness is high. Indeed there can be problems with more brittle systems, since the method relies on a fully plastic ligament and stable crack growth.

J-FRACTURE TOUGHNESS OF POLYMERS AT SLOW SPEED

G.E. HALE and F. RAMSTEINER

1. INTRODUCTION

This paper concentrates on the application of a multiple specimen resistance curve approach as the basis for measuring the J-fracture toughness of plastics, at slow speeds, i.e. displacement rates of typically 1mm/min.

Hence, the principal objectives of this paper are therefore:

1. to outline the theoretical background of the J-integral method;
2. to identify and discuss potential problem areas which must be considered when applying a J-based fracture mechanics approach to plastics;
3. to show how the testing protocol works in practice, by summarising results from a series of interlaboratory comparison test programmes.

The testing protocol for conducting J-crack growth resistance curve tests on plastics is given as an appendix at the end of this paper.

2. BACKGROUND TO THE J-INTEGRAL METHOD

To characterise crack instability and crack growth in materials, which deform in an elastic plastic manner, Rice [1] introduced the J-Integral method. He showed that the difference between the external work and the change of the internal potential energy within the area surrounded by an integration line Γ can be expressed by the line integral along this line:

$$J = \int_{\Gamma} w dy - \int_{\Gamma} T \frac{du}{dx} ds \tag{1}$$

where:

w	energy density
T	stress vector at the outer side of the integration line Γ
u	displacement
x, y	co-ordinates
s	arc length along the integration line

As long as there is no energy dissipation within the area surrounded by Γ, the value of the integral J is zero. Any externally applied work is stored elastically in the material. However if the integration line encloses a propagating crack, the value of the J-Integral accounts for the work done per unit area of crack growth. On the basis of this model, J-values are defined by:

$$J = (1/B)dU/da \tag{2}$$

where:

B specimen thickness
a crack length
U total external work

For practical applications, this leads to equation (1) in 'Introduction to Elastic-Plastic Fracture Mechanics'.

Based on the above definition of the J-Integral for characterising crack growth, it is possible to derive a plane strain condition. Hence this method can only be applied to specimens where:

- the deformation in the third direction can be ignored; e.g. the minimum thickness of the specimens must be large in relation to the yield zone; also the crack length and the distance from the initial notch tip to the specimen surface must exceed minimum values as given in the J-protocol;

- no kinetic effects must be included - this is not critical at low speeds;

- the material temperature remains constant during testing. If significant thermal energy is generated, then Rice's theory is no longer strictly valid as it is based solely on mechanical energy considerations. Hence, values of J would no longer be geometrically independent, since thermal conductivity and local heating are affected by the specimen shape;

- the applied load must increase steadily; the material should not be unloaded during the test.

Since it is experimentally difficult to observe directly the onset of crack instability, i.e. the point at which the critical value of J_C is defined, extrapolation procedures have been developed to evaluate this value. Begley and Landes [2] used a compliance technique in which several specimens with slightly different crack lengths are deformed. J-values and crack lengths can be evaluated since the compliance of the specimens varies as a function of the crack length. In addition, Landes and co-workers [3] developed a single specimen unloading compliance method for plastics. Finally Landes and Begley [4] used the correlation as given in equation (2) to develop a multiple-specimen resistance curve method, which forms the basis of the ESIS protocol for measuring the J-fracture toughness of plastics (see Appendix).

3. PROBLEMS ASSOCIATED WITH THE DEVELOPMENT OF THE TEST PROTOCOL

3.1 Crack Growth Measurement

Early work by the ESIS group indicated that measurement of crack growth Δa is one of the major factors which controls the effectiveness of this test method. While the initial crack length (a_o) can be determined relatively easily, accurate definition of the final crack front position is more difficult. Experimental work by a variety of laboratories demonstrated that two inter-related factors have a major effect on crack length identification after testing:

- accurate interpretation of features on the fracture face arising as a result of the various deformation mechanisms which occur in different polymers;

- an awareness of features introduced during the breaking-open operation.

Individual laboratories used their own preferred method for marking the crack front after testing. These include:

- cooling the tested specimen in liquid nitrogen or solid carbon dioxide before fracturing at either normal loading rates or at high impact velocities;

- high rate impact fracture at ambient temperature;

- fatigue cycling after the test at either ambient or lower temperatures;

- injection of ink into the crack to mark the front;

- measurement of crack length from polished sections under load (see Fig B2 in the appendix).

All of these methods have advantages and disadvantages. Cooling in liquid nitrogen may result in the specimen shattering when it is reloaded. In some plastics (e.g. certain grades of HDPE (high density polyethylene)), additional crescent features have been observed close to the crack front with both liquid nitrogen and/or high rate impact loading. This resulted in some laboratories reporting abnormally high values of crack growth. Conversely, impact at room temperature is quick. Ink injection techniques are dependent on using a low viscosity fluid so that it penetrates the crack completely. In materials which craze such as polypropylene with its crystalline lamellae and spherulites, the ink may be drawn into the craze and the precise crack front is then difficult to define. Fatigue cycling gives a definite change in fracture morphology, but it is time consuming especially as the frequency has to be limited to around 10Hz. Sectioning can be very effective, but it is also time-consuming if used as the only measurement method. Finally, heat-tinting (which is the preferred method for metallic materials) is not effective with thermoplastics.

However, it has been demonstrated that many of the potential problems associated with crack length measurement can be resolved if a limited number of sections perpendicular to the fracture face are prepared and correlated directly with features on the fracture face. This is the approach recommended in the protocol as illustrated schematically in Figure B2.

3.2 Definition of an Initiation Toughness Parameter

There is experimental evidence that some form of crack tip blunting occurs in certain thermoplastics as mentioned in 'Introduction to Elastic-Plastic Fracture Mechanics'. For instance, HDPE exhibits blunting at room temperature, whereas at –20°C, the crack propagated by voiding and fibril formation, i.e crazing (Fig. 1). Corresponding J - Δa plots are shown in Fig. 2. At 23°C, the notch simply deforms by yielding and the J-values follow a ($2\sigma_y\Delta a$) blunting line (in agreement with equation (3) in 'Introduction to Elastic-Plastic Fracture Mechanics'). In contrast, at –20°C, experimental data at higher values of Δa clearly deviate from the blunting line.

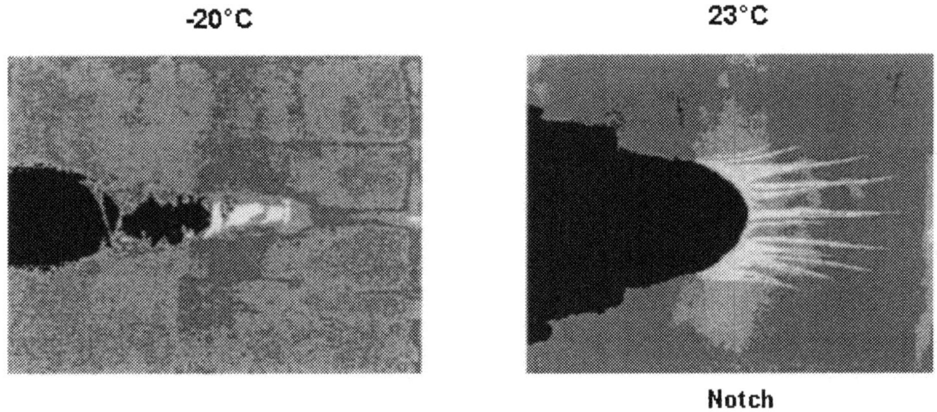

Fig.1 Crack tip in a CT-specimen of HDPE (side view) after deformation at −20°C (crazing) and 23°C (blunting)

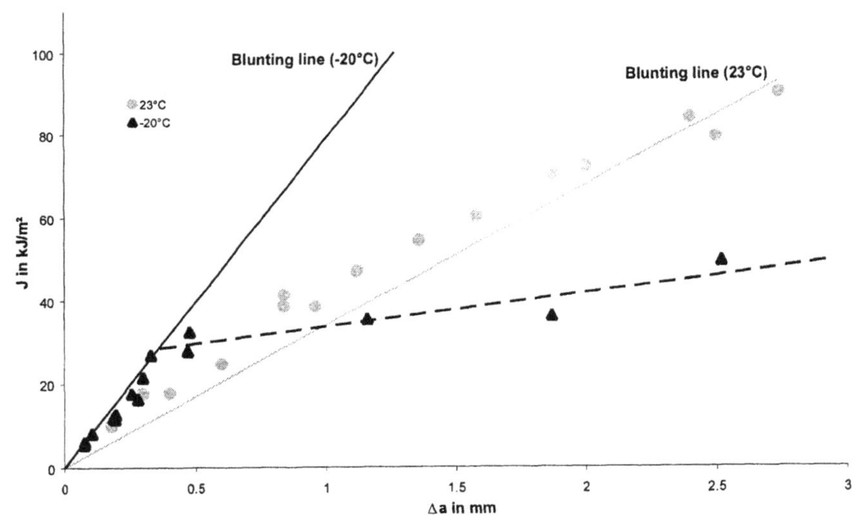

Fig.2 J - Δa diagrams of HDPE, including blunting lines, after testing at −20 and 23°C

In many cases, experimental observations indicated however that the initial slope of the resistance curve did not coincide with the calculated blunting line or the point of intersection between the blunting line and the remainder of the resistance line was not well defined. Rather, it has been established that the entire J-crack growth resistance curve should be described by a power law $J = A(\Delta a)^N$ as discussed in 'Introduction to Elastic-Plastic Fracture Mechanics'. This approach was used in the round robin tests described below.

With this approach, the critical J_C-value has been replaced, in most instances, by a pseudo-initiation value $J_{0.2}$, which defines crack resistance at 0.2 mm of total crack growth. A value of 0.2 mm is small enough to be close to the point at which the notch becomes unstable and crack initiation commences, but large enough to be experimentally measurable. To allow for situations where some form of blunting or multiple cracking does occur, the protocol defines initiation toughness as the lower value of a $J_{0.2}$ parameter or a J_{BL} value (specified as the J-value at the intersection of the blunting line with the J-Δa curve).

4. ROUND-ROBIN (INTERLABORATORY COMPARISON) TESTS

4.1 Participants and Materials

Representatives from academic and industrial organisations in Europe and the USA (see Table 1) have collaborated to assess the effectiveness of this test procedure by a series of round robin tests on different polymers.

A transparent ABS (acrylonitrile-butadiene-styrene copolymer) and a modified PVC (polyvinylchloride) were used as amorphous materials. A polyamide blend (rubber modified PA (polyamide) with PPO (polyphenylene oxide) addition), HDPE (high density polyethylene) {with a density ρ =0.944 g/cm^3 corresponding more closely to a MDPE (medium density polyethylene)}, a PP (polypropylene) homopolymer, and a random ethylene/propylene copolymer were used to study semi-crystalline polymers. Further details of the manufacturer of these thermoplastics, original form and dimensions are given in Table 2.

Table 1 Participants in the ESIS TC4 round-robin programmes on J-testing of plastics

Organisation	Contributors
TWI	G E Hale
EMPA	P Flueller, B Wagner
Solvay	D Adem, A Goldberg
Politecnico di Milano	A Pavan, M Rink
Pipeline Developments	P Marshall, Morley, Hepburn
Imperial College	J G Williams, W N Chung, H McGillivray, Chan, Breda,
ICI	D R Moore
Eastman Kodak	E Moskala
EAHP	R Schirrer
DSM	E H Gaalman, H Bos
Du Pont	H E Daeniker
Cranfield University	I Partridge
BASF	F Ramsteiner
BP Chemicals	M Cawood, A J Hemingway, A Gray
GKSS	E Reese

Table 2 Materials used in the round-robin tests

Material	Manufacturer/ Supplier	Product Form	Size (mm)	Yield strength in tension (MPa) (1)
ABS	BASF	Injection moulded disc	185 diameter x 16 thick	33
Polyamide blend ®Ultranyl	BASF	Injection moulded disc	185 diameter x 16 thick	47.4
Modified PVC	Pipeline Developments	Sheet produced by flattening pipe in a heated press	450 x 300 x 9.5	43.9
Polypropylene	BASF	Extruded sheet	500 x 250 x 15	29.9
HDPE	BP Chemicals	Compression moulded plaques	500 x 500 at 19 and 23 thick	23.1
Ethylene/ Propylene random copolymer	ICI	Centre-gated injection moulded discs	150 diameter x 12.5 thick	31.6 (2)

(1) All tests carried out at a loading rate of 1mm/min and at 23°C, unless stated otherwise
(2) Tested at 2mm/min

As the investigations progressed, it became apparent that the thermoplastics studied may be grouped into three categories:

- Easy to characterise, e.g. ABS, polyamide blends, modified PVC.
- Plastics where additional crack face features complicate interpretation, e.g. HDPE, polypropylene
- Thermoplastics exhibiting fine-scale fracture face features which make crack growth difficult to measure, e.g. ethylene/propylene copolymers

4.2 Easy to characterise materials

ABS is a good example of a thermoplastic where the absence of sidegrooving results in substantial crack front bowing which compounds the presence of a large stress whitened plastic zone ahead of the crack tip. Fig.3 shows the development of a stress-whitened zone in compact tension (CT) non-sidegrooved specimens of ABS at various points on the deformation curve. Since the material was transparent, it was possible to look at the development of the crack through the sides of the testpiece. At point 4 (see Fig. 3), the specimen was broken open so that the side view of the growing crack in the unbroken testpiece and the actual fracture surface from the same specimen could be compared. The actual crack front is substantially less extended than indicated by the position of the stress-whitened zone. The overall effect is to hamper definition of the precise position of the crack front. This resulted in considerable data scatter between different laboratories. Hence, the importance given to the use of a combination of fracture face measurement and sectioning to ensure that an optimal approach to crack growth measurement is adopted.

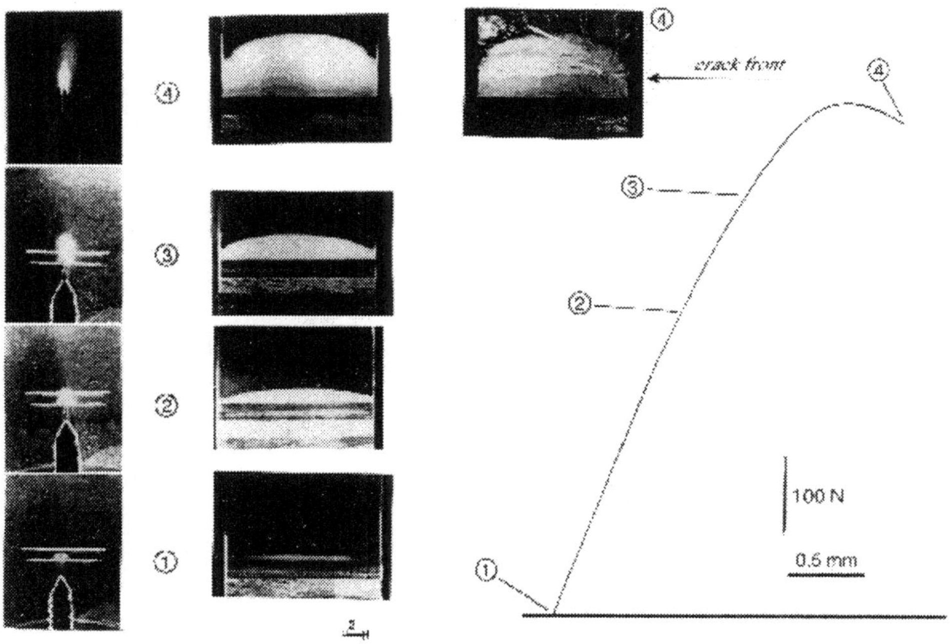

Fig.3 Crack growth development in non-sidegrooved CT specimens of transparent ABS at different points on the deformation curve.
At points 1,2 and 3, the crack was photographed through the sides of the transparent specimen.
At point 4, the specimen was also broken open so that the fracture surface could be compared to the side view of the developing crack.

Considerably less scatter was observed in sidegrooved SENB specimens of ABS (Fig. 4) and the data can be approximated by the power law:

$$J = 7.35\,\Delta a^{0.49} \quad (3)$$

where J is in kJ/m² and a is measured in mm.

In this case, $J_{0.2} = 3.35$ kJ/m².

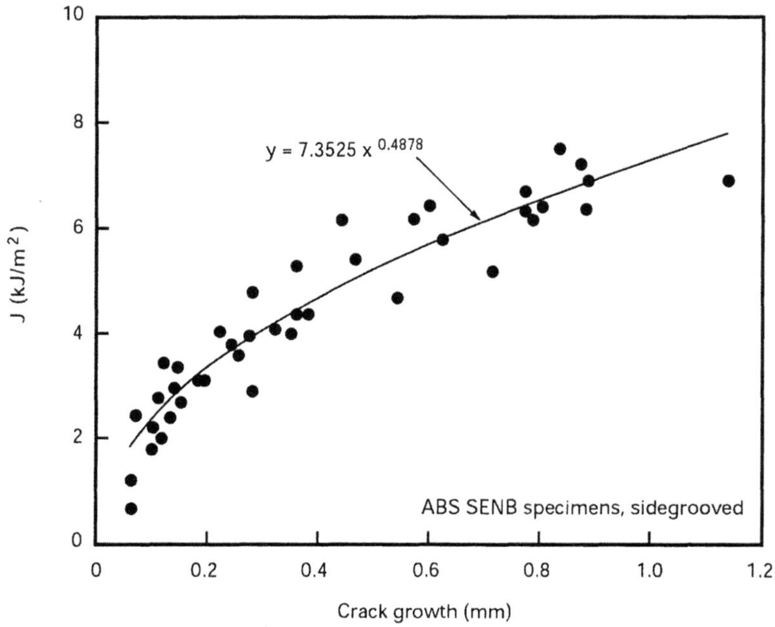

Fig.4 Power law fit to ABS data (side grooved SENB specimens) generated by six laboratories,

Table 3 contains the A and N parameters and $J_{0.2}$ values for six laboratories.

Table 3 Compilation of J R-Curve Data for ABS sidegrooved SENB specimens

Organisation	Notch sharpening method	$J_{0.2}$ (kJ/m^2)	Curve-fitting parameters		Comments
			A	N	
1	Tapped scalpel blade	2.78	7.62	0.626	
2	Sliding carpet knife blade	2.55	8.12	0.72	Very sharp sidegrooves
7	Tapped razor blade	3.86	8.665	0.502	
8	Fly cutter	4.12	8.04	0.415	
10	Tapped razor blade	3.35	6.94	0.45	Only 7 specimens used
14	Sliding razor blade	3.3	7.32	0.493	

Experimental work on a modified PVC indicated that sidegrooving appeared to have a less substantial effect than observed in the studies carried out on ABS. At low levels of crack growth, the amount of scatter between different laboratories is small (Fig. 5). After assessing all the experimental data for PVC, an average value for $J_{0.2}$ is 10.2 kJ/m^2.

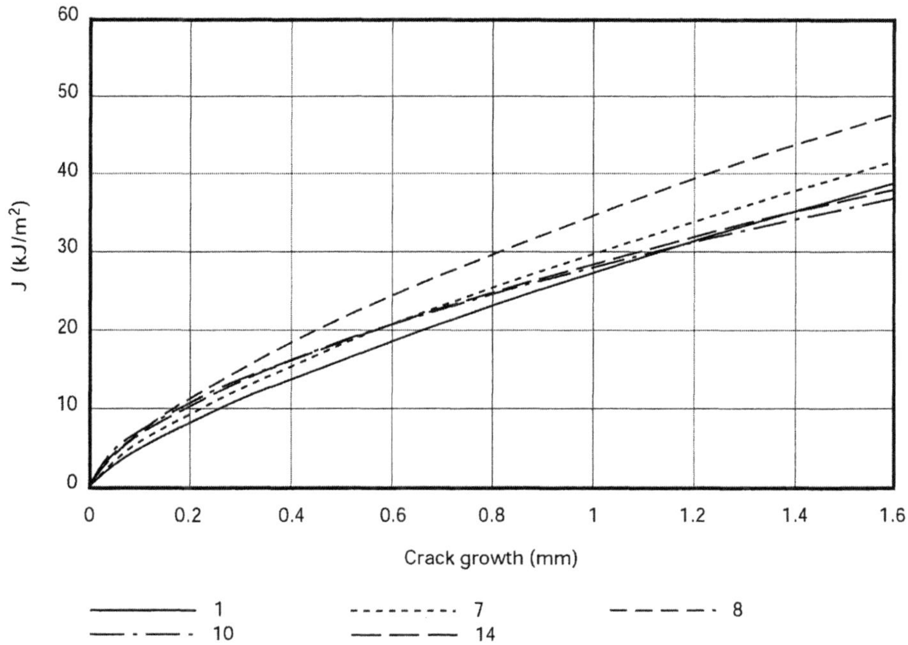

Fig. 5 Comparison of J R-curves for modified PVC (sidegrooved SENB testpieces) measured at five laboratories.

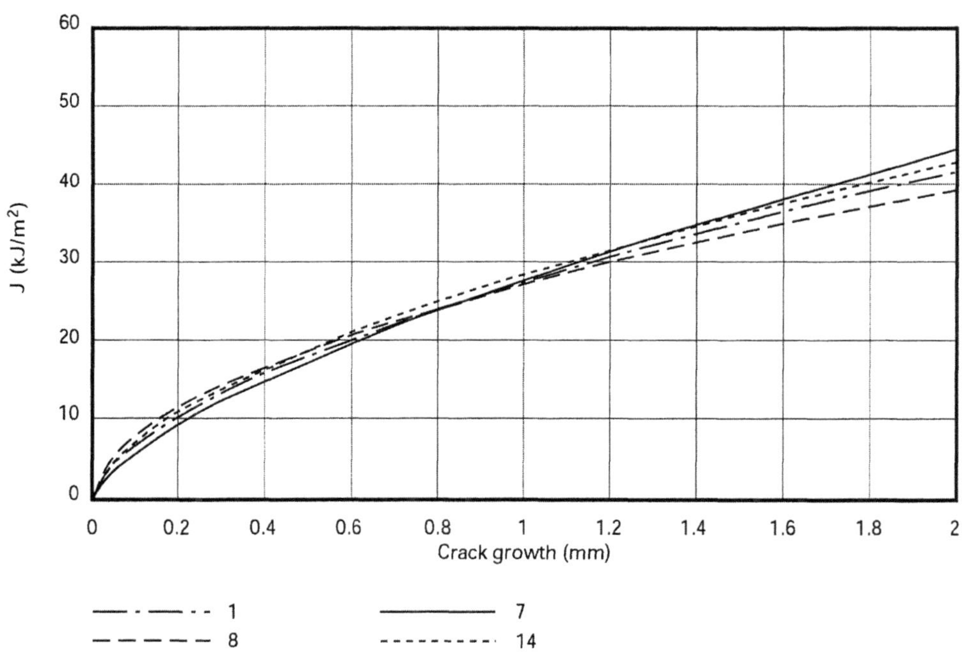

Fig.6 Compilation of J R-curves for sidegrooved SENB specimens of a polyamide blend, measured at four laboratories.

Excellent comparative data were obtained between four laboratories after carrying out tests on SENB specimens extracted from 16mm thick discs of a polyamide blend (Fig. 6). Table 4 summarises the data generated by each laboratory. In this case, the mean value of $J_{0.2}$ is 10.6 kJ/m².

Table 4 Compilation of J R-Curve Data for a polyamide blend (sidegrooved SENB specimens)

Organisation	Notch sharpening method	$J_{0.2}$ (kJ/m^2)	Curve-fitting parameters	
			A	N
1	Tapped scalpel blade	10.54	27.64	0.599
7	Fly wheel cutter and razor tap	9.43	28.01	0.676
8	Fly cutter	11.50	27.35	0.538
14	Sliding razor blade	11.00	28.60	0.592

4.3 Plastics where additional crack face features complicate interpretation

In some instances, e.g. with some grades of HDPE, abnormally large crack growth values were reported by a number of laboratories.

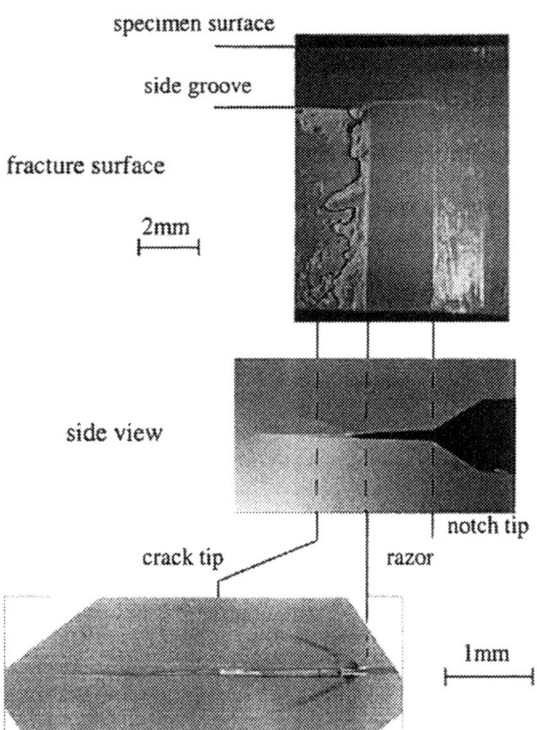

Fig.7 Fracture surface and side view section taken perpendicular to the direction of crack propagation in a CT-specimen of HDPE tested at –20°C, with J = 30kJ/m², Δa ≈ 1mm

If a comparison is made between the fracture surface and a section taken perpendicular to the fracture from a cracked but unbroken testpiece, it becomes clear that the crack front (or what is thought to be the crack front) is very wavy in this material. Fig.7 compares the fracture surface itself and a side view from a section cut perpendicular to the fracture surface at the midplane of the specimen before it was broken open. While there is good correspondence between the crack tip on the fracture surface and the side view, precise definition of the final crack length is not a simple task. This situation is further complicated if the notch propagates by multiple cracking, as shown in Fig. 8 for polypropylene.

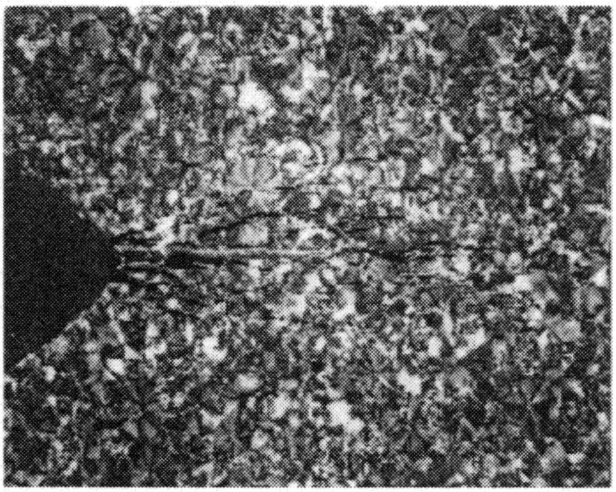

Fig.8 Side view of a section (taken under polarised light) from a specimen cut perpendicular to the direction of crack growth which illustrates the development of multiple cracking in polypropylene

4.4 Thermoplastics exhibiting fine-scale fracture face features

Provided that experimental conditions are carefully controlled and the requirements of the test protocol are followed, it is possible to obtain consistent fracture toughness data for polymers which exhibit fine multiple cracks ahead of the primary crack tip.

For instance, Fig. 9 shows data obtained by three laboratories on specimens extracted from a 12.5mm thick random ethylene/propylene copolymer which have been treated as one homogeneous set of data. The results regress to a power law fit:

$$J = 16.27\ \Delta a^{0.54} \tag{4}$$

In this case, $J_{0.2} = 6.8$ kJ/m².

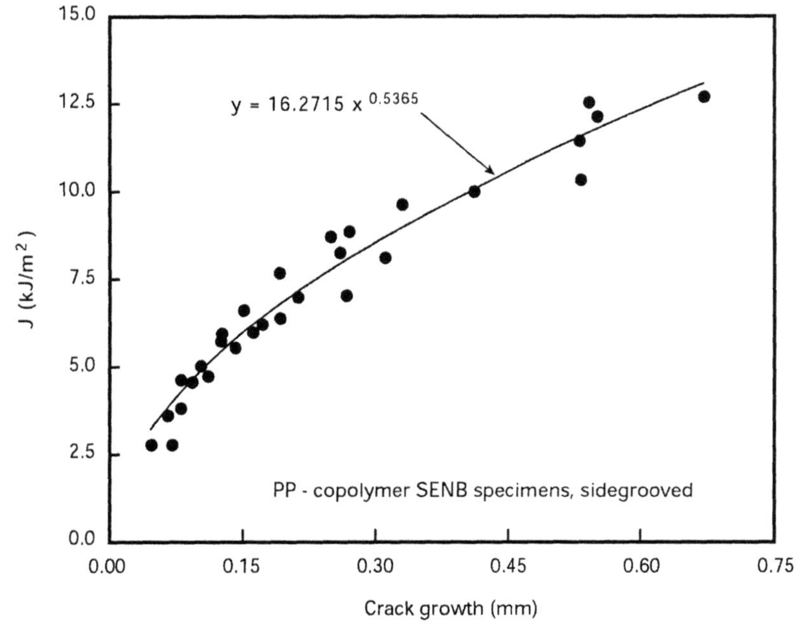

Fig.9 Power law fit to data generated for a random ethylene/propylene copolymer (sidegrooved SENB specimens; data obtained from three laboratories)

Good correlation was obtained between the crack opening observed on sections and measurements on the fracture faces themselves. Examination indicated that partial fracture, i.e. cracking and voiding, also occurs ahead of the crack tip in this type of semi-crystalline polymer as shown in Fig. 8 for polypropylene. In many instances, the microstructural inhomogeneity present in such semi-crystalline polymers can make crack tip measurement insufficiently precise and hence two other options should be considered:

(1) a full J R-curve may be a more effective way of describing the fracture resistance of this type of thermoplastic;

(2) it may be feasible to apply a blunting line approach to polymers which deform in this manner.

Data determined in one laboratory (Fig. 10) illustrate that a blunting line can be appropriate.

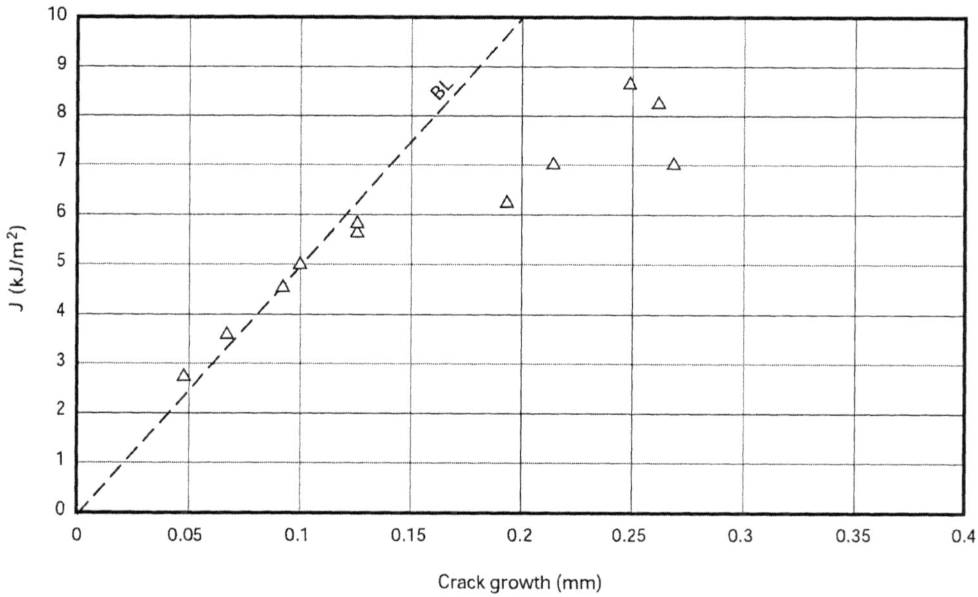

Fig.10 J - Δa data for an ethylene/propylene copolymer illustrating potential crack tip blunting in the early stages of deformation at 23°C and a loading rate of 1mm/min, using SENB specimens.

It should be noted however that four other laboratories were unsuccessful in characterising this material by a J R-curve approach, as they were unable to detect any significant crack growth over a range of test conditions. Even in the three laboratories that did generate valid experimental data, the maximum measured crack growth was only 0.67mm, although these figures were correlated by comparing fracture face measurements and those determined via sectioning.

Similar work on polypropylene indicated that crack propagation was strongly influenced by its spherulitic structure which resulted in partial splitting of the crack front (Fig.8).

Overall, experimental evidence indicates that the use of a static J - crack growth resistance curve approach to characterise thermoplastics which exhibit this form of multiple cracking (with a spacing of 0.1 to 0.3mm) on the fracture face, is not readily applicable.

5. CONCLUSIONS

This paper summarises work undertaken by laboratories across Europe (plus one laboratory in the USA) to assess the applicability of a J-testing procedure developed by ESIS to determine the crack growth resistance behaviour of plastics.

The major conclusions are:

1. The elastic-plastic fracture mechanics parameter J can be used successfully to characterise the fracture toughness of ductile thermoplastics at slow speeds by employing a multiple

specimen approach to determine a complete J-crack growth resistance curve.

2. The controlling factor in obtaining consistent results is accurate measurement of the amount of crack growth which occurs during the J-test. It was found that two inter-related factors have a major effect on crack length measurements after testing:

   - the method used to break-open the testpiece;

   - interpretation of features on the fracture face arising as a result of the various deformation mechanisms which occur in different polymers.

3. A pseudo-initiation parameter - $J_{0.2}$ (i.e. J determined at 0.2mm of total crack growth) - can be determined to characterise crack instability

4. A J R-curve approach is no longer applicable in cases where:

   - the fracture face exhibits fine-scale features (typically of the order of 0.1 to 0.3mm) which result in apparent uneven crack growth and the presence of multiple cracking ahead of the principal crack tip;

   - the material is very inhomogeneous as a result of a coarse grained structure;

   - the material is extremely tough and yielding occurs at the crack tip rather than actual crack propagation.

6. RECOMMENDATIONS

The testing protocol should be adopted as "best practice" for determining the J-crack growth resistance behaviour of thermoplastics, until such time as ISO or CEN standards become available.

7. REFERENCES

1. Rice, J. R. (1968). *Trans ASME, Journal of Applied Mechanics*, Vol. 35, June pp. 379-386.
2. Begley, J. A. and Landes, J. D. (1972). The J Integral as a fracture criterion, In: *ASTM STP 514, Fracture Toughness, Proc. 1971 National Symposium on Fracture Mechanics*, Urbana, Illinois, USA, 31 Aug - 2 Sept 1971, Part II, published by ASTM, Philadelphia, USA, pp 1-20
3. .Landes, J.D. and Zhou, Z. (1993). Application of load separation and normalisation methods for polycarbonate materials, *Int. Journal of Fracture*, Vol. 63, No. 4, 15 October, pp 383-393
4. Landes, J. D. and Begley, J. A. (1974). Test results from J-Integral studies: An attempt to establish a J_{Ic} testing procedure, In: *ASTM STP 560 , Fracture Analysis, Proc. 1973 National Symposium on Fracture Mechanics*, Univ. of Maryland, Maryland, USA, 27-29 Aug 1973, Part II, published by ASTM, Philadelphia, USA, pp 170-186
5. Hashemi, S. and Williams, J.G., (1986). Fracture toughness study on low density and linear low density polyethylenes, *Polymer*, Vol. 27, No. 3, March, pp 387-392

8. ACKNOWLEDGEMENTS

The authors wish to thank all the members of the ESIS Technical Committee on Polymers and Composites who have contributed their data and expertise during the development and evaluation of the testing protocol. Thanks are due to BASF, BP Chemicals, Du Pont, ICI, Pipeline Developments and Solvay for the provision of materials at various stages of this project, without which this study could not have been undertaken. GEH wishes to acknowledge the funding provided by the UK Government Department of Trade and Industry and the Industrial Members of TWI.

APPENDIX

A TESTING PROTOCOL FOR CONDUCTING J-CRACK GROWTH RESISTANCE CURVE TESTS ON PLASTICS

June 2000

Prepared on behalf of the
ESIS Technical Committee on
Polymers and Composites

NOMENCLATURE

Specimen dimensions

a_o	initial crack length (i.e. machined notch plus razor sharpened tip)
a_f	final crack length
b_o	original uncracked ligament ($W-a_o$)
B	specimen thickness
B_N	net thickness of sidegrooved specimens
S	span of single edge notch bend specimen
W	specimen width

Material Properties

E	axial modulus of elasticity (Young's modulus)
σ_f	flow stress
σ_u	rupture strength
σ_y	yield strength

Fracture Parameters and Related Quantities

Δa	crack growth
Δa_{max}	validity limit for J-controlled crack growth
P	load
v	displacement
C	elastic compliance ($\Delta v/\Delta P$)
J_o	fracture resistance not allowing for crack growth
J_c	fracture resistance at crack initiation
J_{BL}	fracture resistance where the blunting line intersects with the J-Δa curve
$J_{0.2}$	fracture resistance at 0.2mm of total crack growth including crack tip blunting
J_{max}	validity limit for J
J_{Ic}	critical value of J, at the onset of stable crack extension
η	calibration factor
U	the area under the load versus load-point displacement record up to the line of constant displacement corresponding to the termination of the test

A TESTING PROTOCOL FOR CONDUCTING J-CRACK GROWTH RESISTANCE CURVE TESTS ON PLASTICS

June 2000

This protocol has been prepared by Dr. Geoff Hale, TWI Ltd and incorporates information gained from a series of round-robin exercises conducted by the ESIS Technical Committee on Polymers and Composites. Valuable comments and suggestions have also been provided by the ASTM task group who were responsible for the preparation of ASTM D6068-96 [1]. Any comments and enquiries should be addressed to Geoff Hale at TWI Ltd.

1. INTRODUCTION

J-fracture toughness testing of ductile metallic materials is currently covered by ASTM E1820-99 [2] and an ESIS procedure, ESIS P1-92 [3]. These documents and their predecessors were used as the basis for developing an equivalent testing procedure for plastic materials.

For polymers, stable crack growth is either measured using several specimens loaded to different displacements (multiple specimen technique) or it may be estimated from the elastic compliance measured during unloading-loading cycles using a single specimen (single specimen technique). The text of this protocol is concerned with a multiple specimen approach. It provides information on the conduct of the tests, those thermoplastics for which the protocol works well and those where care is needed to avoid problems.

2. SPECIMEN CONFIGURATION AND SIZE

Compact tension (CT) or three point bend (SENB) specimens may be used (Fig. 1). As in the case of the LEFM standard [4], the specimen thickness is usually taken as the maximum available, i.e. the sheet thickness B. If the material surface is uneven, it may be skimmed, but as little material as possible should be removed so that the greatest possible sheet thickness is maintained. The width, W, is generally equal to twice the specimen thickness, i.e. W = 2B, and this configuration maximises the opportunity to obtain plane strain conditions at the crack tip. This is the approach used conventionally for metals and it has been employed successfully when carrying out LEFM tests on plastics.

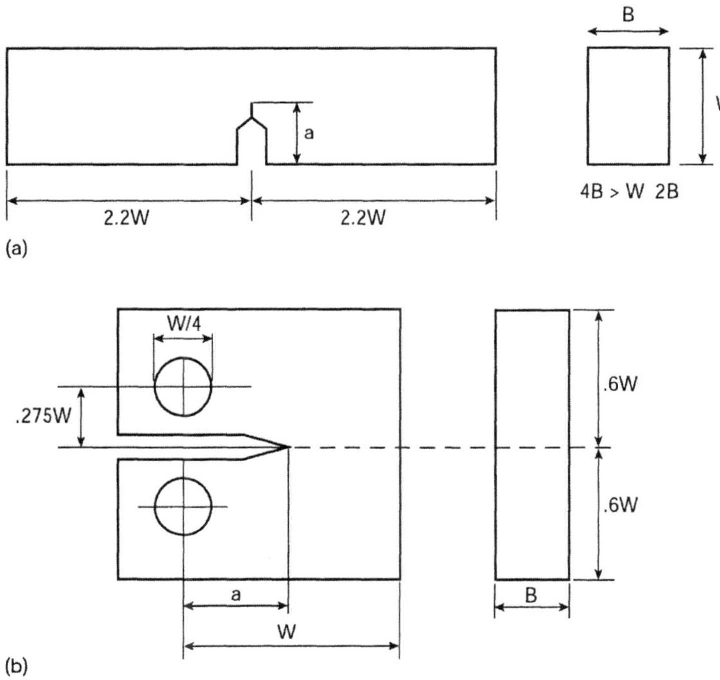

Fig. 1 Specimen configurations
(a) Three point bend specimen (SENB)
(b) Compact tension configuration (CT)

3. NOTCHING

The crack tip should be as sharp as possible and it is suggested that specimens are precracked by sliding or tapping a razor blade into the root of a machined notch, which is ideally produced using either a broach or a single point flycutter so that the tip radius $\rho \leq 20\mu m$. Alternatively, fatigue precracking can be employed. (Note, to avoid hysteretic heating in polymers, it may be necessary to fatigue at very low frequencies, <4Hz in some instances). Pressing the razor blade into the crack tip is **not** permissible.

The length of the precrack (a_o) for both SENB and CT specimen geometries should satisfy the requirement:

$$0.55 \leq a_o/W \leq 0.65$$

Furthermore, the initial pre-crack lengths should not vary by more than 0.05W.

4. SIDEGROOVING

Specimens must be sidegrooved whenever possible to promote growth of a straight ductile crack front so that measurement of total crack growth at the end of the test is simpler. The sidegrooves should be equal in depth and have an included angle of 45° ± 5° with a root radius of 0.25 ±

0.05mm. The total reduction in thickness shall not exceed 0.20B. Past experience suggests that 10% sidegrooving on each face with a cutter having an included angle of not less than 45° and a minimum root radius of 0.25mm is likely to be sufficient. (If the side-grooves are too sharp, then crack growth initiates from the root of the sidegrooves and not from the centre of the testpiece.) If plane-sided specimens are used, the difference between the mean crack growth and any of the measurement points should be less than 30%. If this figure of 30% is exceeded, then sidegrooved specimens **must** be employed.

5. TEST CONDITIONS

Since plastics are viscoelastic materials, it is essential that the temperature and loading rate are recorded. In the first instance, to aid comparison, it is recommended that all tests are conducted at a temperature of 23°C and a loading rate of 1mm/min. These conditions are known to yield consistent crack growth resistance curves for those thermoplastics which exhibit stable and well-defined crack growth. If a material exhibits unstable crack growth during the test (i.e. short arrested brittle cracks, sometimes described as 'pop-ins'), then either a slower loading rate (probably 0.1mm/min) or a higher test temperature must be used. Further advice is given below.

6. MEASUREMENT OF DISPLACEMENT

In order to calculate J, the displacement must be measured on the load line.

For a CT specimen, this can be achieved by an extensometer (i.e. a LVDT or a clip gauge) placed in the notch on the load line, or the displacement can be calculated from the crosshead displacement rate of the test machine corrected by the compliance of the fixtures and indentation.

For a SENB specimen, it is not as easy to measure the real displacement at the load line. Corrections should be made to account for deformation due to roller indentation, pin penetration etc. in a similar manner to the procedure described in the LEFM standard [4]. Alternatively, a comparator bar may be used. Further details of each approach can be found in Annex A.

7. LOADING RIGS

For SENB specimens, a rig with moving rollers of sufficiently large diameter to minimise plastic indentation is recommended (Fig. 2). For the CT geometry, loading is via pins through the holes (Fig. 1).

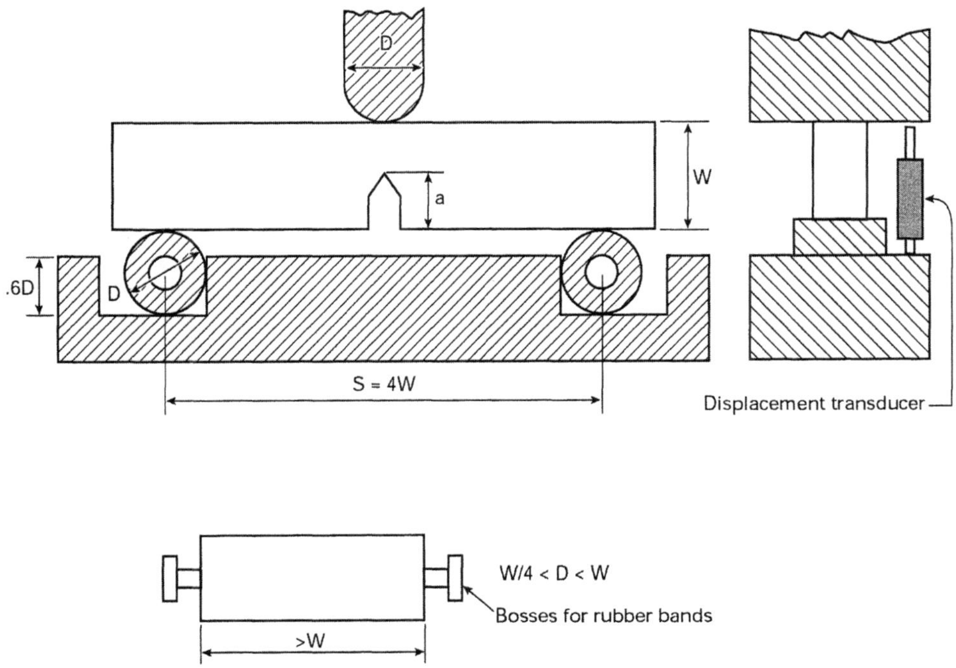

Fig. 2 Bending rig

8. TEST PROCEDURE

A series of specimens are loaded to different displacements and the amount of crack extension which occurs during testing is determined.

With the first specimen, it is good practice to just exceed maximum load so that subsequent displacement levels can be estimated more accurately using this initial record. To control crack growth beyond maximum load, displacement or clip gauge control should be employed.

Break open the specimen and determine the amount of crack growth following the procedures described in Section 9.

Evaluate J using the procedure given in Section 11.

The test procedure is repeated for at least six further specimens choosing displacements such that the crack growth satisfies the validity requirements given in Section 12.

9. CRACK LENGTH MEASUREMENT

The final crack growth resistance curve is strongly influenced by the accuracy of the crack growth measurements made once testing is complete. These measurements must be carried out directly on the fracture face, but certain precautions are necessary to minimise the risk of incorrect measurements. While the initial crack length (a_o) can be determined relatively easily, greater care is required when measuring the final crack length.

Measurement of the final crack front from the fracture face is clearly a subjective matter and is therefore prone to error. Furthermore, there has been some evidence that, in certain plastics, the 'breaking-open' operation may cause additional crack growth. Therefore, while direct measurement from the fracture face is the primary approach adopted, this must be supplemented by sectioning **at least two** specimens through the centre. One specimen should exhibit a small amount of crack growth, e.g. around 0.1 to 0.2mm. For the second specimen, crack growth should be measured at or close to maximum load. Comparisons of the crack growth measured on the sectioned face and that of an equivalent testpiece broken open by one of the methods described in Annex B can be made. With this approach, it is possible to confirm which feature on the fracture face corresponds to the actual crack growth that occurred in the test itself. Once established, all subsequent measurements can be made on the fracture face up to the defined feature.

Information on crack length measurement directly from the fracture face is given in Annex B together with advice on sectioning.

The initial and final crack lengths (a_o and a_f respectively) are measured using either a travelling microscope or a high power optical microscope.

There is generally no difficulty measuring the initial crack length (a_o) at the tip of the razor-sharpened region. Three equidistant measurements are used as shown in Fig. 3. The initial crack length (a_o) is the average of the three measured values.

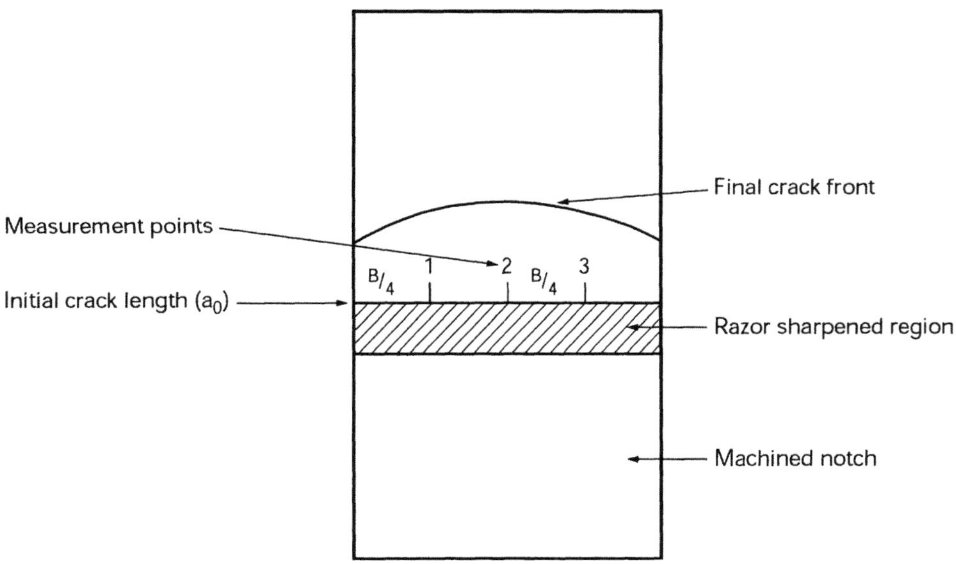

Fig. 3 Measurement of initial crack length (a_o)

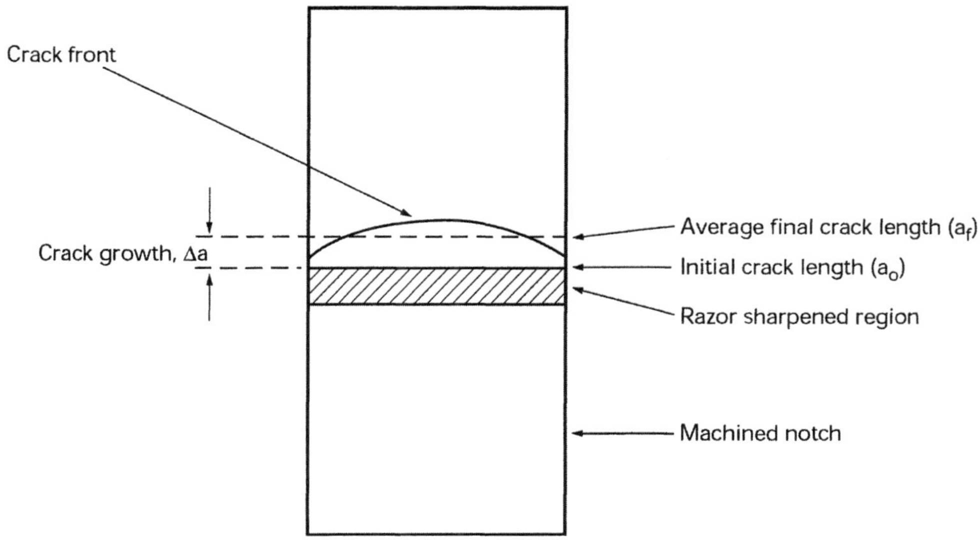

Fig. 4 Measurement of final crack length (a_f)

The final crack length (a_f) is determined by establishing a 'subjective' average crack growth (Fig. 4). This is obtained by positioning the microscope cross-hair at the centre of the curved crack front. (Note: *This is less rigorous than a true average, but more practical when a large number of specimens are to be tested. The previous protocol used a nine-point average as adopted by the metals community*).

Illumination of the fracture face can cause difficulties in interpreting the limit of ductile crack growth. One suggestion is to light the specimen from the side but other options may be explored.

10. RESPONSE OF DIFFERENT THERMOPLASTICS

The crack growth behaviour exhibited by thermoplastics can be described as:-

a. Easy to characterise. Distinct markings visible on the fracture face which correlate with the measurements made on sections cut at 90° to the fracture face. Typical examples include: rubber-toughened PMMA, ABS, polyamide blends etc.

b. Additional features such as half-moons may appear on the fracture face (see Fig.B1 in Annex B). These can be mistaken for actual crack growth unless a section is taken as described above. Examples include high density polyethylenes.

c. Crack growth obscured by an uneven and rough fracture face. This often occurs in thermoplastics which exhibit crazing at the crack tip, e.g. polypropylenes. This may result in thin slivers of material overlaying the main fracture face. Once again, sectioning perpendicular to the fracture provides additional confirmation of the region of crack growth.

d. It is recognised that the fracture faces of some semi-crystalline thermoplastics which exhibit small-scale inhomogeneous structures (typically in the range of 0.1 to 0.3mm), such as polypropylenes and some polyethylenes, are extremely difficult to characterise because of uneven crack growth and the presence of multiple cracking ahead of the principal crack tip. In these materials, it can be difficult to decide if crack growth has taken place or not. Alternative approaches should be employed, if possible, to measure the fracture toughness of the material, e.g. LEFM testing at lower temperature.

If sectioning perpendicular to the fracture does not provide a positive answer, then characterisation of the material's resistance to stable crack growth is not likely to be successful.

11. ANALYSIS PROCEDURE

11.1. Fracture Resistance J

There are several formulae available for determining the fracture resistance J. The formula given here is considered the most appropriate for multiple specimen testing. It has the advantage of avoiding the need to partition the area U under the load displacement record into elastic and plastic components as used in the ASTM methods. Furthermore, over the allowable a/W range, the formula is virtually identical to those which partition U.

Calculate J_o for each specimen using the relationship

$$J_o = \frac{\eta U}{B_N (W - a_o)} \tag{1}$$

where

J_o fracture resistance not allowing for crack growth

η $2 + 0.522\,(1-a_o/W)$ for compact tension specimens,
2 for single edge notch bend specimens

B_N net thickness of sidegrooved specimens

U is the area under the load versus load-point displacement record up to the line of constant displacement corresponding to the termination of the test (Fig. 5).

If plane-sided specimens are used replace B_N with B in the above formula.

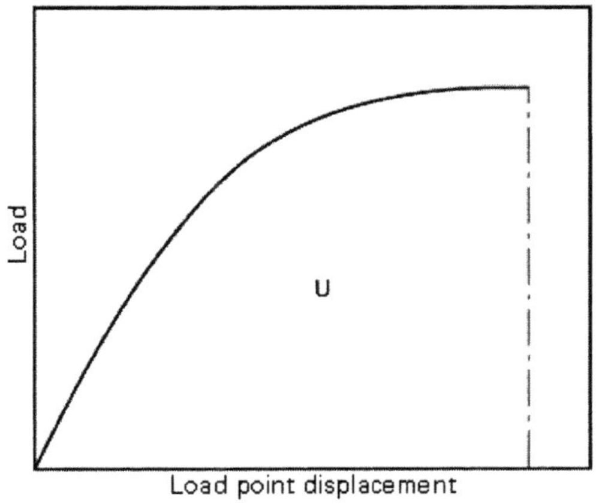

Fig. 5 Definition of absorbed energy U

11.2. Crack Growth, Δa

Δa is the difference between the final crack length (a_f) and the initial crack length (a_o) measured from the fracture faces of each specimen.

12. CONSTRUCTION OF VALID CRACK GROWTH RESISTANCE CURVES

Using the data determined in Section 11, a plot of fracture resistance against crack growth is constructed.

Since small amounts of crack growth are difficult to measure and hence are subject to error, a 0.05mm exclusion line parallel to the J-axis is used (Fig. 6). Only data beyond this exclusion line are used to determine the best fit curve.

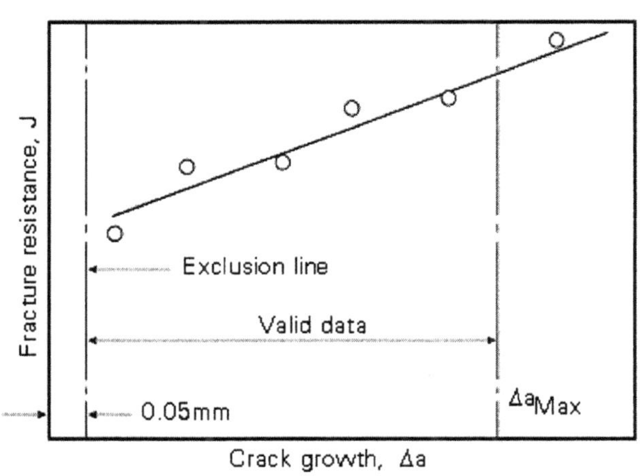

Fig. 6 Data points to be used for curve fitting

Determine the magnitude of the parameter, Δa_{max} where:

$$\Delta a_{max} = 0.1(W - a_o)$$

where $(W-a_o)$ is the initial uncracked ligament.

Some of the earlier work on the use of J-fracture toughness for polymers indicated that a 6% limit on uncracked ligament size was too restrictive for plastics [5,6] and therefore a higher value of 10% has been adopted. Above this limit (i.e. Δa_{max}), J may not necessarily be a valid characterising parameter.

There are insufficient data available at present to establish whether the limit on uncracked ligament size (Δa_{max}) should be greater than 10%.

A second exclusion line parallel to the J-axis is drawn through Δa_{max} (Fig. 6).

To ensure that the data points are evenly spaced, the interval between the 0.05mm and Δa_{max} exclusion lines (see Fig.7) is divided into four equal sections. At least one (J,Δa) data point is required in each quadrant. However, since there is an interest in determining some form of fracture initiation parameter, a further two specimens are required in the first section so that the position of the J-Δa curve, in the vicinity of $\Delta a = 0.2$mm, is more clearly defined. At least seven specimens must fall between the 0.05 and Δa_{max} exclusion lines for a J-Δa curve to be valid.

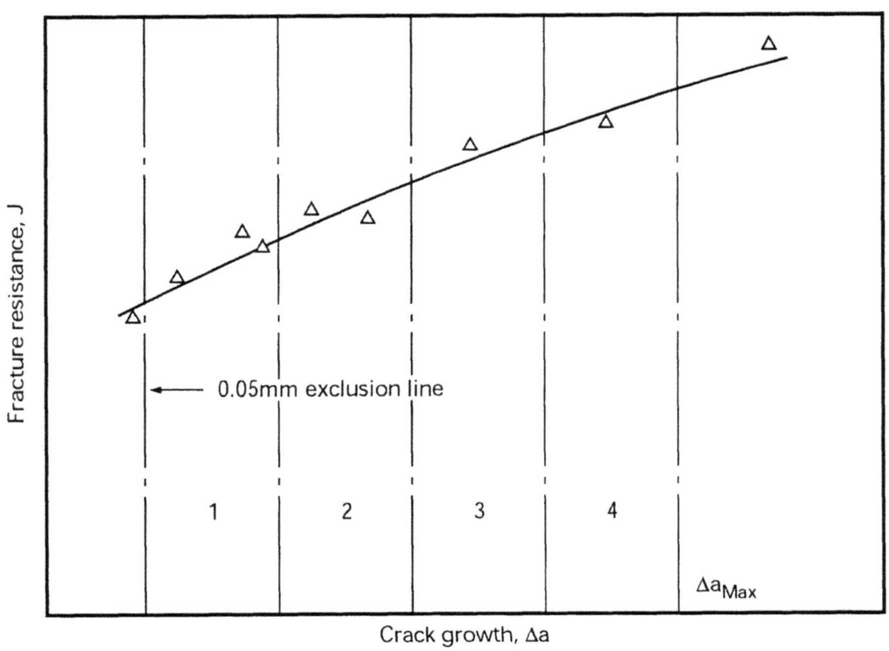

Fig. 7 Data spacing requirement

Users should state in their report whether the requirements outlined above can be satisfied or not. If not, they should still continue with the curve fitting routine described below but they should make it clear in their report where they failed to meet the validity and spacing requirements.
A best fit curve may now be plotted through **all** the data points falling between the 0.05mm offset exclusion line and the Δa_{max} exclusion line, using a simple power law of the form:

$$J = A\Delta a^N \tag{2}$$

where A and N are constants.

and $N \leq 1$

If $N > 1$, then the validity of the data should be questioned and it must be rechecked.

If the maximum amount of crack growth achieved in the test is less than Δa_{max}, then all the data points between the 0.05mm exclusion line and this value should be used for the curve-fitting routine. However, in most instances, the crack growth measured in a specimen taken beyond the maximum load position on the load-displacement trace will be greater than Δa_{max} and, as noted earlier, this procedure should be employed with the first testpiece to obtain a reference point.

In the metals field, the concept of a J-controlled crack growth resistance curve is widely used and certain validity limits are defined to establish the region within which a J-controlled crack growth process occurs. There is insufficient experimental evidence available to indicate if the same limits can be applied to plastics and hence the concept of a region of J-controlled crack growth has not been pursued at this point in time.

13. DETERMINATION OF INITIATION TOUGHNESS, J_C

The initiation toughness is defined as the lower value of a $J_{0.2}$ parameter (i.e. J measured at 0.2mm of total crack growth {including crack tip blunting}) or a J_{BL} value (specified as the intersection of the blunting line with the J-Δa curve) as shown in Fig. 8.

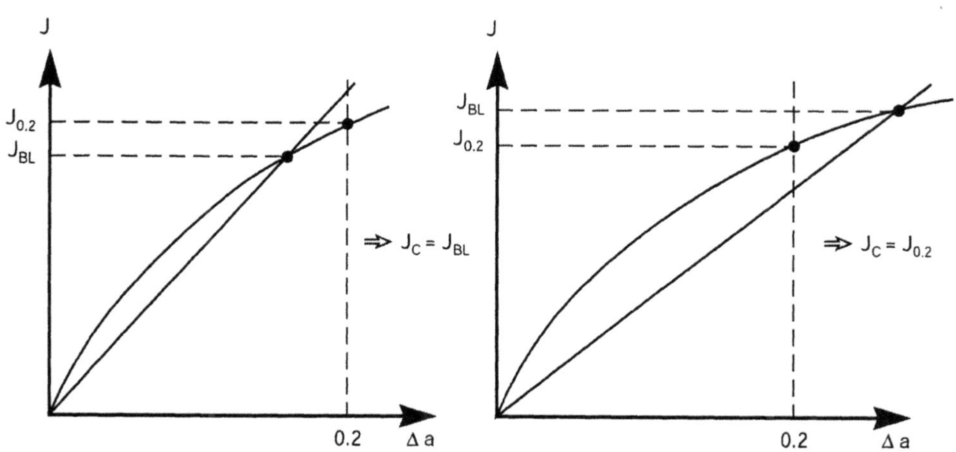

Fig. 8 Definition of initiation parameter (J_c)

$J_{0.2}$ is considered to be valid if:

(i) at least one J-Δa point falls between 0.2 and 0.4mm of crack growth

(ii) $J_{0.2} \leq J_{max}$

where J_{max} is the smaller of $(W-a_o)\sigma_y/20$ or $B\sigma_y/20$

Similarly, J_{BL} must also be less than J_{max}.

In this protocol, the flow stress, σ_f [$=(\sigma_y + \sigma_u)/2$], used in the metals field is replaced by the conventional polymer physics definition of uniaxial yield strength, i.e. the first attainment of the maximum load. To minimise the risk of brittle fracture in tensile tests, it is often necessary to polish the edge of tensile specimens. The yield strength should be measured at 1mm/min (or at a slower displacement rate as defined in Section 5 above).

If any of the specimens exhibit brittle cleavage fracture before maximum load, under the test conditions specified, i.e. temperature and loading rate, this must be stated in the report.

14. REPORTING

An outline reporting form is attached (see Annex C). This should be completed as fully as possible. A copy of the final J-Δa best fit curve should be included. All experimental points and exclusion lines should be shown on this plot. Sketches showing the variation in precrack length (a_o) and final crack length (a_f) would be useful. Photographs from some of the fracture faces should be attached if possible.

15. REFERENCES

1. ASTM D6068-96. Test method for determining J-R curves of plastic materials

2. ASTM E1820-99. Standard test method for measurement of fracture toughness

3. ESIS P1-92. ESIS recommendations for determining the fracture resistance of ductile materials, European Structural Integrity Society, January 1992.

4. ISO 13586. Determination of fracture toughness (G_{Ic} and K_{Ic}) - Linear Elastic Fracture Mechanics (LEFM) approach, 2000

5. Hashemi, S. and Williams, J. G: (1986). The effects of specimen configuration and notch tip radius on the fracture toughness of polymers using J_c, *Plastics and Rubber Processing and Applications*, Vol. 6, No. 4, pp. 363-375.

6. Chung, W. N. and Williams, J. G: (1991). Determination of J_{Ic} of polymers using the single specimen method, In: *ASTM STP 1114, Elastic-Plastic Fracture Test Methods: The User's Experience, Proc. 2nd Symposium on User experience with elastic plastic fracture test methods*, Lake Buena Vista, Florida, USA, 8-9 November 1989, published by ASTM, Philadelphia, USA, pp 320-339

ANNEX A

DETERMINATION OF LOAD-LINE DISPLACEMENT

A.1. Correction Procedure for Extraneous Displacements

As noted in Section 6, values of J must be determined from the area under the load versus load-line displacement diagram. To accurately determine load-line displacement, it is necessary to take account of indentation effects, pin penetration, machine stiffness etc. These extraneous displacements are additive, so that measurements derived from machine crosshead displacement, or relative displacements between the specimen and the testing machine will overestimate the true load-line displacement. The degree of overestimate will vary with material, temperature, loading rate, specimen dimensions, loading fixtures and test machine.

A test configuration as shown in Fig. A.1a or A.1b using identically prepared, but unnotched, samples is used to generate a load-displacement correction curve. This correction curve is then **subtracted** from the load-displacement curve obtained during the actual fracture test with notched samples. This subtraction is performed by subtracting the correction curve displacement from the fracture test displacement at corresponding loads.

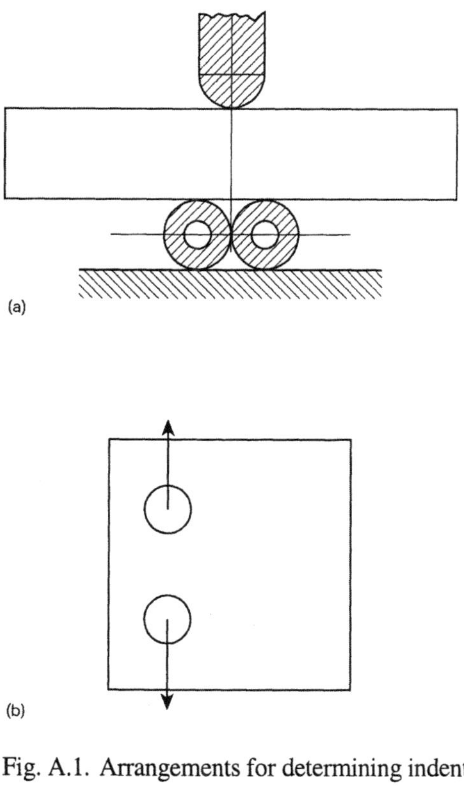

Fig. A.1. Arrangements for determining indentation displacement
(a) SENB specimen
(b) CT testpiece

In practice, a linear correction curve can usually be obtained (up to the maximum loads recorded in the fracture test). Use of a linear correction simplifies the displacement correction. Any initial non-linearity due to penetration of the loading pins into the sample is observed during both the calibration test and the actual fracture test, so a linearisation of the near-zero correction data and the fracture test data can effectively correct for this initial non-linearity.

The indentation tests should be carried out at the same loading rate as the fracture test, i.e. normally 1mm/min.

The corrected energy U for each specimen is calculated from integrating the appropriate load versus load-point displacement diagram as illustrated in Fig. A.2a and correcting for indentation, using the curve in Fig. A.2b.

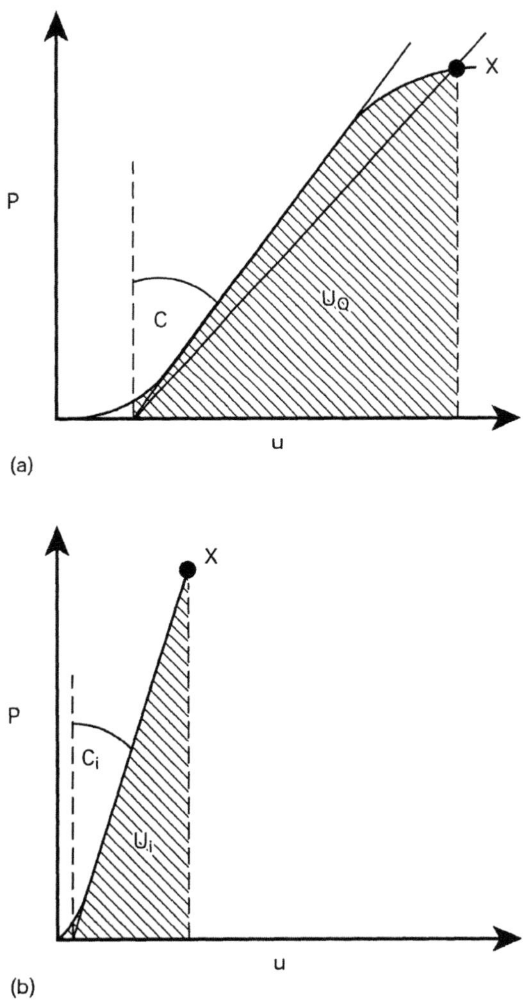

Fig. A.2. Method of correcting for indentation
(a) Load - deflection: fracture test
(b) Load - deflection: indentation

The corrected fracture energy is given by:

$U = U_Q - U_i$

Total energy corrections are usually <20%.

The fracture resistance, J, may then be calculated from U using Eq. (1) in the text of this protocol.

Further details of these correction procedures are given in the LEFM standard for plastics [Ref.4 in protocol above].

A.2. Use of a Comparator Bar

The only way to obtain load-line displacement directly in a SENB specimen is to measure the relative movement of appropriate points on the specimen. For example, one can measure the vertical displacement of the notch tip relative to a horizontal line that is a fixed distance from the undeformed edge of the specimen near the outer loading points. This type of direct measurement may be obtained by using a horizontal comparator bar [Ref. A.1.] and determining the vertical displacement of the bar relate to the notch tip or notch mouth as shown in Fig. A.3.

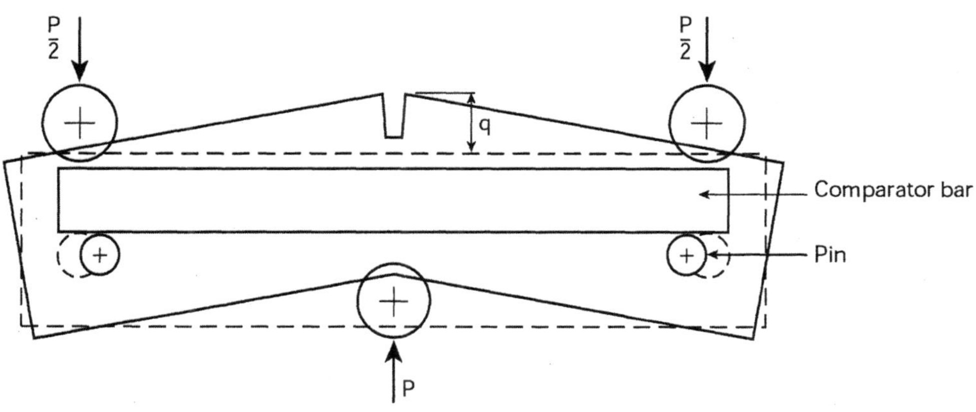

Fig. A.3. Principle of comparator bar measurement

A.3. References

A.1. Dawes, M. G: (1977). Elastic-plastic fracture toughness based on the COD and J-contour integral concepts, In: ASTM STP 668, *Symposium on Elastic-plastic Fracture*, p. 607.

ANNEX B

CRACK LENGTH MEASUREMENT

B.1. A major source of error with these tests can occur with measurement of the final crack front on the testpiece. There is a need to establish unambiguously which region on the fracture face is characteristic of actual crack growth in the test.

B.2. There are a number of methods available for breaking open a specimen so that the amount of crack growth that occurs during the test can be determined more precisely. These are:

i. high speed impact with or without prior cooling (in either solid carbon dioxide {-70°C} or liquid nitrogen), to produce a brittle fracture;

ii. high rate fatigue cycling - interference with any existing craze/process zones should be considered:

Useful results can be obtained if the sample is fatigued over a load range between 25 and 75% of the highest load achieved in the fracture test. Generally, room temperature cycling is adequate. However, for tougher materials, it may be useful to fatigue the specimen whilst it is cooled below ambient temperature. The crack should be propagated by at least 0.2mm by fatigue, although greater amounts of fatigue crack growth are acceptable. To finally break the testpiece open, it should be reloaded at the same rate as the original fracture test (i.e. typically 1mm/min) to give a smooth fracture face.

iii. injection of inks or dyes into the crack before breaking open - surface tension may be a problem here;

iv. two-step cooling - initially in solid carbon dioxide and later in liquid nitrogen.

B.3. Problems which have been identified are:

i. Direct immersion in liquid nitrogen for 5 minutes prior to breaking open may, in some cases, cause the specimen to shatter. This can often be avoided by making a sawcut through the back face opposite to the crack to a depth of approximately 5mm. The sample is then cooled using solid carbon dioxide (i.e. to around -70°C) for 5-10 minutes, after which it can be reloaded in the test rig and broken open. A further refinement is to fit a spacer (e.g. a broken piece of sawblade) into the sawcut before reloading so that the back face cannot close up and the amount of bending in the remaining ligament is reduced. With both these approaches, the compressive strain at the back face is reduced and this tends to minimise the risk of the material shattering.

ii. The appearance of additional features, e.g. half-moon shapes on the fracture face, has been noted (Fig. B.1.). These are probably a function of the cooling and breaking open cycle given to the specimen.

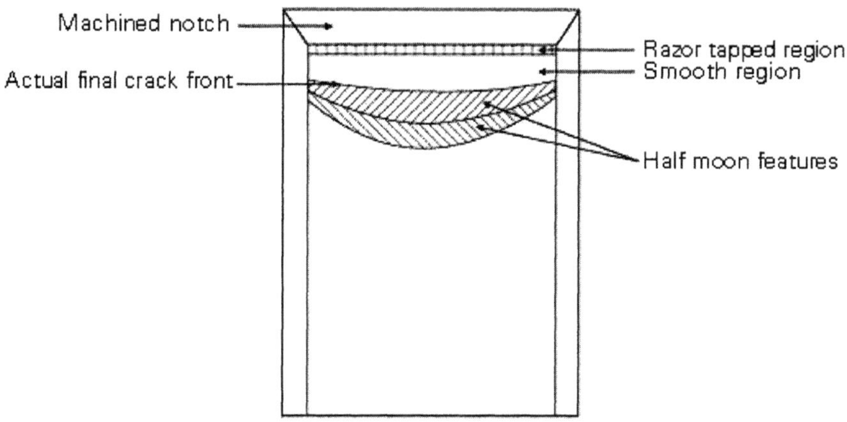

Fig. B.1. Possible features observed on polymer fracture faces after cooling in liquid nitrogen

iii. Additional features are likely if the material under test tends to exhibit multiple cracking ahead of the principal crack tip, e.g. polypropylene, and this can make interpretation of the fracture face more difficult.

To assess the influence of both the latter points, an additional precracked but untested sample should be broken open. This will help to indicate which features are characteristic of the cooling/breaking open operation for each polymer.

B.4. Where possible, optical or low magnification SEM photographs of the fracture faces should be taken and included with the report.

B.5. It is suggested that a fine bandsaw is employed to section a specimen midway through its thickness (Fig. B.2.). The cut face is then polished on successively finer grades of emery or silicon carbide to at least a 600 grit finish.

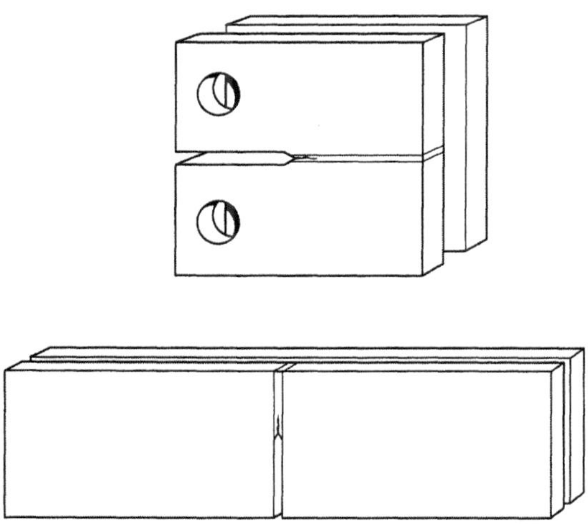

Fig. B.2. Diagram to show sectioning procedure for compact tension and single edge notch bend specimens

B.6. A reliable measurement of crack growth on a sectioned specimen can only be obtained if the crack is wedged fully open. This can be done by inserting a wedge into the crack mouth (Fig. B.3) after loading the testpiece to at least 80% of the total displacement reached in the test.

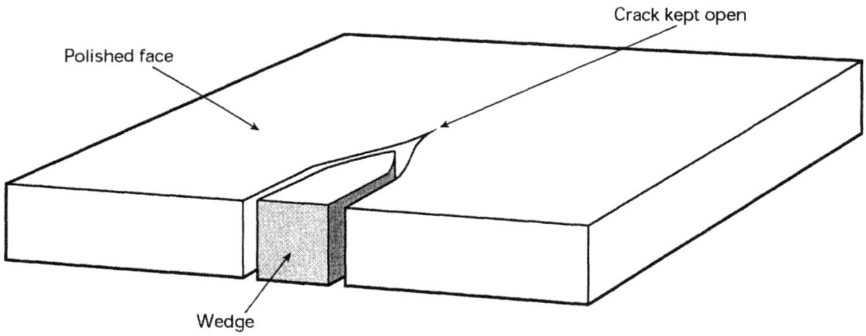

Fig. B.3. Use of a wedge to open crack tip

B.7. Alternatively, a small three-point loading rig can be used in which the load is applied by tightening a screw which forces the crack faces apart. (Further information on this type of rig can be obtained from Phil Marshall of Pipeline Developments or Roy Moore at ICI).

B.8. The crack growth on the section should be measured using either a travelling microscope or a high power optical microscope.

ANNEX C

Reporting form for J R-curve tests on plastics

Date(s) tests carried out				
Name				
Organisation				
Description of material				
Supplier				
Specimen type (CT or SENB)				
Mean specimen dimensions (mm)				
W	B	B_N	a_o	a_o/W
Sidegrooving - Y or N				
Included angle at root of sidegroove		Root radius of sidegroove (mm)		
Notching method used				
Root radius of crack tip (if measured)				
Yield strength (MPa)				
Loading rate in tensile test (mm/min)				
Young's Modulus E (MPa) - not essential				
Loading rate in J-test (mm/min)				
Crack length measurement method				
Resolution of measuring technique				
Δa_{max} (mm)		J_{max} (kJ/m^2)		

Specimen No.	1	2	3	4	5	6	7	8	9	10
J (kJ/m^2)										
Δa (mm)										
Used for curve fitting (Y/N)?										
Brittle fractures or 'pop-ins' (Add Y against relevant specimen)										
No. of specimens used for curve fitting										
Were the spacing requirements satisfied (Y/N)?										

Power law constants for $J = A\Delta a^N$	A		N	
$J_{0.2}$ (kJ/m^2)		Valid (Y/N)?		
J_{BL} (kJ/m^2)		Valid (Y/N)?		

J - FRACTURE TOUGHNESS OF POLYMERS AT IMPACT SPEED

H. J. MACGILLIVRAY

1. INTRODUCTION

It is often necessary to determine the fracture toughness of a material at loading rates higher than those used for conventional quasi-static tests. With the continual extension of the use of tough, ductile polymers into service applications where they encounter high loading rates, a simple, reliable method is particularly desirable. A further reason for testing at higher loading rate can be to obtain valid fracture toughness conditions, when the samples available are too small to provide sufficient constraint. Testing such samples at impact rates where the yield stress may be raised, may allow valid results to be obtained.

The objective of this test procedure is thus to extend the use of the ESIS elastic-plastic J crack growth method [1] described in the previous paper, to cases where impact loading is required, in a similar relationship to the static and dynamic K_c & G_c procedures discussed in 'Determination of Fracture Toughness...' and 'Determination of the Impact Fracture Toughness...'. This paper gives details of the particular difficulties which are met when testing at higher loading rates, and how they have been overcome in the ESIS annex on Impact J Testing. The development of the protocol is explained, and results from the comprehensive round-robin test programmes are discussed.

2. BACKGROUND TO THE PROCEDURE

A number of practicable impact J test methods have been developed over the last 25 years, and were reviewed in [2]. These were developed primarily for use with pressure vessel steels, but can be readily adapted to use with plastics [3].

One of the simplest to perform is the multi-specimen 'low-blow' or 'reduced velocity' test, and this has been selected as a basis for the ESIS TC4 procedure. Only the simplest designs of pendulum or dropweight testing machines are required to perform this test, preferable but not necessarily fitted with a force-instrumented striker. The machine must be of appropriate energy capacity and constructed so that the weight or pendulum can be released from any desired height, and the span should be variable. To perform the test, the energy of the impactor is deliberately limited so that the initial crack is extended but the specimen is not completely broken during the test. A series of nominally identical notched specimens is impacted at increasing energies to produce a range of crack extensions. The energy input to cause crack growth is either calculated from the potential energy available in the striker, or measured by integrating the force-displacement record. The specimens are then cooled in liquid nitrogen or some other medium, broken open and the ductile crack growth Δa during impact is measured. A plot is made of J against Δa, and the value of J at 0.2mm growth is found. The power law fit to the data is also determined.

This method has a number of advantages, including the possible use of simple non-instrumented testing machines, and uses identical specimens to the ESIS quasi-static J test. This allows a direct comparison between static and impact results. Among the drawbacks of the method are that it is generally suitable only for relatively low impact velocities, the initial velocity generally varies from test to test, and reduces during each test to a final value of zero, inevitably giving variable crack growth rates. The impact velocity is not independently variable unless the mass of the striker can be altered in small increments.

Alternative test methods were considered which could overcome some of these objections. The testing machine may be fitted with stop blocks to arrest the movement of the striker after a pre-determined travel: by adjusting the stop position in small increments, multiple specimens can be impacted to produce a range of crack extension. This method has the distinct advantage of permitting any desired impact velocity to be used, but loads on the machine are high, the striker must be instrumented and there is a possibility of some crack extension continuing under the inertial loading of the sample's own mass. The Chipperfield method [3] uses a specially-shaped specimen with shoulders machined to varying width. This allows the specimen to slip through the supporting anvils after a pre-determined amount of crack extension; since the load on the specimen is removed, there can be no further inertial crack growth. However, the striker must be instrumented and the specimens require careful preparation.

Single-specimen dynamic key curve test procedures have been successfully used to produce R-curves for polymers [4]. Work is also in progress to detect ductile crack initiation directly, using sensors positioned near the crack tip [5]. Although relatively simple experimentally and requiring only one or two specimens, thus offering advantages for production testing, the data evaluation required for these procedures is relatively complex and the results obtained would not be directly comparable with those from the ESIS quasi-static method. Single-specimen methods are however under continual development and will certainly become the future method of choice. However, none of these alternatives in their current form rivals the simplicity of the low-blow test.

The testing protocol is given in full in the following section. Where the provisions are identical to the main quasi-static method, a note refers to the Parent Protocol.

3. A TESTING PROTOCOL FOR CONDUCTING J - CRACK GROWTH RESISTANCE CURVE TESTS ON PLASTICS UNDER IMPACT LOADING: DRAFT 2 - OCTOBER 1996

This appendix extends the ESIS document 'A Testing Protocol for conducting J-Crack Growth Resistance Curve Test on Plastics: May 1995' [1] [referred to as the *Parent Protocol*] to impact loading conditions. The testing procedure used is the 'low-blow' or 'reduced velocity' method, where the energy of the impactor is deliberately limited so that the specimen is not completely broken during the test. A drop weight or pendulum testing machine of appropriate capacity is required to perform these tests, preferably with a force instrumented striker.

The recommendations of the Parent Protocol are to be followed as far as possible, and only the variations required are considered here. The section numbering used is identical for ease of reference.

Draft 1 - October 1993 was produced by Barry Crouch of DuPont, Wilmington. USA: this draft has been produced by Hugh MacGillivray of Imperial College, London, UK. Experience gained from the first round-robin test programme on toughened polyamide is included.

Introduction
It is often of interest to measure the toughness of a material at impact loading rates. However, particularly in tough polymers, it may not be possible to obtain a valid LEFM fracture toughness because samples of sufficient thickness are not available. An obvious alternative is to conduct an Impact J-Crack Growth Resistance test.

Specimen Configuration and Size
Refer to Parent Protocol. At present the impact test method is applied only to three point bend (SENB) specimens.

Notching
Refer to Parent Protocol.

Side – Grooving
Specimens must be side-grooved whenever possible.

Test Conditions
The test temperature, normally 23°C must be recorded. The procedure produces a variable loading rate during the test as the impactor gradually slows from its initial velocity to zero, then rebounds. The initial impact velocity may also vary between specimens, when different drop heights or pendulum angles are used to generate the require initial energies. This situation can be avoided if the mass of the impactor is adjusted between tests. [see section 8]

The impact velocity or range of velocities should be reported.

Measurement of Energy
The load line displacement is not measured directly.

Two options are available to estimate the energy input into the specimen at maximum displacement.

i) Calculate the potential energy of the falling weight or pendulum using:

$$U = mgh \qquad (1)$$

where m is the mass of the weight, g is the acceleration due to gravity and h is the drop height.

This method is simple and avoids the need for any instrumentation of the test. However it ignores any energy losses due to friction in the test equipment and during impact. Critically it also excludes the striker indentation correction. Thus any direct comparison with J results obtained using the Parent Protocol is subject to errors which may be difficult to quantify. For these reasons, this option is less preferred, but may be used for material ranking or quality control purposes where absolute accuracy is not required.

ii) Record the force - displacement signal during the impact. Determine U from the area under the curve at maximum displacement U_Q as shown in Fig 1. Correct this energy to allow for energy dissipation around the impact point and the supports. This is done by first measuring the (assumed linear) contact stiffness K of the material by performing impacts on fully supported un-notched specimens. At least three repetitions should be made, and the average value of K determined. If the specimen supports are not moved together, as shown in Fig 2a, a fixture such as that developed for the impact K_c - G_c protocol ('Determination of Fracture Toughness..) and shown in Fig 2b may be used.

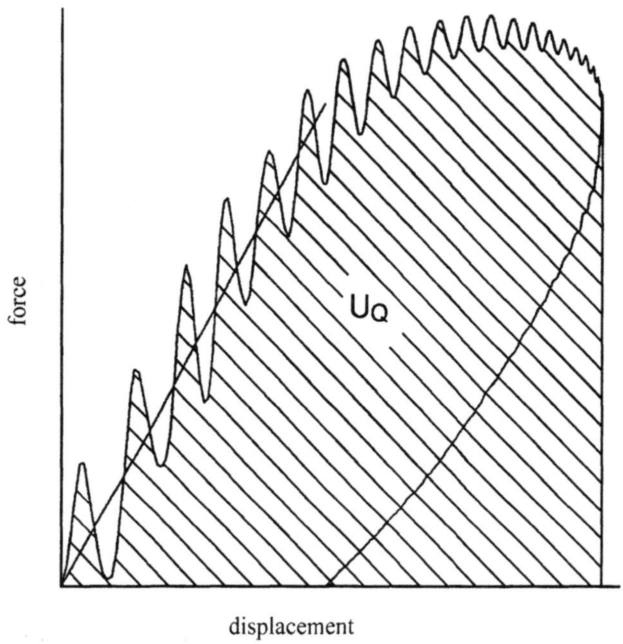

Figure 1. Typical force-displacement record for an impact-J test, showing energy at maximum displacement

Then the estimated energy dissipation due to indentation U_i is given by:

$$U_i = 0.5P^2 / K \qquad (2)$$

where P is the force at maximum displacement and:

$$U = U_Q - U_i \qquad (3)$$

Loading Rigs
A rig with moving rollers as shown in Fig 2 of the Parent Protocol is not normally used for impact testing. Fixed rollers or standard flat anvils with minimum 1mm radius corners should be used.

Test Procedure
A series of specimens is impacted with different energies to generate different amounts of crack growth. J is calculated from the energy U at maximum displacement. Either the same impact weight is used for all specimens, and the impact energy is altered by changing the drop height in the falling weight tower or pendulum, or the drop height may be held constant and the weight varied. An appropriate striker mass m can be estimated from the expected toughness of the material and the desired range of velocities. For example, if the maximum

applied J needed to generate sufficient crack growth can be estimated (J_{max}) and the dimensions of a typical specimen are known, then the required value of U can be estimated.

This defines the striker mass for a given maximum impact velocity V as:

$$m = J_{max} B (W - a) / V^2 \qquad (4)$$

After the impact the specimens are broken open and the amount of crack growth generated is measured. Cooling the specimens in liquid nitrogen or using dry ice may aid the identification of crack growth. Tests are continued until a spread of data points that satisfies the validity requirements has been obtained.

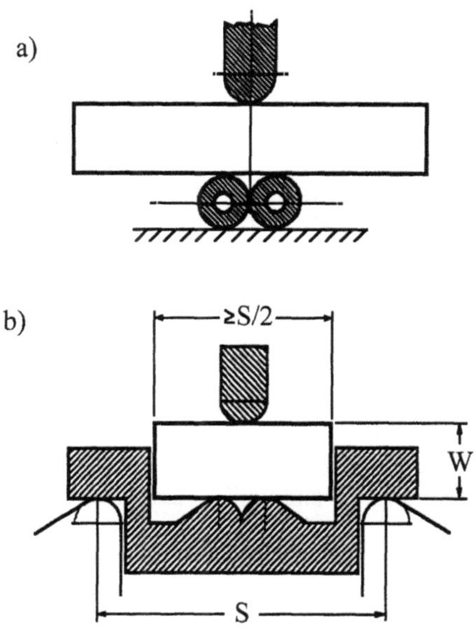

Figure 2. Arrangements for the indentation correction test
(a) rollers
(b) support fixture for use with fixed anvils

Crack Length Measurement
Refer to the Parent Protocol.

Response of Different Thermoplastics
Refer to the Parent Protocol.

Analysis Procedure
Refer to the Parent Protocol. U is the energy discussed in Section 3.6.

Construction of Valid Crack Growth Resistance Curves
Refer to the Parent Protocol. Specimen size requirements based on the yield stress of the material can be conservatively estimated using yield stresses measured at low strain rate. However these may be unnecessarily restrictive, and the dynamic yield stress at the same displacement rate (±50%) should then be used.

Determination of Fracture Parameters
Refer to the Parent Protocol.

Reporting
Refer to the Parent Protocol.

4. USING THE PROCEDURE; TEST RESULTS FROM THREE ROUND-ROBIN PROGRAMMES

A series of round-robin test programmes has been conducted to explore the practicality and repeatability of impact J testing using the protocol. Eleven European and one US laboratory have contributed their results. Testing began in 1994 and continued until 1997, and three materials with widely-varying properties have been evaluated. The first was a toughened polyamide, then a compression moulded acrylonitrile-butadiene-styrene terpolymer (ABS), and finally a high density polyethylene (HDPE). Test specimens were machined by the individual laboratories.

The results of these three programmes are presented, with references to individual laboratories removed, in the following tables and figures. Specimen dimensions, whether or not sidegrooved, and the dropweight striker mass, or pendulum energy, and velocity range are reported. In each case, the first table shows the data as supplied, and the second after elimination of any data sets not strictly complying with the protocol, or any individual data points which were outside the prescribed Δa_{min} and Δa_{max} limits. This has resulted in some slight differences between the original and final $J_{0.2}$ and power law fits.

The test protocol allows two variations of the test procedure, using either measured force [designated P] or non-instrumented potential energy methods [designated PE]. In addition, test specimens may be side-grooved to a depth of 10% each side. The objectives of the test programmes thus included determination of the effects of these allowed variations, in addition to the level of scatter inherent in the basic method. All laboratories used small pendulum or dropweight machines, impact velocities in the range 0.5 - 2 m/s, and some were able to vary the mass of the striker and thus conduct all their tests at a constant impact velocity.

Toughened polyamide was chosen for the first round-robin because it was expected to have a relatively high initiation J value and cause no problems with the measurement of ductile crack growth. Seven laboratories took part and the 11 sets of results they supplied are shown in Table 1; any values unreported are shown as [-] in the tables. The valid results are shown in Table 2, divided for clarity into PE and force groups.

In one case, the pendulum release angle was only variable in 10° increments, which caused problems in achieving the required energy distribution. Labs 5 and 7 were able to compare the force and PE methods directly, and lab 5 additionally tested both sidegrooved and non-sidegrooved samples under otherwise identical conditions.

Table 1: Impact-J results as reported from all first round-robin tests on toughened polyamide

Lab	B [B_N] (mm)	W (mm)	Span (mm)	Mass/Energy (kg/J)	Vel (m/s)	Side groove	PE / Force	$J_{0.2}$ (kJ/m²)	Power law fit
1	6.1	12.0	-	0.418	0.2/0.6	N	PE	18.9	49.2 x $^{0.610}$
2	6.1	12.0	48	varies	2.0	N	PE	18.7	50.3 x $^{0.616}$
3	[4.95]	12.0	-	0.61	0.5/1.0	Y	P	11.3	37.1 x $^{0.741}$
4	10.0	20.0	70	0.48	1.0/2.0	Y	PE	23.8	64.0 x $^{0.616}$
5a	6.1	12.0	48	25J	0.2/0.6	N	PE	23.8	76.8 x $^{0.743}$
5b	[5.08]	12.0	48	25J	0.2/0.6	Y	PE	23.0	60.7 x $^{0.733}$
5c	6.1	12.0	48	25J	0.2/0.6	N	P	20.4	66.2 x $^{0.753}$
5d	[5.08]	12.0	48	25J	0.2/0.6	Y	P	20.6	50.9 x $^{0.563}$
6	6.2	12.0	100	1.68	0.5/1.0	N	P	21	23.1 x $^{0.065}$
7a	6.1	11.9	-	0.955	0.5/1.5	N	PE	15.9	49.9 x $^{0.710}$
7b	6.1	11.9	-	0.955	0.5/1.5	N	P	14.1	45.7 x $^{0.730}$

Table 2: Final impact-J results from first round-robin tests on toughened polyamide

Lab	Side groove	PE / Force	$J_{0.2}$ (kJ/m²)		Power law fit
1	N	PE	18.4		49.0 x $^{0.607}$
2	N	PE	18.2		49.1 x $^{0.617}$
5a	N	PE	23.3		75.5 x $^{0.731}$
4	Y	PE	23.4		64.6 x $^{0.632}$
5b	Y	PE	23.4		55.8 x $^{0.540}$
		mean [PE method only]:	**21.3**	mean fit	**56.0 x $^{0.616}$**
		SD:	**2.8**		
3	Y	P	11.1		38.9 x $^{0.778}$
5c	N	P	20.3		65.7 x $^{0.729}$
5d	Y	P	20.6		51.1 x $^{0.565}$

The data shows a fairly wide scatter, but with the results for side-grooved specimens well within the general band. This indicates that the specimen size criteria have been achieved, and indeed the calculated B or [W-a] requirement is between 6mm and 8mm, depending on the value of σ_{y_2} used. The reported $J_{0.2}$ values are close to or slightly above the calculated J_{max} of 17kJ/m^2.

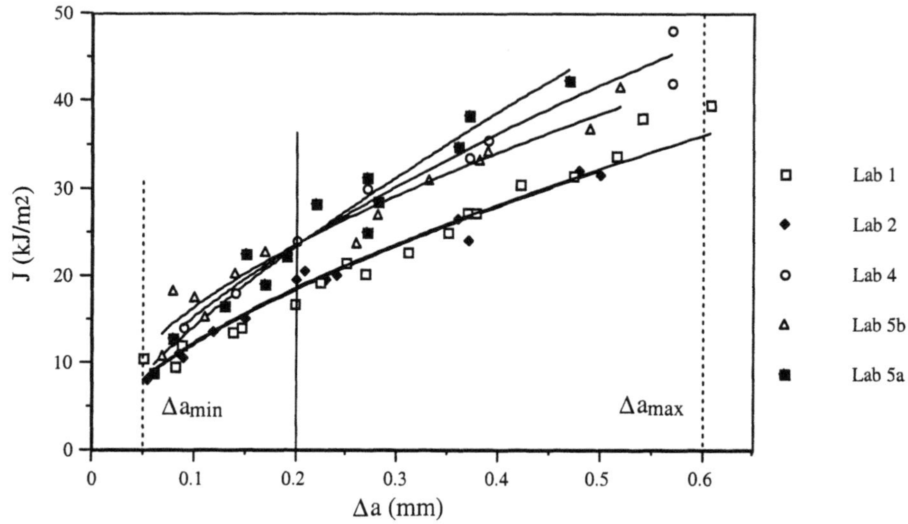

Figure 3. Individual R-curves for the five toughened polyamide data sets obtained using the potential energy method

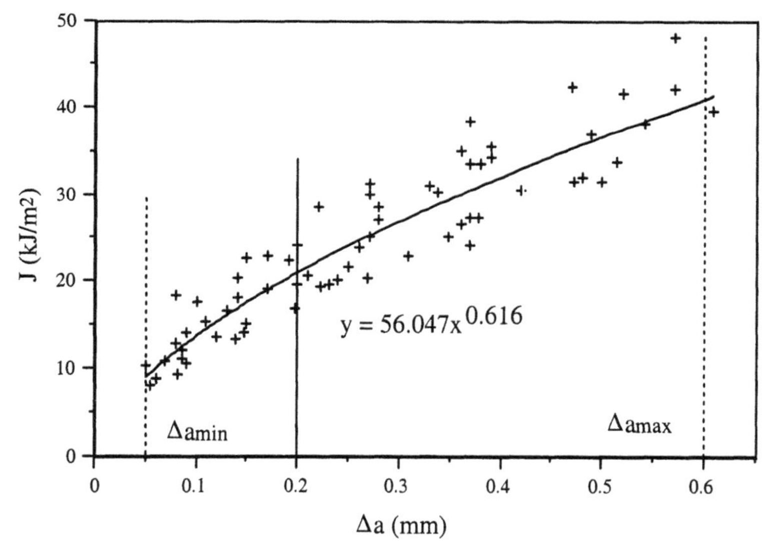

Figure 4. Toughened polyamide results for the potential energy method giving a mean figure for $J_{0.2}$ and the R-curve

Figure 3 shows the R-curves for the five data sets obtained using the potential energy method. If this data is fitted as a single set, Figure 4 results which gives a mean value for

$J_{0.2}$ and the R-curve. Figure 5 shows the same data re-plotted on a log-log scale and indicates the scatter clearly.

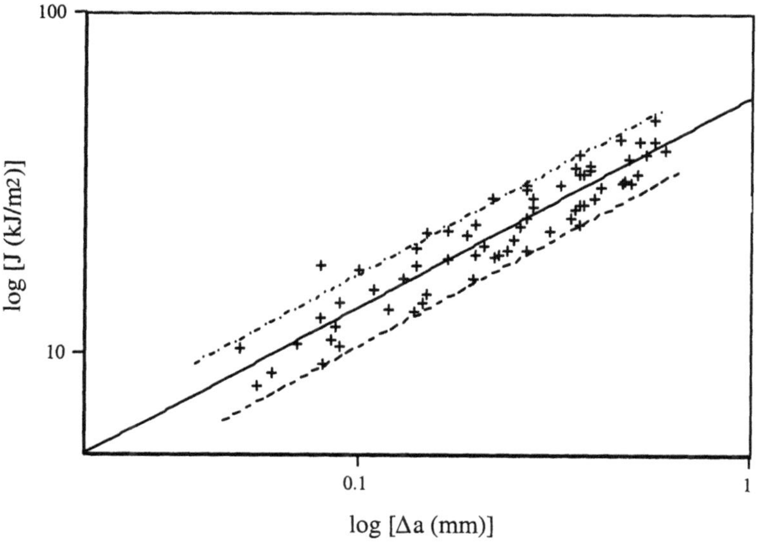

Figure 5. The toughened polyamide data shown in Figure 4 plotted on a log-log scale showing uniform scatter of data points

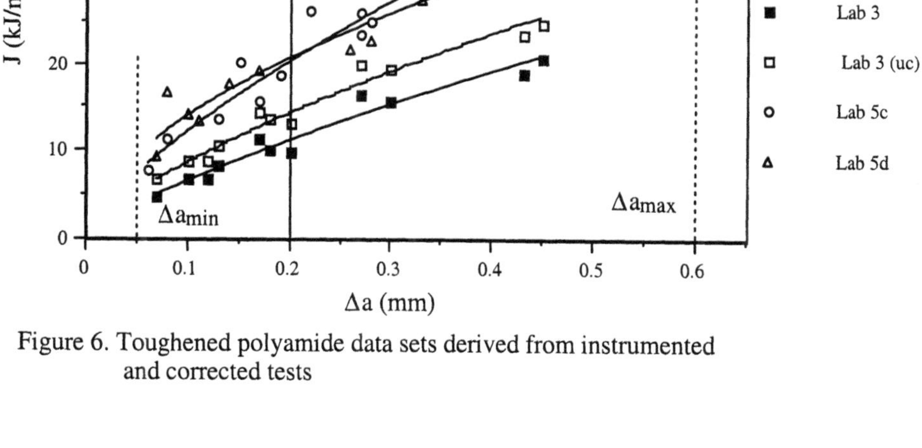

Figure 6. Toughened polyamide data sets derived from instrumented and corrected tests

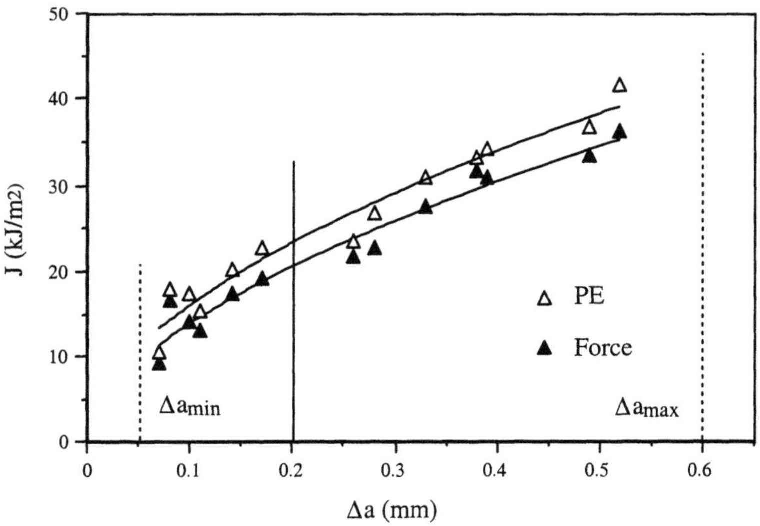

Figure 7. Comparison of R-curves for toughened polyamide obtained by the potential energy [PE] and the measured force methods for sidegrooved specimens from one laboratory

The corrected data sets derived from instrumented tests are plotted in Figure 6. Less data is available and the scatter is greater. There is clearly a significant difference between the Lab 3 and Lab 5 R-curves, which it has not been possible to resolve. The Lab 5 results were obtained from the same specimens as that laboratory's PE data, but using integrated force measurement and indentation correction, and are close to but slightly below, the mean PE curve in Figure 4.

The Lab 3 results are fully self-consistent, and cannot be explained by incorrect load calibration or striker mass. They seem to reflect a genuinely lower energy requirement of about 25%, which may be connected to the method of notching the specimens. This was the only laboratory to report using razor tapping: all others who specified a method used razor sliding. It might however be expected that any such effect would decrease with greater crack growth, but this has not occurred. The difference is further emphasised by the larger indentation correction U_i in the Lab 3 data, in the range 0.2 to 0.4 of the total energy, while the other laboratory reported corrections of around 0.1 - 0.2. The Lab 3 uncorrected curve is shown. It is just possible that Lab 3 was sent a different grade of polyamide in error.

The differences to be expected between R-curves obtained using the potential energy PE and the measured force methods are shown clearly in Figure 7. This is data for the same set of sidegrooved specimens tested by one laboratory. The measured force method gives a $J_{0.2}$ value 2.5 kJ/m^2, or 10.8% lower than by potential energy, and R-curves of generally similar shape which gradually diverge with increasing crack extension.

The second round-robin on compression moulded ABS included 8 laboratories who supplied a total of 15 data sets. Table 3 gives the reported data, with the final valid results presented in Table 4.

Table 3: Impact-J results as reported from all second round-robin tests on ABS

Lab	B [B_N] (mm)	W (mm)	Span (mm)	Mass/Energy (kg/J)	Vel (m/s)	Side groove	PE / Force	$J_{0.2}$ (kJ/m^2)	Power law fit
1	[7.7]	18.3	72	10J	1.2/1.3	Y	P	2.71	$5.36\ x^{0.424}$
2a	6.0	12.0	48	2J	-	N	p	2.8	$6.81\ x^{0.560}$
2b	[4.8]	12.0	48	2J	-	Y	P	2.45	$7.57\ x^{0.680}$
2c	6.0	12.0	48	2J	-	N	PE	3.0	$7.47\ x^{0.563}$
2d	[4.8]	12.0	48	2J	-	Y	PE	2.8	$8.33\ x^{0.688}$
3	6.0	12.0	-	-	0.3/0.7	N	PE	3.7	$6.22\ x^{0.317}$
4	[5.0]	12.0	-	-	0.5/0.6	Y	PE	3.43	$6.26\ x^{0.374}$
5	[5.0]	11.8	-	-	0.3/0.6	Y	PE	5.41	$11.5\ x^{0.470}$
6	[4.8]	12.0	48	0.5J	0.7/0.9	Y	PE	[1.6]	$6.49\ x^{0.880}$
7a	[4.8]	12.0	48	0.038	1.4/2.1	Y	PE	5.4	$9.70\ x^{0.360}$
7b	6.0	12.0	48	0.038	-	N	PE	5.1	$7.76\ x^{0.260}$
7c	6.0	12.0	48	0.038	-	N	PE	5.45	$10.4\ x^{0.400}$
7d	6.0	12.0	48	0.5J	1.2/1.9	N	PE	6.9	$11.6\ x^{0.320}$
8a	[4.8]	12.0	48	0.5J	0.7/1.3	Y	PE	5.47	$11.6\ x^{0.460}$
8b	[4.8]	12.0	48	1J	0.5/0.6	Y	PE	5.06	$9.09\ x^{0.360}$

Table 4: Final impact-J results from second round-robin tests on ABS

Lab	Side groove	PE / Force	$J_{0.2}$ (kJ/m^2)		Power law fit
2d	Y	PE	2.75		$8.33\ x^{0.688}$
3	N	PE	3.73		$6.22\ x^{0.317}$
4	Y	PE	3.43		$6.26\ x^{0.374}$
5	Y	PE	5.4		$11.5\ x^{0.470}$
7a	Y	PE	5.43		$9.70\ x^{0.360}$
7b	N	PE	5.1		$7.76\ x^{0.260}$
7c	N	PE	5.46		$10.4\ x^{0.400}$
8a	Y	PE	5.53		$11.6\ x^{0.460}$
	mean [PE method only]:		**4.60**	mean fit:	**8.53 x** $^{0.391}$
		SD:	**1.11**		
2b	Y	P	2.53		$7.57\ x^{0.680}$

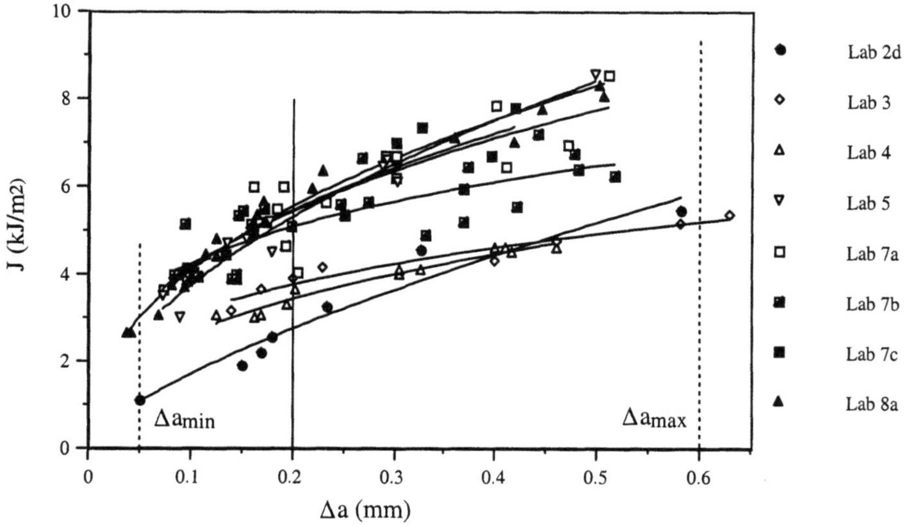

Figure 8. Individual R-curves for the eight ABS data sets obtained using the potential energy method

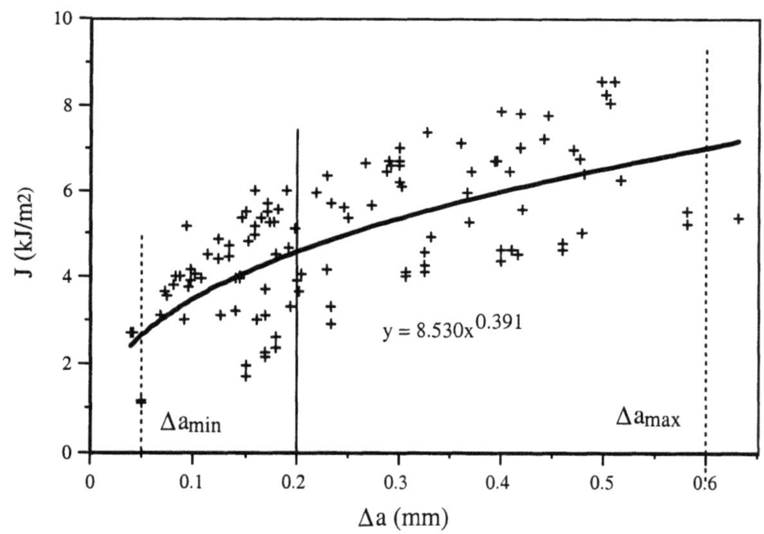

Figure 9. ABS results for the potential energy method giving a mean figure for $J_{0.2}$ and the R-curve

All these results are plotted in Figure 8, and a large scatter is apparent. The data divides almost into an upper and lower group, with a range of nearly 2:1 in $J_{0.2}$, and with power law fits having different slopes. Some effect of hammer mass on the results 8a and 8b was noted, from tests under otherwise similar conditions.

Figure 9 shows the data plotted as a single set, and gives a mean figure for $J_{0.2}$ and the R-curve. The absolute value of scatter at around $\pm$ 1.1 kJ/m^2 at $J_{0.2}$ is smaller than that for toughened polyamide, and may represent the limit to be expected for the test method. The very low energy required to cause crack growth in ABS caused problems for some labs, who were unable to use their instrumentation or could not obtain the smaller crack extensions required. Side grooves have no noticeable effect in this material. Direct comparison between the measured force and potential energy methods is possible for results 2b and 2d, and indicates a value for $J_{0.2}$ which is 0.35 kJ/m^2, or 12.5% lower for measured force than potential energy; this effect is of similar magnitude and sense to that noted for toughened polyamide.

Table 5: Impact J results as reported for third round-robin tests on HDPE

Lab	B [B_N] (mm)	W (mm)	Span (mm)	Mass/Energy (kg/J)	Vel (m/s)	Side groove	PE / Force	$J_{0.2}$ (kJ/m^2)	Power law fit
1a	6	12	-	-	0.4/0.8	N	PE	{8.7}	14.1 x $^{0.299}$
1b	12	24	-	-	0.4/0.8	N	PE	{8.3}	14.3 x $^{0.350}$
2	[5.0]	12	-	-	0.8/1.0	Y	PE	16.2	21.3 x $^{0.170}$
3	[5.3]	12.1	-	-	0.4/0.7	Y	PE	12.1	26.6 x $^{0.490}$
4a	[5.4]	18	72	0.31	1.0/1.6	Y	PE	14.9	21.1 x $^{0.215}$
4b	12	18	72	0.48	1.0/1.6	Y	PE	15.3	22.9 x $^{0.274}$
5a	[4.7]	12.7	24	.25/.42	1.0	Y	PE	13.1	23.8 x $^{0.371}$
5b	[4.7]	12.7	24	.25/.42	1.0	Y	P	11.9	21.9 x $^{0.380}$
5c	[10.1]	23.9	50	1.0/1.7	1.0	Y	PE	13.5	17.7 x $^{0.168}$
6a	[4.8]	12	48	1J	0.9/1.2	Y	PE	10.8	17.0 x $^{0.280}$
6b	[4.8]	12	48	2J	0.7/1.1	Y	PE	12.0	18.1 x $^{0.265}$
7a	[5.5]	12	48	.05/.11	2.0	Y	{P}	9.4	17.0 x $^{0.370}$
7b	[5.5]	12	48	.09/.11	2.0	Y	{P}	13.5	21.3 x $^{0.280}$
7c	[5.5]	12	48	.08/.10	2.0	Y	{P}	{19.2}	42.8 x $^{0.500}$
7d	[5.5]	12	48	.05/.19	2.0	N	{P}	11.3	26.8 x $^{0.540}$
7e	[5.5]	12	48	0.5J	1.6/2.1	N	{P}	10.6	18.8 x $^{0.350}$

The third round-robin was performed on HDPE, with the seven laboratories that took part supplying a total of 16 data sets. Since a size effect could be anticipated for this material, specimens of differing absolute size were tested and most labs sidegrooved their samples. The data supplied is shown in Table 5: note that some data which was very far from the validity requirements is shown thus { }. Lab 7 measured energy from the force-displacement record, but did not make indentation corrections, and the results have been treated as equivalent to PE. The final valid data is presented in Table 6.

It can be noted from Table 5 that sidegrooving has been successful in increasing constraint, since no size effect is visible in the pairs of results from the laboratories that tested two specimen sizes, 4a/4b and 5a/5c. Most of the $J_{0.2}$ results are also below the calculated J_{max} of 16 kJ/m^2.

Table 6: Final impact-J results from third round-robin tests on HDPE

Lab	Side groove	PE / Force	$J_{0.2}$ (kJ/m^2)	Power law fit
2	Y	PE	16.1	20.5 x $^{0.151}$
3	Y	PE	12.1	26.6 x $^{0.490}$
4a	Y	PE	14.9	22.9 x $^{0.266}$
4b	Y	PE	14.7	22.9 x $^{0.274}$
5a	Y	PE	13.0	21.8 x $^{0.323}$
5c	Y	PE	13.5	17.6 x $^{0.164}$
6a	Y	PE	10.8	17.0 x $^{0.280}$
7a	Y	{P}	9.4	17.0 x $^{0.367}$
7b	Y	{P}	13.5	21.7 x $^{0.296}$
	mean [PE method only]:		<u>**13.1**</u>	mean fit: **19.5 x $^{0.246}$**
		SD:	<u>**2.1**</u>	
5b	Y	P	11.7	19.8 x $^{0.326}$

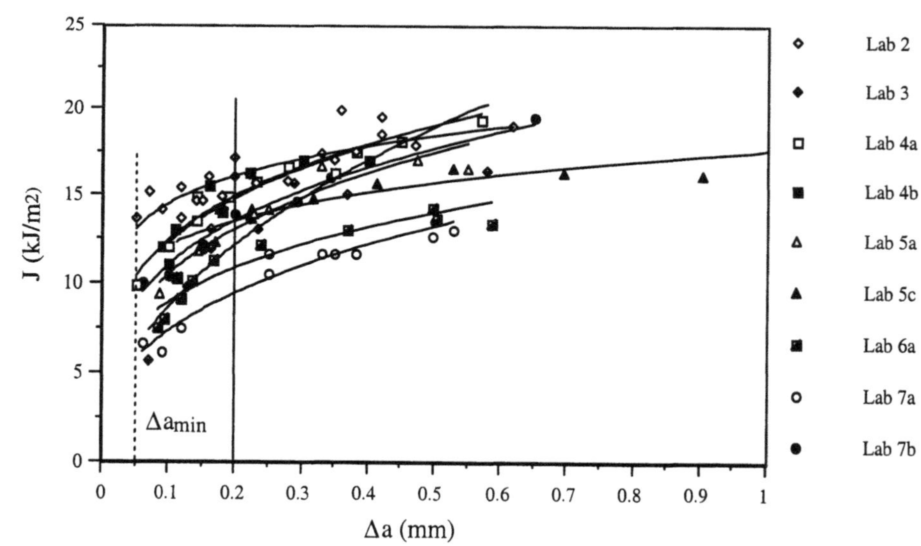

Figure 10. Individual R-curves for the nine HDPE data sets obtained using the potential energy method, replotted and curve fitted.

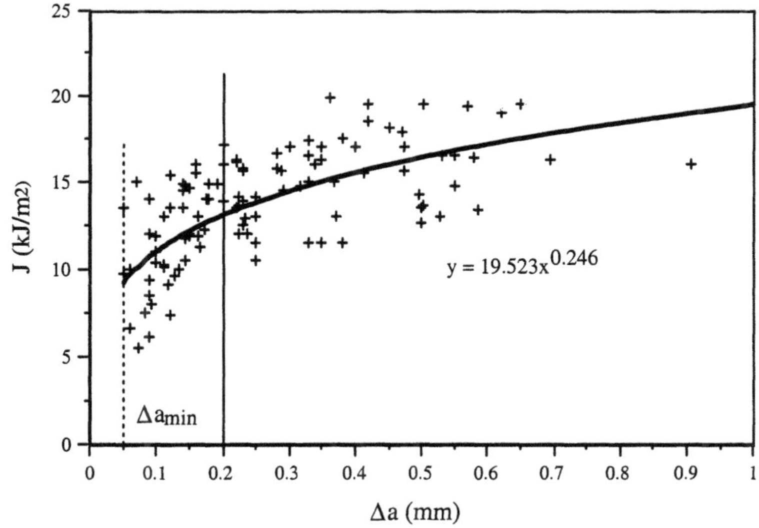

Figure 11. HDPE results for the potential energy method giving a mean figure for $J_{0.2}$ and the R-curve.

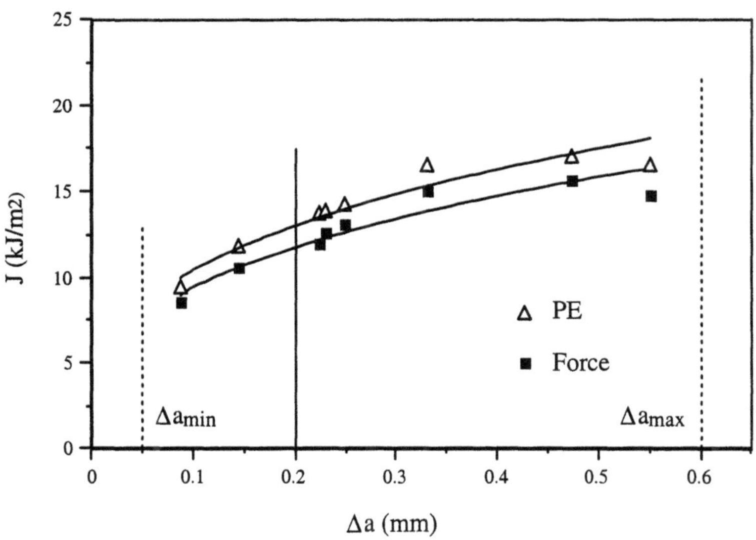

Figure 12. Comparison of R-curves for HDPE obtained by potential energy [PE] and measured force methods for sidegrooved specimens from one laboratory

The individual valid data is plotted in Figure 10, and shows a scatter only slightly lower than that for toughened polyamide. Care was required to avoid confusing the craze zone that develops ahead of the crack tip with actual crack growth, with some laboratories using dye to improve contrast, and others employing the centreline section technique to confirm their measurements.

Figure 11 again shows the data plotted and curve-fitted as a single set. Finally Figure 12 indicates the comparison between R-curves obtained using the potential energy PE and the measured force methods, using results 5a and 5b for the same set of specimens tested by one laboratory. The measured force method gives a $J_{0.2}$ value 1.3 kJ/m^2, or 10% lower than by the potential energy method, and very similar R-curves were obtained.

5. CONCLUDING REMARKS

The test programmes have shown the ESIS method to be practical and able to yield reasonably repeatable results when used with care. The extent of scatter which may be anticipated in test results has been evaluated, and it is clear from the R-curves that uncertainty is greater in the measurement of J than of crack growth Δa. Generally, comparable results were obtained from sets of data measured within individual laboratories, but larger variations were found from one laboratory to another. This suggests that the scatter is mainly caused by calibration errors. In the case of the potential energy technique, errors can occur in release height or angle measurement, in the effective mass of the striker or because of friction in the guides or bearings. Where force-displacement is measured, errors may be introduced by incorrect loadcell calibration or the integration method used to determine energy. For both techniques, machine alignment, and striker/anvil wear or damage could be secondary sources of error. It may be worth considering the development of a calibration method, perhaps using standard test specimens in the manner generally used for validating Charpy pendulum machines for metals.

Some comparisons with the results from earlier TC4 quasi - static J testing round-robin are interesting; impact tests using toughened polyamide showed a standard deviation of 2.8 while the quasi-static deviation was considerably smaller, at about 1.0. For ABS the respective dynamic and static deviations were closer, at 1.11 and about 0.8. The mean values of $J_{0.2}$ for impact were 4.6 kJ/m^2 for PE and 2.53 kJ/m^2 using force, while the mean quasi-static was intermediate at 3.32 kJ/m^2 .

The potential energy technique has been shown to give satisfactory results, and the extra complication needed for force measurement and indentation correction may in practice be unnecessary for most purposes. The effect of the notching method was not specifically investigated in these programmes; generally razor sliding was used successfully. Again it may be useful to run some additional tests. A number of useful suggestions have been received, and will be incorporated in a future draft of the protocol.

6. REFERENCES

[1] A Testing Protocol for conducting J-Crack Growth Resistance Curve Tests on Plastics, ESIS Technical Committee TC4, May 1995.

[2] MacGillivray, H. J. and Turner, C. E., 4th Int. Conf. on Mechanical Properties at High Strain Rates, Oxford 1989.

[3] Huang, D. D., Toughened Plastics 1, Science and Engineering, Advances in Chemistry Series 233, ACS, Washington D.C. 1993.

[4] Seidler, S., Grellmann, R. and Lach, R. Proc. 2nd ESIS Conf. on Polymers and Composites, Les Diablerets, 1999.

[5] Lenkey, G.B., Winkler, S., Major, Z. and Levay, I., Proc. ECF 12, Poitiers 1996.

7. LIST OF PARTICIPANTS

The author wishes to thank all the members of ESIS TC4 Technical Committee on Polymers and Composites, who have generously contributed their expertise to the testing programmes and data evaluation, and given permission for their results to be used.

ICI plc	Dr D R Moore
BASF	Dr F Ramsteiner
BP Chemicals	Dr E Clutton
CEAST Turin	Dr M Grosso
DuPont, Wilmington	Dr B Crouch
Lausanne Polytechnic EPFL	Dr P Beguelin
Imperial College	Prof J G Williams
	Mr H MacGillivray
Martin Luther University, Merseberg	Prof S Seidler
Politechnico di Milano	Prof A Pavan
Rhone Poulanc	Dr G Orange
Shell	Dr Ir A Cervenka
Twente University	Dr P Reed

8. MATERIAL SUPPLIERS

The author wishes to thank DuPont, BASF, TWI and BP for the supply of materials used in the testing programmes.

ESSENTIAL WORK OF FRACTURE

E. CLUTTON

1. BACKGROUND AND AIMS

There is increasing use of the essential work of fracture (EWF) method to determine the toughness of thin plastic films. There are a number of potential areas for use of this technique such as in the measurement of the fracture toughness of thin coatings and paints, the study of packaging properties, etc. The method is attractive in that it deals with the sort of gross ductility that can occur in the plane stress fracture state. For most materials of this nature, it would be inappropriate to take the steps of moving to thicker test specimens in order to suppress crack tip ductility. This would bring about a change in stress state which does not occur in practice.

The EWF approach is a means of partitioning the energy associated with fracture into two parts. One element is specific to the fracture of the material and, as such, is a material parameter. The second element is related to gross plastic deformation and depends upon the geometry of the fractured specimen or component. The basis of this partitioning was originally suggested by Broberg [1]. He proposed that the non-elastic region at the tip of a crack should be sub-divided into a region where the fracture process takes place and an outer region surrounding it where further gross ductility occurs. This idea has been developed by Cotterell, Reddel and Mai for metals [2,3] and more recently by a series of workers [4 -8] for ductile polymers. Cotterell et al. [2] have labelled the crack tip specific work the **essential work of fracture, w_e**, and the work done in the outer region the **non-essential work**. The essential work of fracture has been shown to be a material property for a given sheet thickness and independent of the specimen geometry [9]. The non-essential work, however, depends on the shape of the plastic zone surrounding the crack and is related to the **plastic work dissipation per unit volume of material, w_P.**

The principle of the technique is to measure the load-displacement trace and hence the energy to fracture for a series of fracture specimens, ensuring that plasticity in the ligament (fracture region) is fully developed. In such cases, it is possible to partition the work of fracture into an element taking place along the fracture line and another element taking place in a volume of material surrounding the crack. The former element is proportional to the fracture area and hence the ligament length (l), while the latter element is proportional to the volume of the outer region. For both metals and plastics, it has been observed that the volume of the outer region is proportional to the square of the ligament length. So for any valid set of conditions, the total energy absorbed in fracturing such a specimen, W_f, is given by the expression

$$W_f = w_e.lt + w_p.\beta l^2 t \quad (1)$$

where l is the ligament length, t is the sheet thickness and β is a shape factor associated with the dimension of the plastic zone normal to the crack line.
Normalising by lt, we obtain

$$w_f\ (=W_f/lt) = w_e + \beta w_p.l \quad (2)$$

and the essential work of fracture may then be determined from a graph of w_f plotted against l. A typical plot of w_f versus l is shown in Figure 1 for LLDPE. The essential work of fracture is then derived from a 'best-fit' linear regression analysis of the data.

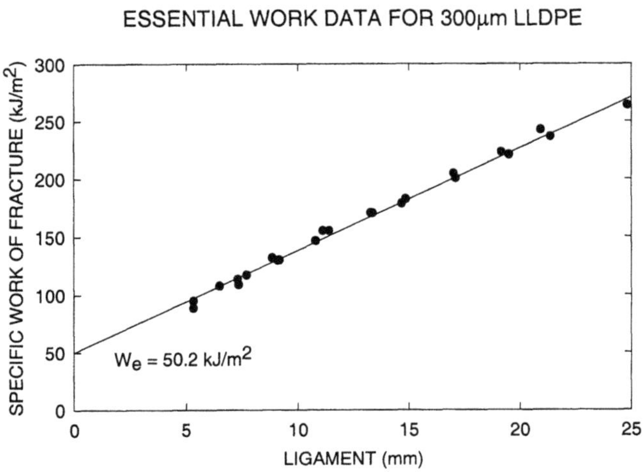

Fig. 1. Typical plot of w_f versus l

It is customary to determine w_e in a state of plane stress, although Saleemi and Nairn [5] amongst others have shown fairly successfully that it is possible to derive a value for plane strain. The measurement of a plane stress value demands that the state of stress in the ligament of each test specimen is one of plane stress. This has implications for the type of specimen used and the range of ligament lengths employed.

Whilst in principle the essential work technique can be applied to any fracture specimen geometry, the constraints on test validity dictate that some geometries are more appropriate than others. It is necessary to have relatively short fracture ligaments to ensure full yielding and this means using deeply notched specimens. Most work in the literature has used either double edge notched tension (DENT) specimens or single edge notched tension (SENT) specimens. In the interests of consistency and reproducibility, the DENT specimen is recommended as the most appropriate specimen type. Details of the nomenclature for this specimen are given under the final test protocol in Figure TP1.

There are suggested constraints on ligament length to maintain a state of plane stress. The suggested minimum ligament is 3 times the thickness, although this is rather arbitrary. The suggested maximum is the minimum of 2 times the plastic zone size and one third of the specimen width. This is based on the need for the ligament to be fully yielded prior to crack propagation.

The aim of the work within the ESIS technical committee (TC4) on the fracture testing of plastics and composites has been the application of the EWF method to measure a plane stress toughness. The committee's attempts have been focused on defining the most appropriate test method to give best reproducibility between laboratories. This paper draws on the experience gained during 7 years of round robin activity involving up to 23 participants.

2. ISSUES AND PROBLEMS WITH THE PROCEDURE

During a seven year period between 1992 and 1999, a series of round-robin exercises were performed under the guidance of the ESIS TC4 committee on the fracture of polymers and composites. A total of 23 laboratories participated in these round-robins (see Appendix 1 for details) and the following aspects of the test were raised and discussed.

Geometry and test speed

From experience in the TC4 round-robin exercises, there is sufficient evidence to conclude that the specimen size (both width and length) does not influence the result, provided the ligament is small compared to the specimen width, say, a maximum of one third of the width. Test speed is somewhat arbitrary, but the current protocol specifies a calculated speed of 0.2 times the gauge length to ensure equivalent strain rates are used.

Shape of the plastic zone

In principle, it is possible to determine the shape of the plastic zone surrounding the fractured ligament and so derive a value for the plastic work energy density, w_p. This might then be compared with the total energy per unit volume required to deform a tensile specimen to failure.

For metals, the zone size tends to be large and easy to measure, but for plastics this is not the case, particularly those that neck before failure. It was found that the shape of the zone could not be measured with sufficient precision and that the definition of the zone boundaries was a highly subjective matter. Attempts to define the zone shape, in order to extract w_p from the data, were abandoned.

Trace coalescence

One of the most noticeable features of the essential work of fracture method is the similarity in shape of the load-displacement diagrams for a specimen set consisting of a standard range of ligament lengths. This is illustrated in Fig. 2 for LLDPE traces from one of the TC4 round-robins.

It was suggested that there might be some way of normalising the traces by factors dependent on the ligament length. It was proposed that there was a factor relating the peak stress and another factor relating the maximum displacement to some characteristic normalised loading curve. In effect, this proposed that the experiments at different ligament lengths could all be reduced to one datum. Unfortunately, when the separate loading curves were normalised and plotted together, it was found that they were similar but sufficiently different to cast doubts on the proposal. There was insufficient evidence to warrant further study along these lines and the approach was abandoned.

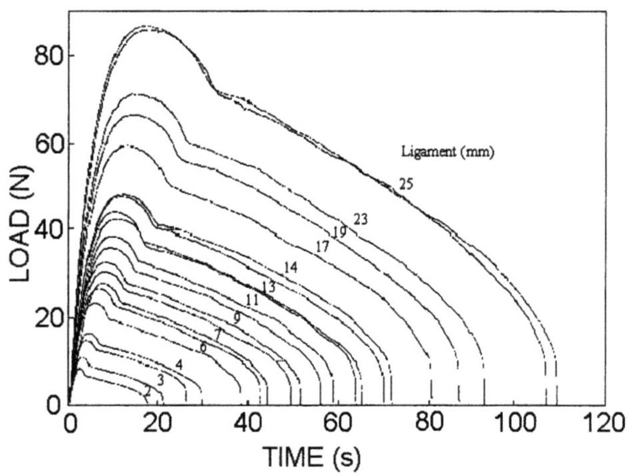

Fig. 2. Typical load - time traces for LLDPE from EWF tests

Number and distribution of data

Initial statistical analysis provided by Imperial College Mathematics Department [10] showed that the most efficient and accurate method of determining the intercept (w_e) and gradient (βw_p) was to perform tests at the limits of ligament length with more points at the lower limit. It was felt that this would not provide any confidence that the data was linear and supported the theory. Therefore, in the initial protocols it was specified that data should cover the entire ligament range, but that there should be a distribution of data according to a certain scheme which gave a weighting towards data at shorter ligaments.

Data from various laboratories showed that data distributed in this manner did not produce any greater accuracy in the determination of the essential work of fracture. More recently the protocol has reverted to a uniform distribution of data.

One laboratory (see [8]) took the LLDPE from one of the round-robins and performed a large number of tests to quantify how the size of the confidence interval on the essential work of fracture was related to the number of data points used. These results were compared with data for zinc and aluminium and this is shown in Figure 3 in the form of $\Delta w_e/w_e$ versus number of data points, where Δw_e is the confidence interval for w_e.

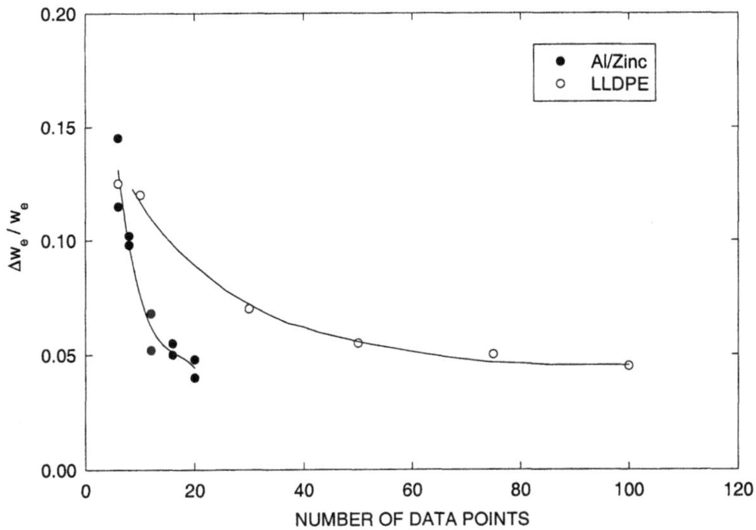

Fig. 3. Level of precision given by the EWF method for LLDPE in comparison to aluminium and zinc.

It can clearly be seen that to achieve the same accuracy in determination of w_e, it is necessary to test many more specimens of LLDPE than is the case for zinc or aluminium. The initial essential work of fracture protocols specified that a minimum of 20 specimens should be tested. This number is vindicated by the above data; 20 data points should lead to an accuracy of better than 10% in terms of $\Delta w_e/w_e$.

Ligament length range

The transition region between plane stress and plane strain is ill defined and appears to vary according to the material studied. For example, in rubber toughened Nylon 66, Saleemi and Nairn [5] observed it at $l = 3t$, whereas in LLDPE Wu and Mai [8] observed it at $l = 14t$, where t is the thickness of the material. Therefore, there is no clearly defined lower bound to ligament length which ensures that data is obtained under plane stress conditions. However, an extensive review of the literature for polymers suggests that imposing a lower bound of max ($3t$, 5mm) is currently the most reasonable position to take.

It is generally advocated that the maximum ligament should be min ($W/3$, $2r_p$), where W is the specimen width and r_p is the plastic zone length. $W/3$ is arbitrarily included to avoid edge effects. The basis for the $2r_p$ is that the ligament should be fully yielded prior to crack growth. However, an extensive search of the literature reveals that the departure from linearity in this upper region bears no relation to these parameters. Indeed, the data is often found to be linear well beyond this limit. Some have suggested that the upper bound be based on observation of whether or not the ligament is fully yielded prior to crack growth, but this is rather subjective and requires a definition of the fully yielded state. Neither is the upper ligament limit well defined. For practical purposes, the most reasonable upper limit is the somewhat arbitrary value of 15mm.

Statistical treatment of the data

Another statistical issue arising was the validity of applying the standard least squares linear regression to the data. This is only strictly valid if the variance in the measured value of specific work is the same for all ligament lengths; i.e. the error in measurement of specific work is independent of ligament length. This aspect was incorporated into round-robins on LLDPE and PET, where laboratories were asked to generate data from replicate specimens at different ligament lengths. There was no clear trend in variance across the ligament length range, with the implication from these data that the conventional unbiased least squares linear regression is applicable to essential work data.

Notch quality

In the round-robin on PET, the various participating laboratories were asked to generate one set of data with their normal method of notching the specimens and, if possible, record with a scanning electron microscope (SEM) the quality of the notch and the notch tip radius. In addition, laboratories were asked to attempt variations to the notching quality and produce second data sets using this variant so that the issue could be quantified.

Several laboratories reported data including a SEM photograph of the notch tip and these results are given in Table1.

Table 1 - PET Data and Measured Notch Radii

Laboratory Number	Notch Type	Estimated Notch Radius	w_e (kJ/m^2)
1	Scalpel	~1 μm	20.1
	Die-punch	50 μm damage zone	58.3
5	Sliding razor	<10 μm	31.2
9	Jig cut + razor	~1 μm	22.2
	Scissors cut		63.3
16	Razor		23.6
17	Razor	25 μm damage zone	40
18	Razor	~5 μm	34.3

It is reasonably clear from the data that lower values for essential work of fracture were obtained from those laboratories that managed to obtain the sharpest notches. In addition, two laboratories were able to produce much higher w_e values from cruder methods of cutting the notches, therefore proving that the sharpness of the notch is very important to the essential work of fracture method.

Stress levels

For the DENT specimen geometry, Hill [11] has performed a fully plastic plane stress analysis, which calculates the maximum stress, σ_{max}, in the specimen to be given by

$$\sigma_{max} = 1.15\ \sigma_y \qquad (3)$$

where σ_y is the yield stress of the material. **This formula was used only as a check for the essential work data.** Most laboratories found that the maximum stresses from their data were of the same order as given by equation (3), but with no consensus on the relative position. It was generally found that, as the ligament length increased, the maximum stresses decreased gradually. There was a steeper decrease in maximum stresses for shorter ligaments. Figure 4 typifies the comparison between maximum stresses and equation (3) for EP copolymer. The dotted line in the figure represents equation (3), where the yield stress was taken as an average of those values measured in the different laboratories.

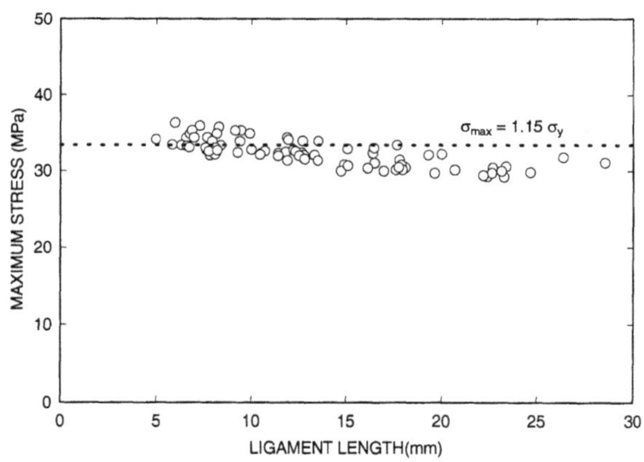

Fig. 4. Comparison of maximum stress with 1.15 σ_y for EP copolymer

Stress criterion

In view of the difficulties in defining the minimum ligament length at which the transition from plane stress to a mixed stress state occurs, it was proposed that a criterion on the maximum stress values might fulfil this purpose. It can be seen in the previous figure that the maximum stress rises for shorter ligaments, which reflects a change in stress state. It was proposed that data for which the maximum stress was above, say, 1.1 times some mean stress should be excluded from the analysis. This criterion could have been based on equation (3), but, because of the discrepancies seen between equation (3) and the data, it was decided to use a mean stress calculated from the essential work data.

In the most recent protocol, an average value of maximum stress is calculated and denoted σ_m. Then the criterion is to reject any essential work data for which the maximum stress is greater than $1.1\sigma_m$ or less than $0.9\sigma_m$. These limits are rather arbitrary, but have been found to be useful in defining the transition to a mixed-mode stress state. In addition, they also exclude any data for which the maximum stress is low, possibly arising from premature crack growth or simply experimental errors.

This criterion has been applied to all essential work data from the TC4 round-robin on 300μm LLDPE and Table 2 illustrates how its application has improved the reproducibility of data from the 11 laboratories involved. Although the average essential work of fracture is virtually unchanged, the standard deviation has been reduced from 9.8 to 6.8.

Table 2 : Essential Work for 300μm LLDPE with and without the Stress Criterion

Laboratory	Original w_e	Revised w_e (after applying stress criterion)
1	41	48.5
2	45	40.3
3	48	59.6
6	71	59.9
7	50	49.5
9	50	47.7
11	60	56.9
12	50	50.2
13	34	47.4
14	43	39.5
21	47	47.6
MEAN	49	49.7
STD DEV	9.8	6.8

Influence of specimen preparation

In one of the round-robins, an exercise was performed to assess the influence of specimen preparation, in which specimens of LLDPE were prepared by ICI using their purpose-built notching device. Each laboratory was issued with a set of specimens for testing and asked to prepare a further set of their own for testing in parallel. The results of this exercise are shown in Table 3 and illustrate a negligible influence arising from specimen preparation.

Table 3 : Comparison of EWF Data for LLDPE : ICI versus 'Own' Specimens

	ICI specimens	Own Specimens
	20	20
	29.3	21.3
	26.2	28.8
	31.4	36.6
	31.8	19
	33	22
	32.1	17.4
	26.2	30
	28.8	32.3
	42.3	23.4
	21.6	25.8
Mean	29.34	25.15
Std Dev	6.05	6.1
SD/Mean	0.21	0.24

Non-linearity

Once the valid data had been defined by applying the stress criterion, many of the individual data sets for the different materials appeared to be linear. However, there were still a number of sets of data suggesting that some non-linearity might still be present. This was the case irrespective of material and is typified by data for PET shown in Figure 5, where a power law relationship has been applied rather than a linear one.

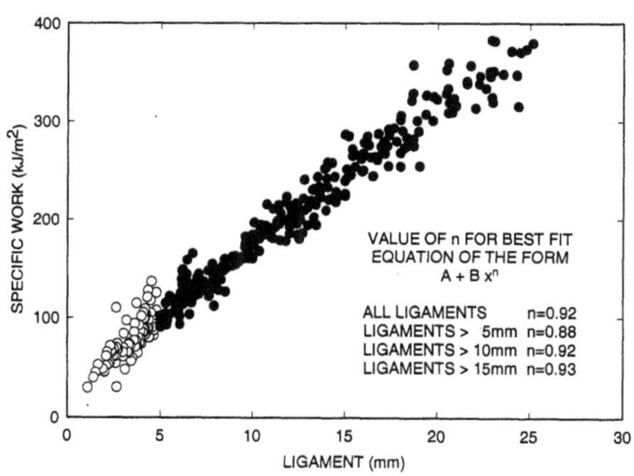

Fig. 5. Analysis of EWF data from all laboratories revealing non-linearity

It is proposed that some of this non-linearity arises from visco-elastic effects due to the differences in time-scale experienced by specimens of different ligament lengths as typified by the data in Figure 2.

If the plastic energy density w_p is time-dependent, then this would translate to a dependence of w_p upon the ligament length, l, and the linear relationship of equation (2) would not apply. Supposing the plastic energy density w_p has time-dependent behaviour which can be described by

$$w_p = w_o \, t_f^{-m} \tag{4}$$

where w_o is a reference value and m is the time-dependent exponent, then it follows that equation (2) becomes

$$w_f = w_e + k\beta w_o \, l^{1-m} \tag{5}$$

where k is a constant.

Figure 5 shows all collated data for PET, together with the results of curve fits of the form of equation (5). It can be readily seen that there is non-linearity in the data and an average value of $(1\text{-}m) = 0.9$, which means a time-dependent exponent of -0.1. This is similar to, but a little higher than, typical exponents found for time-dependent modulus and yield stress.

This proposal was supported by the maximum stress data, which were found to be described well by a power law relationship of the form

$$\sigma_{max} = \sigma_o \, l^{-m} \tag{6}$$

This is illustrated in Fig. 6 for PET and suggests a lower time-dependent exponent of -0.04, which is more in keeping with the likely time dependence of the yield stress.

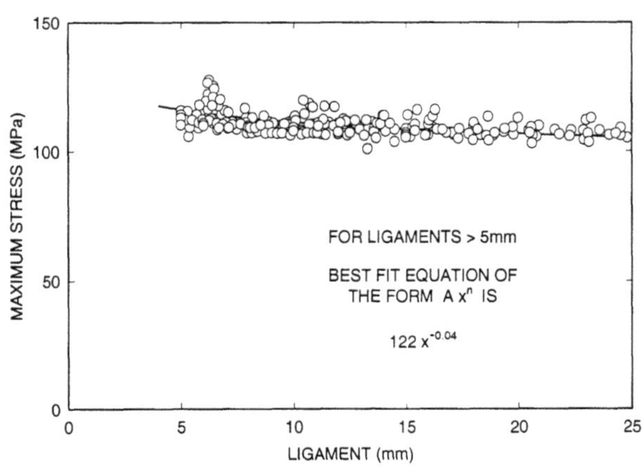

Fig. 6. Maximum stress data for PET showing power law dependence

Further work is required to test this explanation of non-linearity and is underway at the time of going to press.

Reproducibility

The primary aim of the ESIS working group on the essential work of fracture method was to develop a test protocol which would be commonly acceptable and deliver valid, reproducible fracture data. Despite the difficulty in resolving issues such as valid data limits, non-linearity, etc., some progress was made on reducing inter-laboratory variability. This can be shown by reference to table 4, which summarises the data for each of the six different round-robins in terms of the mean, standard deviation and scatter. It should be noted that all of the data has been re-analysed using the most recent protocol.

Table 4 : Inter-laboratory Variation in Essential Work Data

Round robin	Number of data sets	Material	w_e	Δw_e	$\Delta w_e/w_e$
1	6	EP copolymer	28.3	14.1	0.5
2	12	PP 30% rubber	68.8	31.7	0.46
3	13	300μm LLDPE1	54.9	17	0.31
4	8	150μm LLDPE1	41.1	13.9	0.34
5	18	PET	28.4	7.9	0.28
6	24	300μm LLDPE2	27.3	5.9	0.22

It is clearly demonstrated that the scatter, or standard deviation divided by the mean, decreases with successive round-robins. Therefore, the measures taken to tighten the constraints on the method have proved effective in improving reproducibility.

3. REFERENCES

1. Broberg, K.B. (1975). *J. Mech. Phys. Solids* **23**, 215.
2. Cotterell, B. and Reddel, J.K. (1977). *Int. J. Fract.* **13**, 267.
3. Mai, Y.W. and Cotterell, B. (1986). *Int. J. Fract.* **32**, 105.
4. Saleemi, A.S. and Nairn, J.A. (1990). *Polym .Eng. Sci.* **30**(4), 211.
5. Chan, W.Y.F. and Williams, J.G. (1994). *Polymer* **35**(8), 1666.
6. Hashemi, S. (1997). *J. Mater. Sci.* **32**.
7. Karger-Kocsis and Czigany, T. (1996). *Polymer* **37**(12), 2433.
8. Marchal, Y., Walhin, J. and Delannay, F. (1997). *Int. J. Fract.* **87**, 189.
9. Wu, J. and Mai, Y.W. (1996). *Polym. Eng. Sci.* **36**(18), 2275.
10. White, L. (1993). Private communication.
11. Hill, R. (1952). *J. Mech. Phys. Solids* **1**, 19.

4. TEST PROTOCOL

Specimen geometry

This protocol recommends the use of the DENT specimen because of its symmetry and minimal specimen rotation during the test. Figure TP1 defines the important dimensions and their nomenclature.

Figure TP1 : DENT specimen and nomenclature

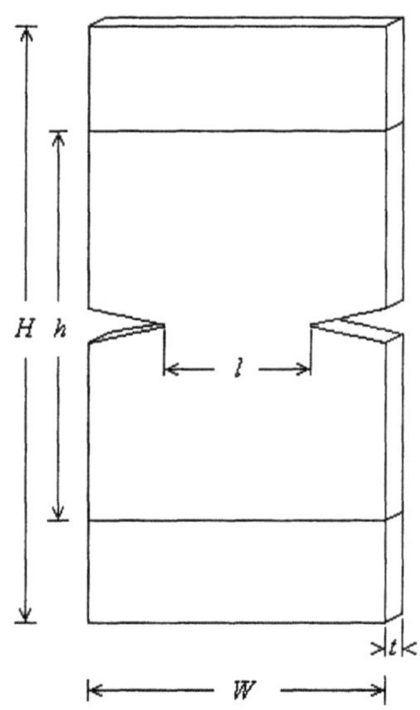

The initial step is to cut rectangular specimens of width W and length H from the test material.

The width W depends on the availability of material, but it is recommended that it is chosen to be at least a factor of 2 times the maximum ligament length used.

The length H includes the gauge length h and the amount of material used in gripping the specimens; again this depends on the availability of material. However, the choice of h is not critical and a typical value of 100mm has been successfully used.

A minimum of 20 valid data points are required and this means that a minimum of 25 specimens per material should be tested.

The maximum ligament length should be 15mm and the minimum ligament length should be the maximum of *3t* and 5mm. For testing of thin polymer sheet or films, for which this method is strictly intended, the usual minimum ligament length will be 5mm. In terms of distribution, the specimens should cover the entire ligament length range. There is no basis for biasing the data towards the intercept, since the confidence limits on the essential work are no smaller in such situations. Therefore, it is perhaps most reasonable to choose, say, 5 ligaments covering the range and perform 5 or more replicate tests at each nominal ligament length.

Notching

For each specimen, the two edge notches should be made by cutting either a shallow angle V-notch or a saw slot finished with a V-notch. Each V-notch should then be extended by a minimum of 1mm so that the final notch tip is sharp. If it is not possible to produce V-notches, the alternative of a square-ended saw slot is acceptable, provided the final notch tip is 3mm ahead of the saw slot. **It cannot be emphasised enough how important it is that the two notches are directly opposite one another. The best aid to achieving this is to draw a line prior to notching across the width of the specimen at its mid-point.** It is also important that the two notches are equal in length and pre-marking is again a useful aid.

There are various techniques that have commonly been used to generate a sharp notch - razor pushing, razor sliding, razor tapping, etc. It is not currently known which is necessary or most appropriate for the essential work test. At this stage, it is suggested that a method be used which is simple and which is thought to give a sharp initial notch. A record of the method used and the quality of notch (via a scanning electron photomicrograph) would be a useful inclusion in the final report.

Having notched the specimens, the thickness t of each specimen should be measured by taking representative readings across the ligament and averaging if necessary.

If the specimens are oriented in relation to any particular process direction, (e.g. 0° to the direction of injection for injection moulded plaques) this should be recorded on the proforma by specifying the direction in which stress is applied.

Testing speed

Polymers are rate dependent and consideration of test speed is important. The speed used must be fast enough to be practicably viable and yet slow enough to promote full yielding of the largest specimen ligament prior to crack growth. In order to ensure comparable strain rates for tests on different specimen gauge lengths (h, in mm), the test speed V (in mm/min) should be chosen to be given by:

$$V = 0.2\,h$$

For example, this gives a test speed of 20mm/min for a specimen gauge length of 100mm.

Each specimen in turn should be gripped in a tensile testing machine and deformed to failure at this calculated constant crosshead speed. The load-deformation traces should be recorded and the total energy to failure, W_f, calculated. The ligament length of each specimen should then be measured using a travelling optical microscope. This is defined as the distance between the tips of the sharp notches.

Check of stress level

Measurement of the peak load, P_{max}, during each test allows the maximum nett section stress to be calculated. According to plasticity theory, e.g. Hill [7], if σ_y is the uniaxial tensile yield stress for the material, then the maximum nett section stress for a DENT specimen in plane stress is $1.15\sigma_y$. A useful check is to plot nett section stress, σ_{max} $(=P_{max}/lt)$, versus l for the series of tests and compare with the horizontal line $\sigma_{max}=1.15\sigma_y$. From experience, it has been observed that this line usually passes through the data. However, the peak stresses observed are higher than predicted by this relationship with specimens having the shorter ligaments and lower than predicted for specimens with the longer ligaments.

Yield stress, σ_y, for the above calculation should be determined in such a way that the time to peak load in the tensile test, i.e. time to yield, is roughly the same as the average time to peak load in the essential work tests. This ensures that the yield stress is derived at an appropriate strain rate. The yield stress determination can be performed on any geometry of tensile specimen, provided the specimens are cut and stressed in the same direction as the essential work specimens. It is recommended that yield stress is determined from an average of at least 3 measurements.

Note that this is merely a check on the stress levels experienced during the test.

Stress criteria

In addition to the above check, it is considered useful to apply a stress criterion (i) to ensure greater likelihood of fracture occurring under plane stress conditions and (ii) to remove data where fracture has occurred prior to full ligament yielding. The following procedure is recommended : For all data, determine an average value for σ_{max} denoted by σ_m. Then apply the criterion that any essential work data, for which $\sigma_{max} < 0.9\sigma_m$ or $\sigma_{max} > 1.1\sigma_m$ be rejected from the determination of $\boldsymbol{w_e}$.

Outlying data criteria

For the data which meets the above stress criterion, values of w_f $(= W_f/lt)$ should be calculated and plotted against l. A least squares regression line should be fitted to the data to provide the slope, the intercept, 95% confidence limits on the intercept and the standard deviation of the data with respect to the regression line. Data for specimens that lie more than 2 times the standard deviation from the best-fit line should be eliminated from the analysis. This procedure of rejecting points should only be applied once to the original data. Having rejected these points, a final least squares linear regression is applied to the remaining data to give the slope, the intercept and the 95% confidence limits on the intercept.

Results Recorded

1. Specimen width *W*, thickness *t*, testing speed, *V*, and test temperature should be quoted.

2. Specimen orientation with respect to any process direction should be quoted.

3. Maximum and minimum values of *l* should be quoted.

4. Dimensions of the tensile specimen, test speed, tensile yield stress, σ_y, and $1.15\sigma_y$ should be reported.

5. The average value of maximum stress, σ_m, should be quoted.

6. The values for w_e (intercept) and βw_p (slope) should be quoted from the final least squares regression of w_f against *l* together with the 95% confidence limits on the determination of w_e.

7. The graph of σ_{max} against *l* should be provided, together with lines to indicate σ_m and the stress criterion limits $0.9\sigma_m$ and $1.1\sigma_m$.

8. The graph of w_f against *l* should be included.

9. A table of values of specimen thickness, ligament length, peak load, σ_{max}, W_f and w_f should be provided. Indication should be made for those data that have been excluded from the essential work determination and the reason for their exclusion.

Note that Appendix 2 provides a proforma for recording of results and Appendix 3 illustrates an example of how the data should be presented.

Appendix 1 : ESIS TC4 Round-robin Participants

ATO - DLO, Holland	Insa, Lyon, France
BASF, Germany	ITMA, Gijon, Spain
BP Chemicals, UK	Kaiserslautern, German
CNRS/ENSMA, France	Politecnico di Milano, Italy
Cranfield Institute, UK	Rhone-Poulenc, France
DRA, Woolwich, UK	Shell Chemicals, Holland
Dresden Polym. Inst., Germany	University of Louvain, Belgium
DSM, Holland	University of Pisa, Italy
EAHP, Strasbourg, France	University of Sydney, Australia
Elf Atochem, France	University of Trento, Italy
ICI, UK	Utah, USA
Imperial College, UK	

Appendix 2 : Proforma for essential work of fracture results

MATERIAL:

ESSENTIAL WORK TEST CONDITIONS

SPECIMEN WIDTH (mm):
SPECIMEN THICKNESS (mm):
TEST SPEED (mm/min):
TEST TEMPERATURE (°C):
SPECIMEN ORIENTATION:
MAXIMUM LIGAMENT (mm):
MINIMUM LIGAMENT (mm):

TENSILE TEST CONDITIONS

TENSILE SPECIMEN DIMENSIONS:
TENSILE TEST SPEED (mm/min):
TENSILE YIELD STRESS, σ_y (MPa):
THEORETICAL DDENT MAXIMUM STRESS ($1.15\sigma_y$):

ESSENTIAL WORK DATA

AVERAGE VALUE OF MAXIMUM STRESS, σ_m:
ESSENTIAL WORK OF FRACTURE, w_e (kJ/m²): ±
AND 95% CONFIDENCE LIMITS

PLASTIC WORK DISSIPATION FACTOR, βw_p (MJ/m³)
SLOPE OF ESSENTIAL WORK PLOT

GRAPHS AND TABLES
GRAPH OF σ_{max} vs LIGAMENT LENGTH
GRAPH OF w_f vs LIGAMENT LENGTH

TABLE OF SPECIMEN THICKNESS, LIGAMENT LENGTH, PEAK LOAD, σ_{max}, W_f AND w_f

Appendix 3 : Example EWF report including graphs and tables

MATERIAL: 300μm LLDPE BLOWN FILM

<u>ESSENTIAL WORK TEST CONDITIONS</u>

SPECIMEN WIDTH (mm): 35
SPECIMEN THICKNESS (mm): 0.3
TEST SPEED (mm/min): 10
TEST TEMPERATURE (°C): 23
SPECIMEN ORIENTATION: 0° (machine direction)
MAXIMUM LIGAMENT (mm): 15.2
MINIMUM LIGAMENT (mm): 4.84

<u>TENSILE TEST CONDITIONS</u>

TENSILE SPECIMEN DIMENSIONS: 0.285mm x 10mm x 150mm strips
TENSILE TEST SPEED (mm/min): 20
TENSILE YIELD STRESS, σ_y (MPa): 9.6
THEORETICAL DDENT MAXIMUM STRESS ($1.15\sigma_y$): 11.04

<u>ESSENTIAL WORK DATA</u>

AVERAGE VALUE OF MAXIMUM STRESS, σ_m: 11.64

ESSENTIAL WORK OF FRACTURE, w_e (kJ/m²): 37.3 ± 7.8
AND 95% CONFIDENCE LIMITS

PLASTIC WORK DISSIPATION FACTOR, βw_p (MJ/m³): 10.2
SLOPE OF ESSENTIAL WORK PLOT

<u>GRAPHS AND TABLES</u>

GRAPH OF σ_{max} vs LIGAMENT LENGTH

GRAPH OF w_f vs LIGAMENT LENGTH

TABLE OF SPECIMEN THICKNESS, LIGAMENT LENGTH, PEAK LOAD, σ_{max}, W_f AND w_f

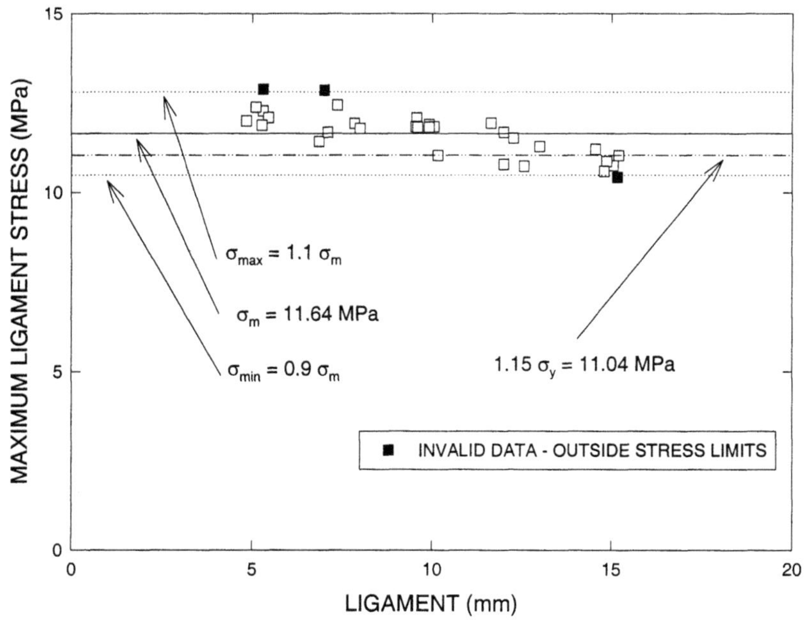

Figure A3.1 : Example EWF Data - LLDPE

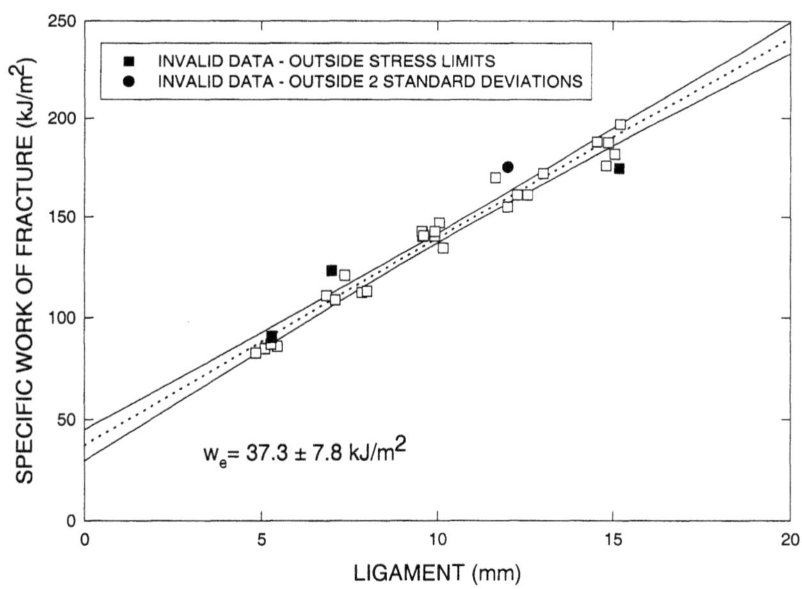

Figure A3.2 : Example EWF Data - LLDPE

Table A3.1 : Example EWF Data - LLDPE

SPECIMEN THICKNESS (mm) t	LIGAMENT LENGTH (mm) l	PEAK LOAD (N) P_{max}	MAXIMUM STRESS (MPa) σ_{max}	ENERGY TO FAILURE (mJ) W_f	SPECIFIC FAILURE ENERGY (kJ/m^2) w_f	INVALID DATA
0.29	5.3	18.9	12.27	139.8	90.98	
0.285	5.1	18.0	12.37	123.5	84.99	
0.285	5.3	19.5	12.88	136.4	90.29	* stress
0.281	5.45	18.5	12.08	131.8	86.03	
0.295	4.84	17.1	11.99	118.3	82.85	
0.285	5.27	17.8	11.87	130.7	87.05	
0.28	7	25.2	12.85	241.4	123.16	* stress
0.286	7.85	26.8	11.92	252.2	112.33	
0.281	8	26.5	11.78	253.9	112.93	
0.286	7.37	26.2	12.43	254.7	120.83	
0.285	6.85	22.3	11.42	216.1	110.7	
0.281	7.1	23.3	11.67	216.7	108.62	
0.29	9.92	34.2	11.89	403.8	140.37	
0.282	10.05	33.5	11.83	416.7	147.04	
0.29	10.16	32.5	11.03	396.1	134.44	
0.285	9.57	32.9	12.07	388.5	142.44	
0.286	9.57	32.4	11.84	383.8	140.24	
0.287	9.92	33.6	11.81	406.1	142.65	
0.281	9.55	31.7	11.83	383.0	142.72	
0.281	9.6	31.9	11.81	379.1	140.55	
0.282	12.27	39.8	11.51	557.7	161.17	
0.29	12.57	39.1	10.73	587.5	161.17	
0.285	12	36.9	10.78	530.7	155.19	
0.29	12	40.6	11.67	609.7	175.2	* 2 stddev
0.285	11.65	39.6	11.93	564.1	169.89	
0.285	13	41.8	11.28	637.1	171.97	
0.285	15.05	46.1	10.75	779.5	181.74	
0.287	14.87	46.3	10.86	800.6	187.59	
0.288	15.17	45.6	10.43	763.3	174.7	* stress
0.29	14.8	45.5	10.59	755.0	175.9	
0.285	14.55	46.4	11.2	779.2	187.9	
0.28	15.2	46.9	11.03	837.8	196.85	

CHAPTER 3

Adhesion Fracture Mechanics

INTRODUCTION TO ADHESION AND ADHESIVES

A. KINLOCH

The practical aspects of adhesion and adhesives have been appreciated and utilised by mankind for many centuries. However, it is only in the last fifty years, or so, that the science and technology of adhesion and adhesives has really progressed significantly and the major advances that have been made may be traced from the middle of the 1940s'. The main reason for this is that the adhesives employed in nearly all the technically demanding applications are based upon synthetic polymers, and such polymers have only been available for about the last fifty years. Synthetic polymers possess the balance of properties that enables them to adhere readily to other materials and to have an adequate strength so that they are capable of transmitting the applied loads from one substrate to the other.

The term *adhesion* is used when referring to the attraction between substances. The level of adhesion forces which are operating across an interface cannot usually be measured by mechanical tests (1,2). For example, the measured energy for interfacial failure is generally orders of magnitude higher than that arising from the *intrinsic adhesion* forces, such as molecular van der Waals' forces or covalent bonds, which may be operating across the interface.

An *adhesive* may be defined as a material which when applied to surfaces of materials can join them together and resist separation. Adhesive is the general term and includes cement, glue, paste, etc., and these terms are all used essentially interchangeably. The materials being joined are commonly referred to as the *substrates*, or *adherends*. There is no universally accepted definition of an *engineering* or *structural adhesive*, but it is generally used to indicate an adhesive which, when hardened, gives a relatively high-modulus, high-strength adhesive so that load-bearing joints may be constructed.

If the various stages in the formation of an adhesive joint, or a bonded laminate, are considered then it is possible to identify three distinct stages. Firstly, the adhesive initially has to be in a 'liquid' form so that it can readily spread over, and make intimate molecular contact, with the substrates. Secondly, in order for the joint to bear the loads which will be applied to it during its service life, the 'liquid' adhesive must now harden. In the case of adhesives used in engineering applications, the adhesive is often initially in the form of a 'liquid' monomer which polymerises to give the high molecular-weight polymeric adhesive. Alternatively, in the case of a flexible laminate, a polymeric layer is typically involved which essentially acts as a hot-melt adhesive. Thus, the laminate is bonded together at an elevated temperature when the hot-melt polymeric layer is in the form of a viscous liquid. This polymeric layer then re-solidifies upon the cooling of the laminate. Thirdly, it must be appreciated that the load-carrying ability of the joint or laminate, and how long it will actually last, are affected by: (a) the design of the joint or laminate, (b) the manner in which loads are applied to it, and (c) the environment which the joint or laminate encounters during its service life.

To understand the science involved, and to succeed in further developing the technology, the skills and knowledge from many different disciplines are required. Indeed, the input from surface chemists, polymer chemists and physicists, materials engineers and mechanical engineers are needed. Thus, the science and technology of adhesion and adhesives is a truly multi-disciplined subject.

In an attempt to bring these different disciplines together, a fracture-mechanics approach to the failure of adhesive joints has been developed. The concepts of fracture mechanics were introduced (3) by A.A. Griffith in the 1920s' whilst working at the Royal Aircraft Establishment, Farnborough. He recognised the importance of flaws in a material or structure. These flaws may be molecular-sized inhomogeneities, or air bubbles, or particles of dirt or dust, or they may be actual cracks. However they arise, Griffith proposed that the strength of a material, or structure, is governed by their presence. From the ideas of Griffith, and later from Orowan (4), one can define a term, the fracture energy, which is the energy needed to propagate a flaw through unit area of the material, or structure. The *fracture energy* (or *fracture toughness*) is given the symbol 'G_c' - where 'G' is for Griffith and the subscript 'c' indicates that it is the critical value for crack growth.

A basic aim of fracture mechanics is to identify fracture criteria, such as G_c, which are independent of the geometry of the cracked body and therefore provide a 'material parameter' for characterising the toughness, or crack resistance, of materials - including interfaces and adhesives. Now, we have been developing methods of fracture mechanics for characterising and predicting the failure of adhesive joints, including flexible laminates. Obviously for adhesive joints or laminates, one can define a value of the *adhesive fracture energy* (also called *adhesive fracture toughness)*, G_c, for either a cohesive failure of the adhesive, or for an interfacial failure along the adhesive/substrate interface.

As noted above, the term *intrinsic adhesion* is often used when referring to the direct molecular forces of attraction between the adhesive and the substrates, to distinguish this phenomenon from the 'measured adhesion'; i.e. from the measured strength or toughness of an adhesive joint. Indeed, even when using a fracture mechanics approach, and interfacial failure of the joint does occur, the value of the adhesive fracture energy, G_c, will usually not be equivalent to the energy associated with rupture of the intrinsic adhesion forces, since the value of G_c will also encompass the energy dissipated in viscoelastic and plastic deformation processes which occur in the vicinity of the crack tip. Indeed, the value of G_c is typically orders of magnitude greater than the energy associated solely with the intrinsic adhesion forces. From a practical standpoint, this emphasises the need not only to establish good intrinsic adhesion across the adhesive/substrate interface but also to develop tough adhesives, where the plastic, or process, zone in the vicinity of the crack tip will be relatively large.

In the present Section two areas are addressed where the 'ESIS TC4' Committee have proposed protocols for fracture-mechanics tests. These areas are (a) for flexible laminates, as might be used in the packaging industry, and (b) for structural adhesives as would be employed in automotive and aerospace applications, for example.

References

1. E.H. Andrews and A.J. Kinloch, Proc. Royal Society, **A332**, 385 (1973).

2. A.J. Kinloch, 'Adhesion and Adhesives: Science and Technology', Chapman and Hall, London, 1987.

3. A.A. Griffith, Phil. Trans. Royal Society, **221**, 163 (1920).

4. E. Orowan, Reports Prog. Physics, **12**, 185 (1948).

PEEL TESTING OF FLEXIBLE LAMINATES

D.R. MOORE and J.G. WILLIAMS

1. BACKGROUND TO PEEL TESTING

Flexible laminates are used in a wide range of industrial and packaging applications. In general, there will be considerable practical importance associated with the adhesive strength between two specific adjacent layers in these laminates. Sometimes it will be important to maximise this adhesive strength, sometimes it will be required to minimise it. Overall, the key requirement will be to control it and to achieve this it is first necessary to measure it.
Adhesion has been considered in terms of peel resistance in a number of application areas. For example, a floating roller test and a T-peel test have been standardised for adhesion to aluminium substrates (1, 2). In both these methods there has been an aim to measure peel strength and this is common to many "adhesion" tests. Peel strength measures the force per unit width of the laminate and as such is a measure of the input energy in conducting the peel; it gives little information on the distribution of the energy in terms of deformation and fracture mechanisms. Therefore, peel strength is a measure of the difficulty involved in pulling the laminate apart. However, what is preferable is a measure of how well the laminate is "stuck" together. It is usually assumed that the measured peel strength is synonymous with this adhesive strength but as we will demonstrate, this is often quite untrue.
The purpose of this protocol is to provide guidance on the measurement of peel strength of the laminate and then to show how the adhesive strength (also known as adhesive fracture toughness or interfacial work of fracture) can be determined from this peel strength and other measurements.

2. ANALYSIS OF PEEL TESTS

The protocol is divided into two parts, one for ***a fixed arm peel test*** and the other for ***a T-peel test***. These geometries are different but there is a common aim in converting the peel strength measurement into an interfacial work of fracture. Some aspects of the analysis of the elastic and plastic deformations will be seen to be similar.

2.1 Analysis of the Fixed Arm Peel Test

Figure 1 shows the peel of a laminate at a peel angle of θ with a force P acting on the peel arm (width B, thickness h). In order to peel one layer of the laminate from the other, it is necessary to provide energy in the form of external work to the laminate. Kinloch et al (3) have shown how this external work (U_{ext}) can be distributed between several deformation and failure processes:-

$$G_A = \frac{1}{B}\left(\frac{dU_{ext}}{da} - \frac{dU_s}{da} - \frac{dU_{dt}}{da} - \frac{dU_{db}}{da}\right) \qquad 1$$

Where U is an energy term and the suffices *ext, s, dt* and *db* refer to external, strain, dissipation in tension and dissipation in bending.
(i) Peel fracture is the breaking of the interfacial bonding forces. This may relate to the work of

adhesion, the work of cohesion or some complicated combination of both. These are all terms that relate to a conceptual macroscopic interpretation of the interfacial fracture processes, because on a microscopic scale there can be no adhesive fracture toughness since the peel crack must run through matter and therefore the peel fracture toughness will always be "cohesive". However, peel fracture is usefully described as the adhesive fracture toughness or as adhesive strength.

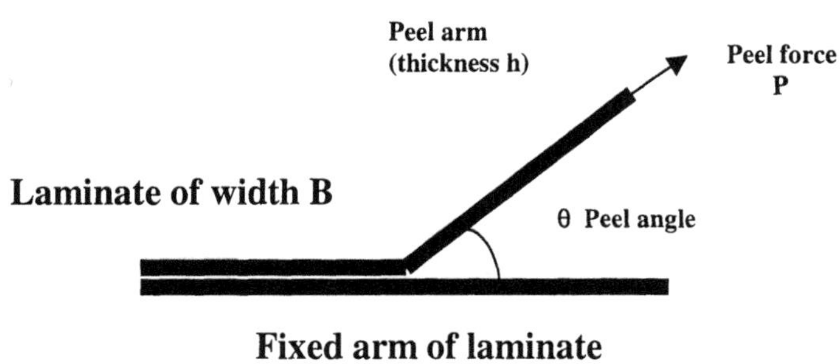

Figure 1 Fixed arm peel test

(ii) Kinetic energy associated with a moving peel fracture. In order to eliminate this term it is necessary to conduct a slow peel test.

(iii) Stored strain energy (U_s) in the peel arm; the thinner the peel arm the higher this energy contribution.

(iv) Dissipated or lost energy $(U_{dt}$ and U_{db} respectively) which can occur in two ways; first as irreversible tensile deformation of the peel arm, second as irreversible bending deformation of the peel arm in the vicinity of the peel crack front.
G_A refers to the adhesive fracture toughness which is a geometry independent property. Although it may be adhesion toughness, cohesion toughness or a combination of both. Kinetic energy is assumed to be zero for this energy balance equation and therefore it will be necessary to accommodate "slow" peel tests.

In an ideal case, there is no strain in the peel arm (known as a peel arm of infinite modulus) and the bending deformations are elastic (ie no irreversible bending deformation) then Kinloch et al (3) have shown that the adhesive fracture toughness is simply related to the peel strength and the peel angle:-

$$G_A = G_A^{\infty E} = \frac{P}{B}(1-\cos\theta) \qquad 2$$

In practice however, this ideal case is seldom realised. If there is tensile deformation in the peeling arm but still no irreversible bending deformation then for a strain ε at a stress of σ , the peel fracture toughness becomes:-

$$G_A = G_A^{eb} = \frac{P}{B}(1+\varepsilon-\cos\theta) - h\int_0^{\varepsilon}\sigma d\varepsilon \qquad 3$$

When there is dissipated energy in bending as well as elastic and irreversible tensile deformation of the peel arm, then the peel fracture toughness becomes:-

$$G_A = G_A^{eb} - G^{db} \qquad 4$$

where

$$G^{db} = \frac{1}{B}\frac{d}{da}\frac{U_{db}}{} \qquad 5$$

G^{db} is a strain energy release rate. It is a complex function but Kinloch et al (3) have determined it in analytical form by using elastic-plastic large displacement theory.
In order to determine G_A without neglecting any of the elastic or irreversible deformations, two experiments are required:

(a) The peel test with a control of the peel angle.

(b) A tensile stress-strain measurement of the peel arm up to fracture. In this analysis, the measured stress-strain curve is approximated to a bilinear form for simplicity, as shown in Figure 2. For many polymeric materials, this approximation will be close to the real material behaviour, and any errors due to this simplification are usually negligible.

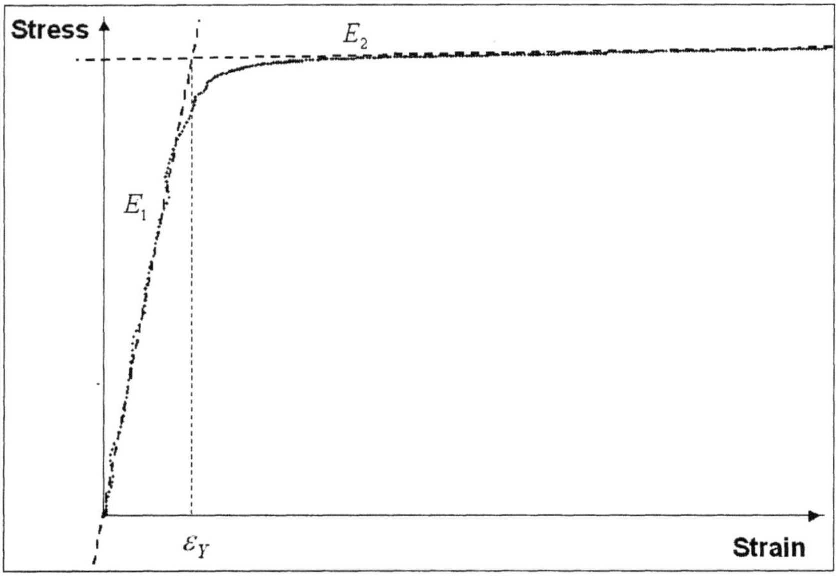

Figure 2 Tensile stress versus strain plot for a peel arm illustrating the definition of E_1, E_2 and yield strain

Data from these tests are then used with the analysis by Kinloch et al (3) in order to calculate the adhesive fracture toughness. A summary of the key equations is given in Appendix 2. For convenience, it is simplest to use a software package in order to process the calculations. This can be achieved in a number of forms but one such package is available on the Imperial College website (http://www.me.ic.uk) which follows the calculations of the Kinloch et al (3) paper.

Adhesive fracture toughness is determined by subtracting the plastic strain energy release rate (G^{db}) from the measured peel toughness (G_A^{eb}). Ideally, the corrections for plastic deformation should not be too large otherwise errors for the determination of adhesive fracture toughness will become significant. The ratio G_A/G_A^{eb} is a helpful guide in this context. For example, if G_A/G_A^{eb} is 1 then plastic deformation correction errors are zero, whilst as the ratio becomes smaller then more attention should be given to possible errors. (These considerations apply equally to the T-peel analysis)

2.2 Analysis of the T-Peel Test.

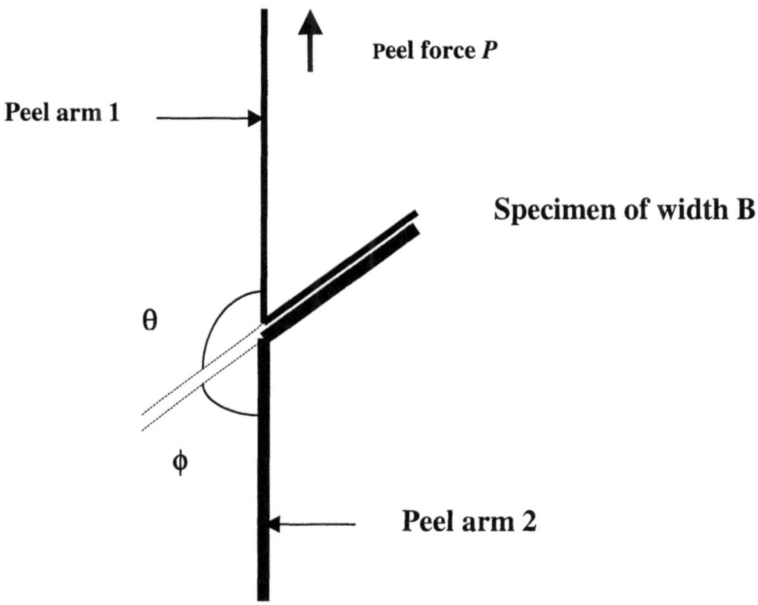

Figure 3 T-Peel Specimen During a Test.

Figure 3 shows the specimen configuration during a T-peel test. When the stiffness of the peel arms are different the peel angles will be ϕ and θ (rather than 90^0). In Figure 3 the stiffer arm is peel arm 2, therefore $\phi<90$ and $\theta>90$. The analysis proceeds along similar lines to that in the fixed arm peel test, except that there are now two peel arms to accommodate. However, only one of the peel angles needs to be considered since $\phi = \pi - \theta$:-

$$G_A^{\,1} = G_A^{\,\infty E} = \frac{P}{B}(1+\cos\phi) \qquad 7$$

$$G_A^{\ 2} = G_A^{\ \infty E} = \frac{P}{B}(1-\cos\phi) \qquad 8$$

Where the super scripts *1* and *2* refer to the two peel arms, respectively.
The peel toughness terms for elastic corrections are then similar to those in equation [3], except that there are two terms. In a similar manner, there will be two terms for the dissipated energy expressions of equations 5. Consequently, there will be two forms for equations [4]:-

$$G_A^{\ 1} = (G_A^{\ eb})^1 - (G^{db})^1 \qquad 9$$

$$G_A^{\ 2} = (G_A^{\ eb})^2 - (G^{db})^2 \qquad 10$$

The adhesive fracture toughness (G_A) from the T-peel test is then the sum of the terms from equations [9] and [10], namely:-

$$G_A = G_A^{\ 1} + G_A^{\ 2} \qquad 11$$

3. A TEST PROTOCOL FOR PEEL OF FLEXIBLE LAMINATES

3.1 Fixed Arm peel Test.

3.1.1 Experimental Procedures in the Fixed Arm Peel Test.

Specimens for conducting peel strength should be in the form of rectangular specimens where the two parts of the laminate have already been adhered but where there is a region of unadhered material (of nominal length 30 mm). The overall dimensions of a peel specimen need not be rigidly defined but for many tests we have found that a length of 100 mm and width 20 mm proves to be quite satisfactory. Three specimens should be tested for each set of conditions.
The choice of peel jig is not unique but the apparatus should incorporate a number of facilities. A successful kit is shown schematically in Figure 4. First, the apparatus should be able to select the peel angle in the range upto 180°. Second, the jig is attached to an Instron or similar universal testing machine such that as peel occurs the peel angle is maintained constant by the jig moving along a low friction linear bearing system. Third, only one side of the laminate is allowed to be the peel arm in the test. Therefore, half of the laminate system is adhered to the peel table thus eliminating a duplication of the deformational factors discussed above. Of course, the laminate can be reversed so changing the peel arm material in another but separate test. Adhering one side of the laminate to the peel table is a critical issue. If this layer can

separate from the table during the test then the energy involved in that process will increase the measured adhesive fracture toughness value to an erroneously high level. The means of gluing the layer to the table should be reported.

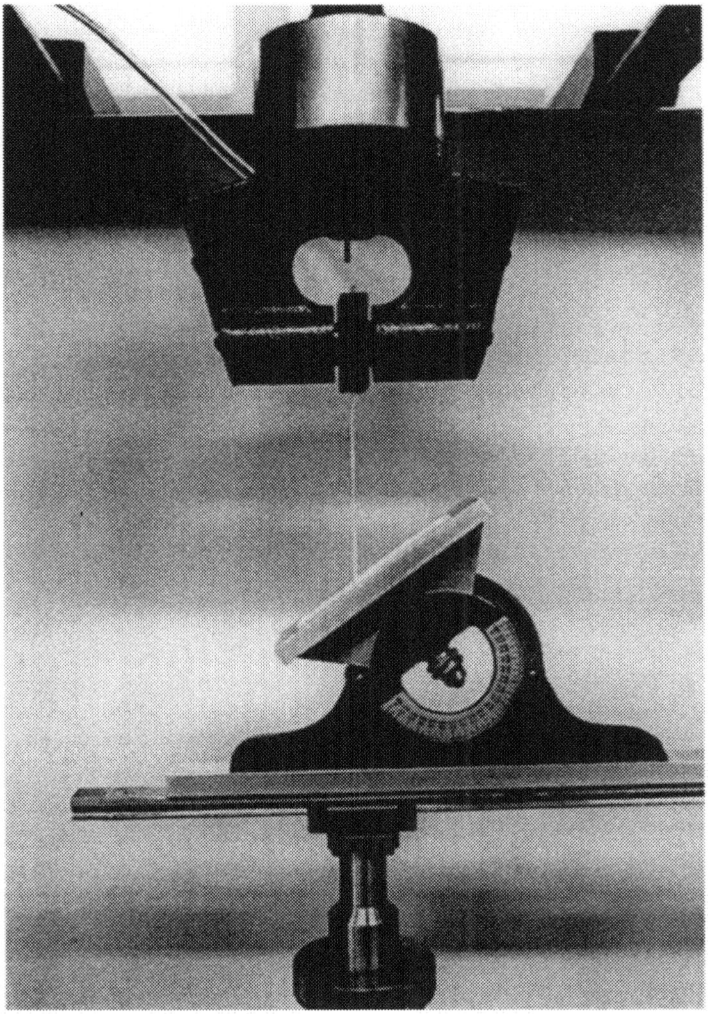

Figure 4 Fixed arm peel fixture with linear bearing system showing a peel angle of 45^0.

The materials of the laminate may or may not be the same. One of these layers should be glued to the peel table of a peel jig; the other layer then becomes the peel arm in the test. It is useful to select the stiffer arm as the peel arm (but not mandatory). The layer to be used as the peel arm should be recorded as should the length, width and thickness dimensions of the specimen. The peel angle needs to be selected on the basis of ensuring that the peel force is large compared with the resolution of the load cell and the frictional forces in the test. For purposes of the protocol it will be necessary to conduct the test at a range of peel angles between 60^0 and 150^0 but the peel angle 90^0 should always be one of the angles. A peel test speed of 10 mm/min will be used. The force versus displacement curve to initiate and propagate a peel fracture should be recorded. At least 30 mm of peel fracture should be established. The lowest steady propagation peel force should then be used to determine the adhesive peel strength or G_A of equation 1.

Of course, the procedures in this protocol maybe used more widely when investigators are conducting research. For these occasions it might be useful to correct or normalise to a common peel crack velocity ($\Delta a/\Delta t$) by recognising that for negligible strain in the peel arm:-

$$\frac{\Delta a}{\Delta t} = \frac{\Delta \delta}{\Delta t}(1 - \cos \theta) \qquad 6$$

Where $\Delta\delta/\Delta t$ is the crosshead speed.

In order to conduct the corrections summarised in equations 2-5 it is necessary to obtain a tensile stress - strain plot on the material of the peel arm. Such a plot is shown in Figure 2 from which values of E_1, ε_Y and α can be obtained. This tensile test should be conducted at the same test speed as the peel test and in order to obtain sufficient accuracy the test specimen should be a rectangular strip of width 10 mm and length 100 mm. In addition, an extensometer will be required to measure the strain deformations necessary to define E_1. The extensometer, ideally, should be of a non-contacting type. The tensile test should continue to such deformations that enable a clear measure of E_2 to be defined. ie a steep slope measurement of E_2 near to the onset of plastic deformation would be inappropriate.

With reference to Figure 2, which includes the actual data from a stress-strain test in tension, it can be seen how the two straight line regions are drawn onto the plot. The elastic modulus (E_1) is defined by using accurate strain at small deformations. The plastic modulus (E_2) is drawn once it is clear that a steady slope has been defined after yield. The extension of the specimen should not be stopped until this has occurred. It is then possible to define the plastic modulus. The intercept of the two lines defines the yield strain.

If it transpires that the "work hardening" portion of the stress-strain curve exhibits a negative slope, then values of α and E_2 should be made zero (and not negative). [It may also be helpful to explore the shape of this curve in terms of a true stress-strain plot, in order to define better the value of yield strain and E_2. This would be particularly helpful for large deformations prior to fracture. Such a curve can be constructed by defining true stress as net stress multiplied by (1 + elongation) and then plotting true stress versus strain.]

A software package is then to be used in order to calculate the various other versions of G_A as indicated in reference 1 and equations 2-5.

3.1.2 Test Report for the Fixed Arm Peel Tests

The following information is required in the test report.

Sample and Laboratory details	Fixed arm peel tests.
Name of Laboratory	
Description of laminate	
Material of peel arm	
Adhesive used for gluing one arm to the peel table	
Test equipment	

Specimen Number---	1	2	3	4	Average
Data from tensile test on the peel arm material:-					
Test speed (mm/min)					
Low strain modulus (E_1) (GPa)					
Strain at yield (%)					
High strain modulus (E_2) (GPa)					
Alpha (E_2/E_1)					
Data from the peel test:-					
Peel angle (^{o})					
Test speed (mm/min)					
Specimen dimensions L (mm)					
B (mm)					
peel arm thickness h (um)					
Peel strength (P/B) (N/mm)					
Derived results by calculations:-					
$G_A{}^{eb}$ (J/m^2) (Equation 3)					
G^{db} (J/m^2) (Equation 5)					
G_A adhesive fracture toughness (Eq 4) (J/m^2) ie $G_A = G_A{}^{eb} - G^{db}$					

Table 1 Summary of experimental details and results for the determination of adhesive fracture toughness of a laminate in the fixed arm peel test.

In addition, it is necessary to include a plot of the peel curve (i.e. force versus displacement in the peel test). The length of peel growth should be marked on this curve together with a clear indication as to how the peel force used to determine the peel strength is derived from the plot.

3.2 T-Peel Test

3.2.1 Experimental Procedures in the T- Peel Test.

Specimens for conducting peel strength should be in the form of rectangular specimens where the two parts of the laminate have already been adhered but where there is a region of unadhered material (of nominal length 30 mm). The overall dimensions of a peel specimen need not be rigidly defined but for many tests we have found that a length of 100 mm and width 20 mm proves to be quite satisfactory. Two specimens should be tested for each set of conditions.

The specimen is mounted into a universal test machine where peel arm 1 is attached to the load cell and peel arm 2 is attached to a fixed clamp. The materials of the laminate may or may not be the same. The test machine should have the usual capabilities for sufficient resolution of the peel force and monitoring the peel force displacement curve (again the curve in Figure 5 could be typical). Once the peel arms are in tension, the specimen configuration will be similar to that show in Figure 3. During the course of peeling it is necessary to measure one of the peel angles (ϕ or φ) and at least three measurements should be made throughout the 30 mm peel fracture; one near the start, one in the middle and one near the end. The average value for the peel angle can then be determined and used in the calculations.

A peel test speed of 10 mm/min will be used. The force versus displacement curve to initiate and propagate a peel fracture should be recorded. At least 30 mm of peel fracture should be established. The average propagation peel force should then be used to determine the peel strength or G_A of equation 11, as shown in Figure 5.

The peel test should be conducted twice in the configuration shown in Figure 3 and then two additional tests should be conducted with the specimen revolved through 180^0 ie with the top peel arm being peel arm 2. The above procedure should be followed and in this configuration particular attention should be given to any difference in the peel angles.

In order to conduct the corrections summarised in equations 7-11, it is necessary to obtain tensile stress - strain plots on the material of the peel arms. (Two tensile stress-strain plots will be required if the materials of the peel arms are different) Such a plot is shown in Figure 2 from which values of E_1, ε_Y and α can be obtained. This tensile test should be conducted at the same test speed as the peel test and in order to obtain sufficient accuracy the test specimen should be a rectangular strip of width 10 mm and length 100 mm. In addition, an extensometer will be required to measure the strain deformations necessary to define E_1. Ideally, this extensometer should be of a non-contacting type. The tensile test(s) should continue to such deformations that enable a clear measure of E_2 to be defined. ie a steep slope measurement of E_2 near to the onset of plastic deformation would be inappropriate.

With reference to Figure 2, which includes the actual data from a stress-strain test in tension, it can be seen how the two straight line regions are drawn onto the plot. The elastic modulus (E_1) is defined by using accurate strain at small deformations. The plastic modulus (E_2) is drawn once it is clear that a steady slope has been defined after yield. The extension of the specimen should not be stopped until this has occurred. It is then possible to define the plastic modulus. The intercept of the two lines defines the yield strain.

If it transpires that the "work hardening" portion of the stress-strain curve exhibits a negative slope, then values of α and E_2 should be made zero (and not negative). [It may also be helpful to explore the shape of this curve in terms of a true stress-strain plot, in order to define better the value of yield strain and E_2. This would be particularly helpful for large deformations prior to fracture. Such a curve can be constructed by defining true stress as net stress multiplied by (1 + elongation) and then plotting true stress versus strain.]

A software package is then to be used in order to calculate the various other versions of G_A as indicated in reference 1 and equations 7-11.

3.2.2 Test Report for the T-Peel Tests

The information required in a test report is outlined in Table 2. Clearly, if there are significant differences in the peel angles and peel fracture toughness values for the two specimen configurations, then the data should not be averaged but instead they should be quoted as two sets of results.

In addition, it is necessary to include a plot of the peel curve (i.e. force versus displacement in the peel test). The length of peel growth should be marked on this curve together with a clear indication as to how the peel force used to determine the peel strength is derived from the plot.

Sample and Laboratory details	**T-peel tests**
Name of Laboratory	
Description of laminates: Peel arm 1	
Peel arm 2	
Test equipment	

Table 2 Summary of Test Results from T-Peel Test

Table 2 ...continued

T-Peel test Specimen Number----->	**1**	**2**	**3**	**4**	**Average**
Specimen configuration	**Peel arm 1 top**		**Peel arm 2 top**		
General laminate Data					
Test speed (mm/min)					
Specimen dimensions L (mm)					
b (mm)					
Data from the T-peel test					
Peel force (N)					
Peel arm 1					
Peel angle θ (^{o})					
Peel arm thickness h (um)					
E_1 (GPa)					
α					
Yield strain (%)					
G_A^{infE} (J/m^2) Equation 7					
$G_A{}^{eb}$ (J/m^2)					
G^{db} (J/m^2)					
$G^1{}_A = G_A{}^{eb} - G^{db}$ (J/m^2) Equation 9					
Peel arm 2					
Peel angle ϕ (^{o})					
Peel arm thickness h (um)					
E_1 (GPa)					
α					
Yield strain (%)					
G_A^{infE} (J/m^2) Equation 8					
$G_A{}^{eb}$ (J/m^2)					
G^{db} (J/m^2)					
$G^2{}_A = G_A{}^{eb} - G^{db}$ (J/m^2) Equation 10					
G_A adhesive fracture toughness (Equation 11) (J/m^2) $G_A = G^1{}_A + G^2{}_A$					

Table 2 Summary of Test Results from T-Peel Test

4. PRESENTATION OF RESULTS AND THEIR DISCUSSION

4.1 Materials

Two laminate systems were used in this study:-

(i) A lamellar structure with six layers based on polyethylene (PE), aluminium (Al) and paper, which was only used for fixed arm peel tests. This is designated Aluminium/PE because the peel arm was a polyethylene system (thickness approximately 0.2 mm) and the fixed arm was a structure composed of PE, aluminium and paper (thickness approximately 0.3mm).

(ii) A lamellar structure with five layers based on polypropylene (PP), an adhesive, an ethylene vinyl alcohol (EVOH) layer another adhesive and another polypropylene layer (Total thickness was 0.12 mm). This is designated PP/adh/EVOH/adh/PP (PP laminate for short) and was used in both fixed arm peel tests and T-peel tests. The peel arms were PP/adh (approximate thickness 50 μm) and EVOH/adh/PP (approximate thickness 70 μm).

4.2 Fixed Arm Peel Test.

4.2.1 Fixed Arm Peel Test on the Aluminium/PE Laminate

The Aluminium/PE laminate provided an opportunity for nine laboratories to conduct a 90^0 fixed arm peel test at a test speed of 10 mm/min. In addition to peeling the PE arm away from the aluminium/paper/PE structure (which was fixed|), each laboratory also measured the tensile stress-strain characteristic of the PE structure in order to derive the three parameters just discussed. Results are summarised in Figure 5 where two sets of results are shown; the peel toughness (G_A^{infE} from equation 2) and the adhesive fracture toughness (G_A from equation 5).

The peel toughness (G_A^{infE}) is a function of a direct experimental measurement and does not include terms derived from the tensile test results and other analysis. It can be seen that the agreement between laboratories for this measurement is good.
The adhesive fracture toughness (G_A) shows more scatter between laboratories. This term does involve further analysis based on parameters from the tensile test. Therefore, both peel and tensile test measurements are critical in the determination of the interfacial work of fracture.
The accuracy in the determination of the yield strain can be seen to be critical, particularly when the ratio of E_2:E_1 is relatively small.

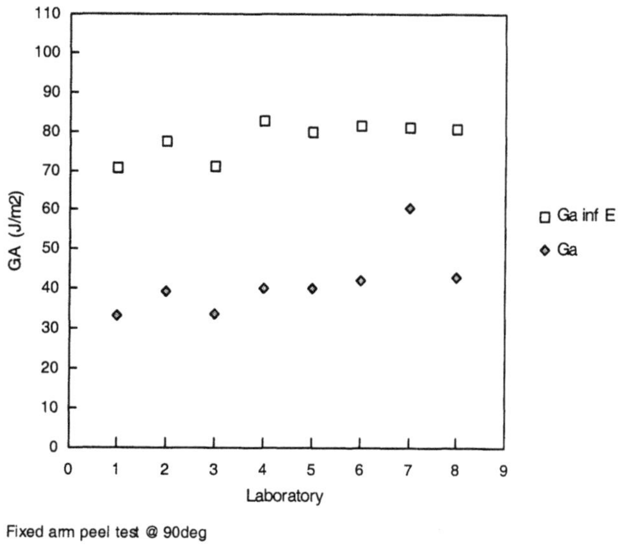

Figure 5 Peel Results for Al/PE Laminate in a Fixed Arm Peel Test

The results depicted in Figure 5 for peel toughness (G_A^{infE}) depend on a correct interpretation of the peel force versus displacement signal. This is illustrated by the results from one of the laboratories as shown in Figure 6.

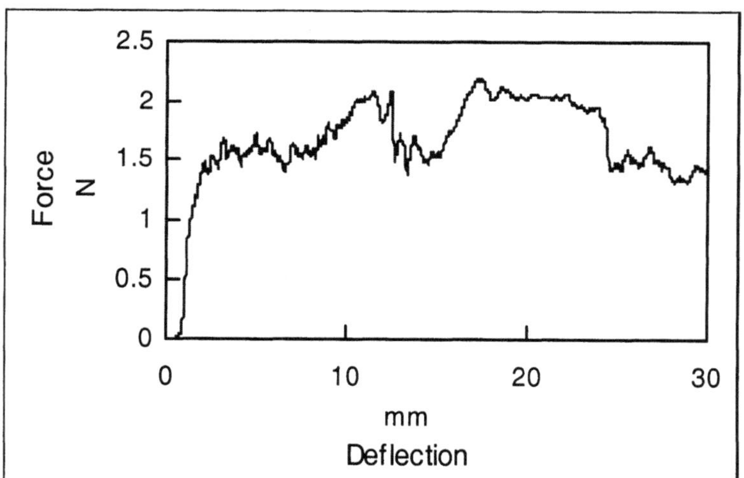

Figure 6 Peel force versus deflection in a fixed arm peel test. The higher peel forces relate to cohesive fracture; the lower peel forces relate to adhesive fracture

The data in Figure 6 show that there are two levels for the peel force. A high level which is interpreted as a cohesive fracture in the peel arm and a low level which is interpreted as an adhesive fracture between the peel arm and the aluminium/paper/PE substrate. In order to achieve consistency for the results in Figure 5 for all of the laboratories, it is important to use only the adhesive peel force from Figure 6. However, it was noted that not all of the laboratories observed this bi-functional feature in the peel curve, although many did.

4.2.2 Fixed Arm Tests on the PP Laminate

Five laboratories obtained peel data at a 90° peel angle. They provided an overview of the results for the PP laminate and this demonstrates that peel toughness, without use of the correction factors, again enables good agreement to be obtained, as shown in Figure 7. However, the values of $G_A{}^{db}$ (from equation 5) and G_A (from equation 4) show some scatter. This is as a direct consequence of the definition of the "plastic" modulus slope E_2 . The variation of the ratio E_2:E_1 is shown in Figure 8 and the scatter is due to variations in the determination of E_2. In turn, variations in E_2 will cause scatter in determination of the yield strain. Of course, the definition of the "elastic" modulus will also influence determination of the yield strain, but all laboratories reported values within 10% of the mean value.

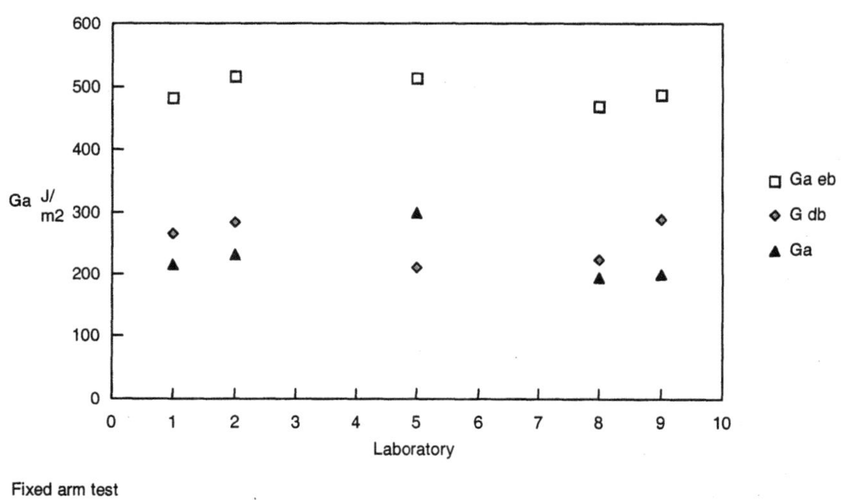

Figure 7 Peel Toughness from Various Laboratories for 90° Peel Angle Tests Using a Fixed Arm Test on PP/adh/EVOH/adh/PP Laminates.

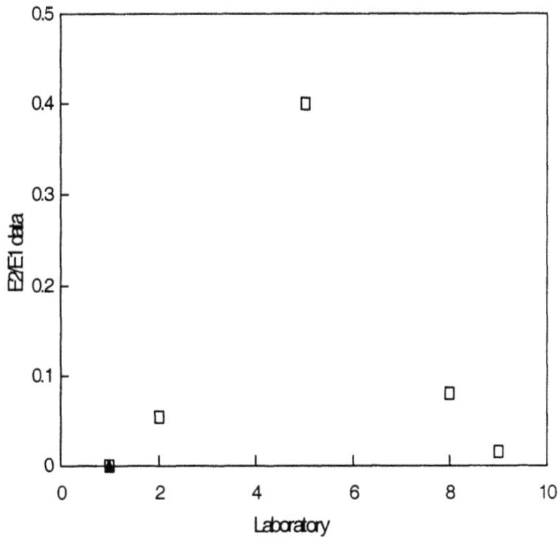

Figure 8 The ratio E_2/E_1 from Several Laboratories in Tension Tests on PP/adh Peel Arms

A number of laboratories conducted multi-angle fixed arm peel tests on the PP laminate. A full set of results is shown in Figure 9 according to analysis by equations 3, 4 and 5.

These results show the dependence of peel toughness on peel angle (G_A^{eb}) but after full analysis they show the independence of adhesive fracture toughness (G_A) on peel angle.
Results for the determination of adhesive fracture toughness (G_A) from a number of laboratories is shown in Figure 10 and it can be concluded that the agreement is reasonable.

Although there is scatter in the data for each laboratory, the mean values between laboratories are in good agreement. For example, the mean value for the multi-angle determination of adhesive fracture toughness (G_A) is 225 J/m^2, whilst the mean value for the 90 ^{0} data only (from five laboratories) is 229 J/m^2.

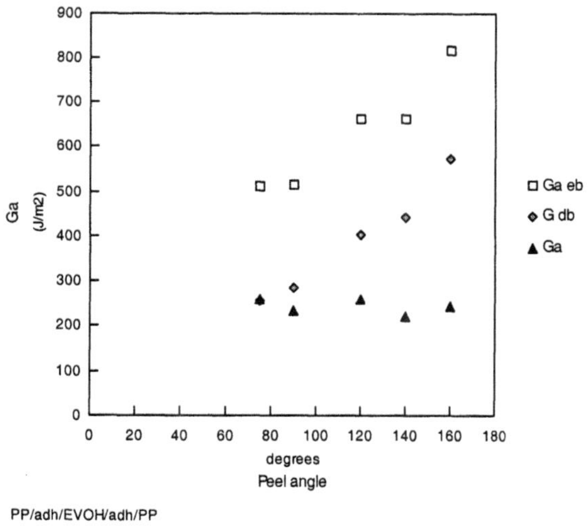

Figure 9 Peel Toughness versus Peel Angle for PP Laminates in a Fixed Arm Peel Test

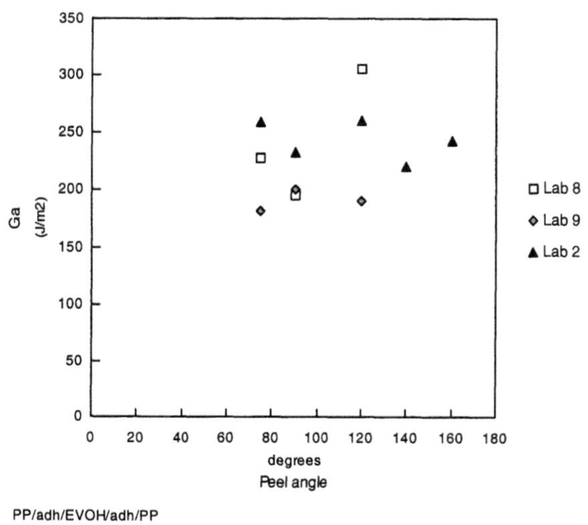

Figure 10 Adhesion Fracture Toughness versus Peel Angle by Three Laboratories Conducting Fixed Arm Peel Tests on PP Laminates

The range of values for the results in Figure 10 is however quite large, namely 180 J/m^2 to 300 J/m^2.

4.3 T-Peel Results on the PP Laminate.

The PP laminate was used in a T-peel test, where peel arm 1 was PP/adh and peel arm 2 was PP/adh/EVOH. Peel arm 2 was the stiffer peel arm with a thickness of 75 μm; peel arm 1 had a thickness of 51 μm. Two configurations we explored, where peel arm 1 was at the top (designated configuration **A**) and where peel arm 2 was at the top (designated configuration **B**). Three laboratories conducted independent experiments using both configurations, **A** and **B**. The average measurements for the smaller peel angle for configuration **A** and configuration **B**, respectively, were 55^0 and 41^0. This was considered to be a real difference, although there was some scatter in the data. Therefore, we can conclude that, at least for flexible laminates, that the location of the stiffer peel arm will influence the peel angle. As suggested in the protocol, such features should be monitored during the testing.
Participants also measured the variation of peel angle during the tests, for both configurations **A** and **B**. Once the peel process has started, that is after tension in the peel arms, then the smaller peel angle (in either configuration) will decrease by a few degrees during 30 mm of peel. This arises because of a dominance from the stiffer peel arm, although the effect is not large. Nevertheless it is worthwhile measuring the peel angle at three different times during the peel process (as recommended in the protocol).
It was only possible to obtain peel arm 1 in an undeformed state in order to conduct the tensile test from which the following parameters were obtained:-
E_1 =0.8 GPa, E_2 = 0.001 GPa, ε_Y= 2.4%
These properties were assumed to apply to both peel arms and this was confirmed to be a conservative assumption by measurement of these properties for peel arm 2. Preliminary measurements of the interfacial work of fracture from the T-peel tests are summarised in table 3 for three participating laboratories.

Specimen Configuration	Range of values for θ^0	Range of values for ϕ^0	Number of tests	Average G_A (J/m^2) (from equation 11)
A	120	60	1	184
A	116-130	64-50	2	211
A	114-140	64-40	4	183
B	36-37	144-143	2	154*
B	44-49	131-136	3	191

*** Based on two values of 120J/m^2 and 184J/m^2**

Table 3 Summary of Interfacial Work of Fracture Results from T-Peel Tests from Three Laboratories.

The results for adhesive fracture toughness in Table 3 show reasonable agreement; there is only one value that would seem to be low and that is based on an average that includes a particularly low value. There would also seem to be agreement for adhesive fracture toughness for either of

the specimen test configurations, even though the peel angle depends on specimen configuration.
Finally, there is a resemblance in the values for adhesive fracture toughness between the T-peel and fixed arm peel tests. The data certainly lie within the same scatter bands. Therefore, this adhesive strength can be argued to be independent of test geometry.

5. CONCLUDING COMMENTS

The aim of this work has been to adopt an analytical approach, already described in the literature, in order to convert peel strength measurements into adhesive fracture toughness for flexible laminate systems. This has been achieved for two types of peel test, namely fixed arm and T-peel tests. At the same time, it was intended to establish a test procedure.
It has been shown that with proper use of the ESIS protocol for measurement that a large group of laboratories can achieve agreement for peel toughness. In particular it is important to interpret the peel force in terms of an adhesive fracture and to measure with accuracy the "elastic" and "plastic" moduli for the tensile deformation of the peel arm.
Analysis of the peel strength can lead to an objective determination of the adhesive fracture toughness. This can be shown to be independent of test parameters such as peel angle, test geometry and the bending and tensile deformations that are input in the test procedure.

6. REFERENCES

1 ASTM D 3167-97 *Standard Test Method for Floating Roller Peel Resistance of Adhesives* 1997
2 ASTM D 1876-95 *Standard Test Method for Peel Resistance of Adhesives (T-Peel Test)* 1995
3 A J Kinloch, C C Lau, J G Williams Int. J. Fract. 66, (1994), p45-70

Appendix 1
Participants in the ESIS TC4 Peel Activity.

ICI plc, UK (D R Moore, R S Hardy)
Imperial College, London (J G Williams)
Elf-Atochem , France (J Pascal)
ATO-DLO, Holland (H Bos)
University of Twente, Holland (P E Reed)
BASF, Germany (F Ramsteiner)
DuPont S.A., Switzerland (S Ducret)
Martin-Luther University, Germany (Grellman)
Politecnico de Milano, Italy (A Pavan)
University of Kaiserslautern, Germany (J Karger-Kocsis)
Tetra-Pak, Switzerland (P Emery).

Appendix 2
Equations and Calculations for the Determination of Adhesive Fracture Toughness.

There are a number of steps necessary to calculate adhesive fracture toughness in a fixed arm peel test. Several terms are required to facilitate the calculations, namely
E_1, ε_y, α, h, B, P and θ.
Calculations proceed through eight steps:-

1. Calculate strain in peel arm

$$\varepsilon = (P/(h * B))/E_1$$

2. Calculate G_A infinite modulus

$$G_A^{\infty E} = \frac{P}{B}(1 - \cos\theta)$$

3. Calculate G elastic maximum

$$G_{max}^{e} = \frac{E_1 \varepsilon_y^2 h}{2}$$

4. Calculate G_A elastic bending

$$G_A^{eb} = \frac{P}{B}(1 + \varepsilon - \cos\theta) - \frac{hE_1\varepsilon^2}{2}$$

5. Perform iteration to determine k_0

Vary k_0 in the following two equations until a value is found which gives the same answer for $f_2(k_0)$

$$f_2(k_0) = \frac{\alpha}{3}\left[1 + 4(1-\alpha)^2\right]k_0^2 + 2(1-\alpha)^2(1+2\alpha)k_0 + \frac{8}{3}\frac{(1-\alpha)^4}{(1-2\alpha)k_0} - 4(1-\alpha)^3$$

$$f_2(k_0) = \frac{G_A^{\infty E}}{G_{max}^{e}} \frac{1-\cos\left(\theta - \left(\frac{4\varepsilon_y k_0}{3}\right)\right)}{(1-\cos\theta)}$$

6. Calculate root angle

$$\theta_0 = \frac{4\varepsilon_y k_0}{3}$$

7. Calculate strain energy release rate for plastic deformation, G^{db}

First calculate $f_1(k_0)$

$$f_1(k_0)=\tfrac{4}{3}\alpha(1-\alpha)^2k_0^2+2(1-\alpha)^2(1-2\alpha)k_0 + \frac{2(1-\alpha)}{3(1-2\alpha)k_0}\left[1-2\alpha+4(1-\alpha)^3\right]-(1-\alpha)\left[1+4(1-\alpha)^2\right]$$

then

$$G^{db} = G^{e}_{max} * f_1(k_0)$$

8. Calculate adhesive fracture toughness

$$G_A = G_A^{eb} - G^{db}$$

Full details are given in reference (1)

FRACTURE TESTS ON STRUCTURAL ADHESIVE JOINTS

B. BLACKMAN and A. KINLOCH

1. INTRODUCTION

The protocol given in Annex 2 is based upon a linear-elastic fracture-mechanics (LEFM) approach and is designed to be used to determine the value of the adhesive fracture energy, G_{IC}, (more generally termed the adhesive fracture toughness) of structural adhesives under Mode I loading. (Although, it should be recognised that if the crack propagates along the interface then both tensile and shear stresses will exist at the crack tip along the fracture plane (1).)

The specimens described in detail in the protocol are bonded joints based upon double-cantilever beam (DCB) and the tapered double-cantilever beam (TDCB) specimens. The DCB specimen shown in Figures 1a-c of the protocol is well-suited for testing joints consisting of an adhesive which is bonding relatively thin sheets of fibre-composite materials, but may also be used when metallic substrates which possess a relatively high yield stress are being employed (e.g. Figure 1c). The TDCB shown in Figure 1d is designed so that, over a large range of values of crack length, the rate of change of compliance with crack length (i.e. dC/da) is constant; and so the value of dC/da is independent of the value of the crack length. (The arms for the TDCB test specimens were accurately contoured using a computer-controlled milling machine.) The rate of change of compliance with crack length being constant is useful since it means that (a) relatively tough adhesives may be tested without plastic deformation of the arms occurring, (b) the substrates may possess a relatively low yield stress, but again no plastic deformation of the arms may be incurred during the test, and (c) the measurement of the adhesive fracture energy, G_{IC}, is independent of the crack length, a.

The details of the testing of the DCB and TDCB adhesive joint specimen, and the calculation of the values of the adhesive fracture energy, G_{IC}, may be found in the protocol given in Annex 2. It should be noted that this protocol has been written so as to be of the same basic format as that for the determination of the interlaminar fracture energy of composite materials under Mode I loading, see 'Mode I Delamination'.

The protocol given in Annex 2 is a development of the ASTM Standard (2), which was based upon the pioneering work of Ripling et al. (3,4). The present protocol has several advantages over the earlier ASTM version. Firstly, it offers a 'corrected beam theory' (CBT) calculation method and this new method incorporates several extra corrections which may be made to the experimental results for the DCB and TDCB tests. These corrections are often particularly important for fibre-composite arms, as may be employed for the DCB test. Thus, the protocol given in Annex 2 readily permits accurate values of G_{IC} to be deduced from DCB and TDCB joints via the 'corrected beam theory' method. Secondly, the protocol provides a direct 'experimental compliance method' (ECM), as well as the beam theory solutions, also from which accurate values of G_{IC} may be determined. The results from these different approaches may then be compared in order to give confidence in the values of G_{IC} so obtained. Thirdly, the possibility of calculating the modulus, E_f, of the substrate arms of the specimen from the DCB fracture-mechanics test results is advanced, and such values may be compared to the independently measured value, E_s. This provides a good cross-check on the accuracy of the results from the DCB test. Fourthly, a check on the linearity of the force versus displacement curve (both upon initial loading and upon final unloading) is incorporated into the present protocol, as well as a correction scheme for the compliance of the test machine. Finally, the protocol addresses the definition of crack initiation in detail; and also considers the determination of crack propagation and 'R-curve' behaviour, in addition to crack arrest.

2. RESULTS OF 'ROUND-ROBIN' TESTS

2.1 Introduction

A total of twelve laboratories took part in the round-robin tests. The joints were all based upon a rubber-toughened epoxy-paste adhesive (cured at an elevated temperature) and the thickness of the adhesive layer was controlled to be 0.4 mm. The joint types which were tested are listed below, and all were manufactured by Imperial College staff.

a. A DCB specimen using uni-directional carbon-fibre reinforced plastic composite as the substrate arms. The composite arms were nominally 1.65 mm thick and 20 mm wide.

b. A DCB specimen using steel (Grade: BS EN32) as the substrate arms. The steel arms were nominally 20 mm thick and 25mm wide.

c. A TDCB specimen using aluminium-alloy (Grade: BS 5083) as the substrate arms. The value of the specimen geometry factor, m, was 2 mm^{-1} and the joint was nominally 10 mm wide.

d. A TDCB specimen using steel (Grade: BS EN32) as the substrate arms. The value of specimen geometry factor, m, was 2 $mm^{-1}$and the joints were nominally 10 mm wide.

Each of the twelve laboratories received four joints of two of the above types to test. The joints all failed by cohesive failure through, approximately, the centre of the adhesive layer. Hence, as intended, there were no effects due to the different degrees of the intrinsic adhesion which are likely to exist between the adhesive and the various types of substrate.

2.2 Effect of Calculation Method

DCB Tests:

As discussed below, no significant 'R-curves' were found for these adhesives joints tested in the round-robin exercise and the mean results for the propagation values represent 'steady-state' propagation values. Thus, the values given in Tables 1 (for CFRP substrates) and 2 (for steel substrates) may be readily used to investigate the effect of the method of calculation upon the value of G_{IC}. In both Tables, the values of G_{IC} deduced from the 'corrected beam theory' (CBT, Equation (5b)) and 'experimental compliance method' (ECM, Equation (9b)) are in excellent agreement. Further, as expected, the values of G_{IC} deduced from the 'simple beam theory' (SBT, Equation (4)) are substantially lower than those calculated employing the CBT and ECM approaches, due to the failure of the SBT method to take into account crack-tip root-rotation effects (5). This aspect is highlighted by the very good agreement of the modulus, E_f, of the arms of the beam ascertained from the CBT method (Equation (8b)) with the independently measured value of the modulus, E_s. (As noted previously (5), if the values of E_f, were to be determined using the SBT method, then they would be found to be a function of crack length, a, and to be in poor agreement with the value of E_s; reflecting the importance of applying the various correction factors embodied in the CBT method.)

Table 1.

DCB-CFRP Substrates: Mean Steady-State Propagation Values of G_{IC} and E_f and Standard Deviations.

Lab.	G_{IC} (SBT) (J/m^2)	G_{IC} (CBT) (J/m^2)	G_{IC} (ECM) (J/m^2)	E_f (GPa)
1	209 ± 14	242 ± 24	238 ± 26	-
3	142 ± 41	175± 56	175 ± 61	137 ± 17
5	142 ± 21	185 ± 63	185 ± 62	148 ± 30
9	174 ± 43	198 ± 69	200 ± 69	171 ± 21
11	190 ± 59	209 ± 61	212 ± 63	169 ± 12
Mean	171 ± 26	202 ± 23	202 ± 22	156 ± 14

Note:

a. Independently measured value of the modulus, E_s, of the CFRP arms was 145 GPa.

Table 2.

DCB-Steel Substrates: Mean Steady-State Propagation Values of G_{IC} and E_f and Standard Deviations.

Lab.	G_{IC} (SBT) (J/m^2)	G_{IC} (CBT) (J/m^2)	G_{IC} (ECM) (J/m^2)	E_f (GPa)
2	519 ± 73	932 ± 48	941 ± 43	219 ± 42
6	688 ± 43	936± 56	935 ± 56	222 ± 22
8	673 ± 30	858 ± 41	850 ± 38	249 ± 56
10	657 ± 69	905 ± 46	910 ± 47	239 ± 55
12	595 ± 38	958 ± 17	960 ± 21	141 ± 13
Mean	626 ± 62	918 ± 34	919 ± 38	214 ± 38

Note:

a. Independently measured value of the modulus, E_s, of the steel arms was 207 GPa.

TDCB Tests:

The values of G_{IC} deduced from the 'corrected beam theory' (CBT, Equation (11)) and the 'experimental compliance method' (ECM, Equation (2)) are in excellent agreement, as may be seen from the results shown in Table 3 for the bonded aluminium alloy TDCB joints. This is reflected in the excellent agreement between the values of dC/da of 2.45×10^{-5} N^{-1} and 2.51×10^{-5} N^{-1} from the ECM and CBT approaches, respectively. It is noteworthy, that the value of dC/da from the 'simple beam theory' (SBT, Equation (3)) is 2.29×10^{-5} N^{-1}, and thus the SBT method gives somewhat lower values of G_{IC}, as may be seen from Table 3. Furthermore, the plots of C versus a from both the CBT and ECM methods are linear, of course, but when extrapolated back to C=0 give a finite, positive, intercept for the crack length, a; unlike the plot from the SBT method which is linear but passes through the origin. As may seen from the protocol given in Annex 2, these observations arise because the simple beam theory expression for G_{IC} incorrectly describes the compliance of the TDCB specimen, since (a) the positions of the loading pins, with their surrounding material, are not taken into account in deriving Equations

(3) and (4), and (b) as for the DCB specimen, the specimen does not behave as a perfectly built-in cantilever beam.

Table 3.
TDCB-Aluminium Alloy Substrates: Mean Steady-State Propagation Values of G_{IC}.

Lab.	G_{IC} (SBT) (J/m^2)	G_{IC} (CBT) (J/m^2)	G_{IC} (ECM) (J/m^2)
2	691 ± 29	756 ± 32	746 ± 28
6	640 ± 53	702 ± 58	678 ± 60
8	666 ± 144	731 ± 158	702 ± 157
10	490 ± 82	537 ± 89	530 ± 114
12	641 ± 104	701 ± 115	684 ± 99
Mean	626 ± 79	685 ± 86	668 ± 82

2.3 DCB versus TDCB Specimen Geometry

The adhesively-bonded steel joints were tested in both the DCB and TDCB test geometry. The values of G_{IC} from the two different test geometries (deduced using either the CBT (Equation (5b)) or the ECM (Equation (9b)) approaches for the DCB test, and either the CBT (Equation (11)) or ECM (Equation (2)) approaches for the TDCB test) were in excellent agreement.

2.4 Initiation Values of G_{IC}

Values for the onset of crack growth for either the original insert or the precrack ascertained via the various options described in the protocol are given, as an example, in Table 4 for the DCB joints prepared using the CFRP substrates. Some general, noteworthy, points emerged. Firstly, the MAX/5% method gives the highest value of G_{IC}(initiation), as would be expected. Secondly, the MAX/5% method does tend to yield the lowest spread of results: the NL and VIS methods allow more judgement of the individual operator to influence the values of G_{IC}(initiation). Thirdly, compared to the steady-state propagation value of 202 ± 22 J/m^2 (see Table 1), then clearly the values of G_{IC}(initiation) indicate that no significant 'R-curve' is seen in these tests.

Table 4.

DCB-CFRP Substrates: Mean Initiation Values of G_{IC} (J/m^2) Calculated via the CBT Method and Standard Deviations. (Values from the Insert and the Precrack).

	Insert			Precrack		
Lab.	NL	VIS	MAX/5%	NL	VIS	MAX/5%
1	183 ± 41	153 ± 33	225 ± 31	-	-	-
3	125 ± 37	147± 54	144 ± 44	163 ± 39	184 ± 32	171 ± 36
5	148 ± 54	-	191 ± 63	202 ± 54	219 ± 57	212 ± 62
9	109 ± 66	117 ± 29	209 ± 104	83 ± 34	117 ± 33	212 ± 102
11	90 ± 20	-	237 ± 72	178 ± 92	-	265 ± 96
Mean	131± 32	139± 16	201± 33	157 ± 45	173± 42	215± 33

Note:

a. NL, VIS, MAX/5% are defined in the protocol, see Annex 2.

b. '-' indicates that the laboratory did not report a value.

2.5 Effect of Substrate Type

The adhesive used in the present tests was selected to always give a cohesive failure through the adhesive layer in all the different types of joints. This was indeed observed. However, the value of G_{IC} does still appear to be dependent upon the substrates employed. For example, compare the results shown in Tables 1, 2 and 3. Subsequent work (6) has revealed that the glass transition temperature, T_g, of this particular rubber-toughened adhesive is very dependent upon the (a) exact heating cycle used to cure the adhesive, and (b) the amount of water present in the CFRP substrates prior to forming the joint. If these factors are controlled to give the same value of T_g for the adhesive layer in all the various types of joints, then the value of G_{IC} is indeed independent of the substrate type (6).

3. CONCLUDING REMARKS

The protocol given in Annex 2 provides a good basis for determining the adhesive fracture energy, G_{IC}, of adhesive joints. It includes several significant changes to the ASTM Standard (2) which: (a) describe 'corrected beam theory'(CBT) solutions, which lead to an improved accuracy for the values of G_{IC} and the stiffness of the specimen, (b) allow a cross-check on the accuracy of the overall approach, (c) enable CFRP, and other fibre-composite, bonded joints to be tested and sound values of G_{IC} to be determined for such joints, and (d) address the issues of crack initiation and 'R-curve' effects.

4. ACKNOWLEDGEMENTS

The authors would wish to thank the ESIS Technical Committee, TC4 and all the laboratories that contributed data to the 'Round-Robin' exercise: Universidad de Oviedo, Universita di Pisa, Imperial College, IFREMER, DSM Research, MERL Ltd, NIST, Rhone-Poulenc Recherches, Politecnico di Milano and BASF. For the supply of materials, we would like to thank ICI plc. We would also wish to thank the EPSRC (Grant GR/L/01626 and Advanced Fellowship AF/992781), NPL and the DTI for financial support.

5. REFERENCES

1. A.J. Kinloch, 'Adhesion and Adhesives: Science and Technology', Chapman and Hall, London, 1987.

2. ASTM 'Fracture Strength in Cleavage of Adhesives in Bonded Joints', D3433-93.

3. E.J. Ripling, S. Mostovoy and R.L.Patrick, ASTM, **STP 360**, 5 (1963).

4. E.J. Ripling, S. Mostovoy and R.L.Patrick, Materials Research and Standards, **64**, 129 (1964).

5. B.R.K.Blackman, J.P. Dear, A.J. Kinloch and S. Osiyemi, J. Materials Sci. Letters, **10**, 253 (1991).

6. B.R.K. Blackman, A.J. Kinloch and M. Paraschi, 'Proceedings of the Adhesion Society', Myrtle. Beach, SC, USA, February 2000, (Adhesion Society, USA, 2000) p.211.

ANNEX 2

Determination of the Mode I Adhesive Fracture Energy, G_{IC}, of Structural Adhesives using the Double Cantilever Beam (DCB) and Tapered Double Cantilever Beam (TDCB) Specimens

1. Scope

This standard specifies a method, based upon a linear-elastic fracture-mechanics (LEFM) approach, for the determination of the fracture resistance of structural adhesive joints under an applied Mode I opening load, using the Double Cantilever Beam (DCB) and Tapered Double Cantilever Beam (TDCB) specimens. The resistance to both crack initiation and propagation are to be determined. The resistance to crack initiation is to be determined from both a non-adhesive insert placed in the adhesive layer and from a mode I precrack. The resistance to crack propagation is to be determined from the mode I precrack. The adhesive fracture energy G_{IC} (also termed the critical strain energy release rate) for applied Mode I loading can be calculated and a resistance-curve (R-curve, i.e. a plot of the value of the adhesive fracture energy G_{IC} versus crack length) can be determined.

2. Normative References

The following standard contains provisions which through reference in this text constitute provisions of this standard. At the time of publication the editions indicated were valid. All standards are subject to revision, and parties to agreements based on this standard are encouraged to investigate the possibility of applying the most recent editions of the standards listed below. Members of IEC and ISO maintain registers of currently valid International Standards.

ISO 291: 1977	*Plastics; standard atmospheres for conditioning and testing.*
ISO 4588: 1991	*Adhesives; preparation of metal surfaces for adhesive bonding.*
ISO 10365: 1992	*Adhesives, designation of main failure patterns.*
ISO 5893: 1993	*Rubber and plastics test equipment; tensile, flexural and compression types (constant rate of traverse); description.*

3 Definitions

For a list of the definitions of symbols and conventions used in this protocol, refer to the central list of symbols in the present book.

4. Principle

This standard uses the Double Cantilever Beam (DCB) specimen, shown in Figures 1a-c, or the Tapered Double Cantilever Beam (TDCB), shown in Figure 1d, for the determination of the adhesive fracture energy, G_{IC}, of structural adhesive joints.

The Double Cantilever Beam (DCB) specimen shown in Figures 1a-c is well-suited for testing joints consisting of an adhesive which is bonding relatively thin sheets of fibre-composite materials, but may also be used when metallic substrates which possess a relatively high yield stress are being employed (e.g. Figure 1c).

The Tapered Double Cantilever Beam (TDCB) shown in Figure 1d is designed so that, over a large range of values of crack length, the rate of change of compliance with crack length is constant and so independent of the value of crack length. This is useful since it means that (i) relatively tough adhesives may be tested without plastic deformation of the arms occurring, (ii) the substrates may possess a relatively low yield stress, but again no plastic deformation of the arms may be incurred during the test, and (iii) the measurement of the adhesive fracture energy, G_{IC}, is independent of the crack length, a. To develop a linear change of compliance with crack length, the height of the specimen is varied by contouring the substrate beam so that the quantity:

$$\frac{3a^2}{h^3} + \frac{1}{h} = m \qquad (1)$$

is a constant, where m is the specimen geometry factor.

5. Apparatus

(1) A tensile testing machine in compliance with ISO 5893, capable of producing a constant cross-head displacement-rate between 0.1 and 5 mm/min in displacement control should be used. The testing machine shall be equipped (i) with a fixture to introduce the load to the pins inserted into the load-blocks, or directly into the substrate beams, or (ii) with grips to hold the piano hinges that both allow rotation of the specimen end, see Figure 1.

(2) The testing machine should incorporate a load-cell which should be calibrated and accurate within ±1% for the chosen load-range (loads are typically expected to be in the range of 100-5000 N). The opening displacement of the test specimen should be deduced from the position of the cross-head. The testing machine shall be equipped with means for recording the complete load versus displacement curves (loading and unloading) during the test. (Note that if extensometry is used to measure the opening displacement of the specimen during the test, then the system compliance correction described in Annex A1 can be neglected.)

(3) The crack length should be measured along the edge of the specimen to an accuracy of at least ± 0.5 mm by either using a travelling microscope or a video camera with suitable magnification.

(4) The thickness of DCB specimen arms should be measured with a micrometer or equivalent with an accuracy of 0.02 mm or better. For measuring the width of the joints, a micrometer or vernier calipers with an accuracy of 0.05mm or better shall be used.

6. Specimens

6.1 Manufacture of adhesive joint specimens

It is not the purpose of this document to give full manufacturing details of the joints to be tested. Such information should be sought from the adhesive manufacturer and or the substrate manufacturer. Appropriate surface treatments may also be determined by reference to *ISO 4588* for metallic substrates. The thickness of the film to be inserted in the bondline during manufacture should be less than 13 microns. A PTFE film is recommended although other release films have been successfully used. If aluminium foil is used, the foil should be coated with releasing agent prior to use. The thickness of the adhesive bondline should be carefully controlled. When fully cured, any excess adhesive should be removed by mechanical means to leave the joint with smooth sides.

6.2 Measurement of specimen dimensions

(1) When DCB substrates have been prepared, the thickness of each substrate should be measured using a micrometer *before* bonding. Measurements should be made at three points along the length of the beam (at 30 mm from either end, and at the mid length) and the average obtained. Thus, a value of the thickness of each substrate, h, is obtained.

(2) If the thickness measurements are repeated on the joint after bonding, a value of the bondline thickness, h_a, may be determined by subtracting the substrate thicknesses, nominally 2h, from the total thickness of the joint.

(3) The width, B, of the DCB or TDCB joint should be measured *after* bonding with vernier callipers or a micrometer at three points along the length of the beam (at 30 mm from either end, and at the mid length) and the average obtained. These width measurements should be made *after* any excess adhesive has been removed from the sides of the beam.

6.3 Preparation of Specimens

(1) Adding a thin layer of typewriter correction fluid ("white ink"), or white spray-paint, on the edges of the sample after conditioning will facilitate the detection of the crack growth. It should be noted that some typewriter correction fluids and paints contain solvents which may be harmful to the adhesive or the laminate matrix material of a composite substrate.

(2) For the measurement of the crack growth, marks should indicate every 1 mm from the tip of the insert or of the Mode I crack for at least the first 10 mm, then marks should be applied every 5 mm. Also, such marks should be applied for every 1 mm for the final 5 mm. For the DCB test specimen the recommended extent of crack propagation is 65 mm, and for the TDCB test specimen the recommended extent of crack propagation is 100 mm.

7. Number of specimens

A minimum of four repeat joints should be tested.

8. Conditioning

The joints should be maintained in normal laboratory conditions for a minimum of twenty four hours prior to testing i.e. they should be maintained at a temperature of 23°C±2°C, and at a relative humidity of 50%±5%.

9. Test Procedure

9.1 Test Set-up and Data Recording

(1) The tests shall be performed under normal conditions in accordance with ISO 291 (23°±2°C, 50%±5% relative humidity) unless prescribed otherwise. After mounting the specimen in the fixture of the testing machine, the end of the specimen may have to be supported in order to keep the beam orthogonal to the direction of the applied load. The load and the displacement signals of the testing machine shall be recorded, either on a paper chart or electronically throughout the test, including the unloading cycle. If using a tensile testing machine with a paper chart recorder, then the following ratios of cross-head speed to chart speed are recommended. When testing joints with metallic substrates, a ratio of cross-head speed to chart speed of about 1:100 is recommended. When testing joints with fibre-composite substrates, a ratio of cross-head speed to chart speed of about 1:10 is recommended.

(2) The crack length should be measured along the edge of the specimen to an accuracy of at least ± 0.5mm by either using a travelling microscope or a video camera with suitable magnification. If unstable crack growth followed by arrest ("stick-slip") is observed during any stage of the test, it should be noted in the report, see Annex B1 for more details.

9.2 Initial loading (the pre-cracking stage).

(1) For testing from the insert (starter film) it is recommended that the specimen should be loaded at a constant cross-head rate of:

(a) 0.1 mm/min for joints prepared using metallic substrates; or

(b) 1.0 mm/min for joints prepared using fibre-composite (with polymeric matrix) substrates.

(2) The point on the load-displacement curve at which the onset of crack movement from the insert is observed on the edge of the specimen should be recorded on the load-displacement curve or in the sequence of load-displacement signals (VIS, Figure 2a).

(3) The loading should be stopped as soon as the crack is seen to move on the edge of the specimen, after which the specimen should be completely unloaded at a constant cross-head rate. Unloading may be performed at up to five times the loading rate. The position of the tip of the precrack should be marked on both edges of the specimen. If the crack lengths a on the edges of the specimen, i.e. the distance between the load-line and the tip of the precrack, differ by more than 2 mm the results should be considered suspect and this be noted in the report.

9.3 Re-loading: testing from the mode I precrack

(1) For testing from the mode I precrack, which has been formed as a result of the above test procedure, it is recommended that the specimen be loaded at a constant cross-head rate of:

(a) 0.1 mm/min for joints prepared using metallic substrates; or

(b) 1.0 mm/min for joints prepared using fibre-composite (with polymeric matrix) substrates.

(2) The point on the load-displacement curve at which the onset of crack movement from the Mode I precrack is observed on the edge of the specimen should be recorded on the plot or in the sequence of load-displacement signals (VIS, figure 2b).

(3) After this, as many crack length increments as possible should be noted in the first 5 mm on the corresponding load-displacement curves, ideally every 1 mm. Subsequently, crack

lengths are noted every 5 mm, until the crack has propagated about 60 mm from the tip of the Mode I precrack for the DCB test and about 95 mm for the TDCB tests. Then again every 1 mm for the last 5 mm of crack propagation.

(4) After this, the specimen should be unloaded at a constant cross-head rate. The unloading may be performed at up to five times the loading rate. A note should be made in the report if the load-displacement curve does not return to its initial point, since this indicates that permanent plastic deformation of the arms of the specimen may have occurred, see Annex A2.

(5) The position of the tip of the crack edges of the specimen, i.e. the distance between the load-line and the tip of the crack should be marked on both edges of the specimen. If the crack lengths a on the edges of the specimen, i.e. the distance between the load-line and the tip of the crack, differ by more than 2 mm the results should be considered suspect and this be noted in the report.

9.4 Measurement of the Machine Compliance

The tensile testing machine with associated grips and pins will not have an infinite stiffness and hence the compliance associated with the machine set-up should be determined and taken into account in the calculations presented in section 10. This is most conveniently achieved by correcting the displacement measured during the DCB or TDCB test to take account of the deflections in the loading system. The procedure for measuring the system compliance is given in the Normative Annex A1 to this document and this should be followed. It is advisable to conduct the system compliance measurement after the fracture tests have been conducted so that the maximum load obtained from the fracture tests is known. This then determines the load range over which the system compliance is measured. For each test, the corrected values of the displacement are then used in the calculations which follow.

10. Data Analysis

10.1 Determining the raw data from the load-displacement trace

The data required for the analysis are the crack lengths a and the corresponding loads P and displacements δ. The values of δ deduced from the load-displacement record should be corrected for system compliance as described in Annex A1 prior to the determination of G_{IC}. Also, any initial non-linearities in the load-displacement trace should be disregarded by extrapolating the linear region of the loading curve back to zero load, as described in Annexes A1 and A2. The following values should be determined from the load-displacement trace.

Initiation values

The crack length for the initiation values from the insert is the distance between the load-line and the tip insert, a_o. The crack length for the initiation values from the precrack is the distance between the load-line and the tip of the precrack, a_p (Figure 1). If possible, the following initiation values, shown in Figure 2, should be determined for testing from the insert (starter film) and from the Mode I precrack for each specimen:

(1) NL, i.e. deviation from linearity: A region of non-linear behaviour usually precedes the maximum load, even if the unloading curve is linear. The point of deviation from linearity (NL in Figure 2) , is determined by drawing a straight line from the origin but ignoring any initial deviations due to take-up of play in the loading system. Experience has shown that it is difficult to reproducibly determine the position of NL on the load-displacement curve. Performing a linear fit on the load-displacement curve starting at 5% of the maximum load and using a consistent criterion for deviation from linearity (e.g the half-thickness of the plotter trace) is recommended.

(2) VIS, i.e. visual observation: This corresponds to the onset of crack growth, i.e. to the first point at which the crack is observed to move from the tip of the insert or of the Mode I precrack on the edge of the specimen (VIS in Figure 2).

(3) 5% or MAX, i.e. 5% increase of compliance or maximum load point: The 5% value corresponds to the point on the load-displacement curve at which the compliance has increased by 5% of its initial value C_0. A best straight line is drawn to determine the initial compliance C_0, ignoring any initial deviation due to take-up of play in the loading system. A new line is then drawn with a compliance equal to C_0 + 5%, and the intersection of this new

line with the load-displacement trace marked. Which ever point occurs first (i.e. max load or 5%) that is the point to be used.

Propagation values

Besides the initiation points (i.e. NL, VIS, 5% or MAX) obtained both from the insert (starter film) and from the Mode I precrack, propagation values (PROP in Figure 2b) should also be determined. These are determined from the Mode I precrack.

10.2 Determining the values of G_{IC}

As the development of this test protocol is still underway, it is recommended that users employ all the methods of analysis shown below for the given test geometry and these values should be quoted in the report. There are three analysis methods for both the DCB test and the TDCB test. If it is not possible for all methods to be used, then it is recommended that the experimental compliance method be used for both DCB and TDCB tests. Since this method, together with corrected beam theory (CBT) method, are considered to be the more accurate methods for determining the values of G_{IC}. However, if unstable ('stick-slip') crack growth occurs then the simple beam theory (SBT) method should be used for both the DCB and TDCB tests, see the Informative Annex B1 for more details.

10.2.1 Double Cantilever Beam (DCB) Tests

Method (1): Simple Beam Theory (SBT)

The value of the adhesive fracture energy, G_{IC}, may be ascertained from:

$$G_{IC} = \frac{P^2}{2B} \cdot \frac{dC}{da} \quad (2)$$

where C is the compliance and is given by displacement, δ/load, P. For thin adhesive layers, it has been shown (References 1 and 2) from simple beam theory that dC/da may be expressed:

$$\frac{dC}{da} = \frac{8}{E_s B}\left(\frac{3a^2}{h^3} + \frac{1}{h}\right) \quad (3)$$

where E_S is the independently-measured flexural or tensile modulus of the substrate. This value of the modulus should be measured from an independent modulus test, or quoted if a standard grade of material is used. Hence, combining Equations (1), (2) and (3):

$$G_{IC} = \frac{4P^2}{E_s B^2}\left(\frac{3a^2}{h^3} + \frac{1}{h}\right) = \frac{4P^2}{E_s B^2} \cdot m \qquad (4)$$

where h is the thickness of the arm of one substrate beam.

For the simple beam theory (SBT) method of calculation, the value of G_{IC} should be calculated via Equation (4).

Method (2): Corrected Beam Theory (CBT)

The simple beam theory expression for the compliance of a perfectly built-in DCB specimen will underestimate the compliance as the beam is not perfectly built-in. A means of correcting for this effect is to treat the beam as containing a slightly longer crack length $(a + |\Delta|)$; and $|\Delta|$ may be found experimentally by plotting the cube root of the compliance $C^{1/3}$, or the cube-root of the normalised compliance $(C/N)^{1/3}$, if load-blocks are being used, as a function of crack length a (Figure 3). The load-block correction N is described below. The extrapolation of a linear fit through the data in the plot yields Δ as the negative x-intercept (References 3 and 4). The propagation (PROP) values only are used for the linear fits i.e. all the initiation values are excluded from the linear fits. The adhesive fracture energy G_{IC} is given by:

$$G_{IC} = \frac{3P\delta}{2B(a+|\Delta|)} \quad \text{... (5a)} \quad \text{or} \quad G_{IC} = \frac{3P\delta}{2B(a+|\Delta|)} \cdot \frac{F}{N} \quad \text{... (5b)} \qquad (5)$$

where P is the load, δ the displacement, a the crack length, and B the width of the specimen. All initiation and propagation values of G_{IC}, if applicable, should be calculated. The load-block correction N is applied if load-blocks are being used, for piano hinges and for loading holes drilled directly though the substrate N = 1. The large displacement correction F becomes important if the displacement δ divided by the crack length a, $\delta/a > 0.4$. The large displacement correction F and the load-block correction N are calculated as follows:

$$F = 1 - \frac{3}{10}\left(\frac{\delta}{a}\right)^2 - \frac{3}{2}\left(\frac{\delta l_1}{a^2}\right) \quad (6)$$

$$N = 1 - \left(\frac{l_2}{a}\right)^3 - \frac{9}{8}\left[1 - \left(\frac{l_2}{a}\right)^2\right]\frac{\delta l_1}{a^2} - \frac{9}{35}\left(\frac{\delta}{a}\right)^2 \quad (7)$$

where l_1 is the distance from the centre of the loading pin to the mid-plane of the arm of the substrate beam and l_2 the distance from the loading pin centre to the edge of the block (Figure 1). Data with large displacement corrections F < 0.9 should be considered suspect and this be noted in the report.

This approach allows the flexural modulus E_f to be calculated as a function of the crack length a by using:

$$E_f = \frac{8(a + |\Delta|)^3}{CBh^3} \quad (8a) \qquad \text{or} \qquad E_f = \frac{8(a + |\Delta|)^3}{\frac{C}{N}Bh^3} \quad (8b) \qquad (8)$$

This calculation is a useful check on the procedure, as a value of the flexural modulus E_f independent of crack length should be obtained. If the maximum variation is more than 10% of the average, the values of G_{IC} should be considered suspect and this should be noted in the report. (The value E_f calculated from Equation (8) should not be quoted as the modulus value and this value should not be used in Equation (4), which requires an independently measured or known value of the modulus to be used.)

Method (3): Experimental Compliance Method (ECM) or Berry's Method

An alternative approach is to plot the logarithm of the compliance C, or of the normalised compliance, C/N, if load-blocks are being used, versus the logarithm of the crack length a as shown in Figure 3. Only the propagation (PROP) values are used for the linear fits, i.e. all the initiation values are excluded from the regression analysis. The slope of this plot, n, can then be used to give G_{IC} as follows:

$$G_{IC} = \frac{nP\delta}{2Ba} \quad \text{(9a)} \quad \text{or} \quad G_{IC} = \frac{nP\delta}{2Ba}\frac{F}{N} \quad \text{(9b)} \qquad (9)$$

with P the load, δ the displacement, a the crack length, and B the width of the specimen. All initiation and propagation values of G_{IC}, if applicable, should be calculated. The same large-displacement correction F and load-block correction N, if applicable, are used as for the corrected beam theory method (see above).

10.2.2 Tapered Double Cantilever Beam (TDCB) Tests

Method (4): Simple Beam Theory (SBT)

The value of the adhesive fracture energy, G_{IC} , may be ascertained from:

$$G_{IC} = \frac{P^2}{2B} \cdot \frac{dC}{da} \qquad (2)$$

For thin adhesive layers, it has been shown (References 1 and 2) from simple beam theory that dC/da may be expressed by:

$$\frac{dC}{da} = \frac{8}{E_s B}\left(\frac{3a^2}{h^3} + \frac{1}{h}\right) \qquad (3)$$

where E_S is the independently-measured modulus of the substrate beam. Hence, combining Equations (2) and (3) and (1):

$$G_{IC} = \frac{4P^2}{E_s B^2}\left(\frac{3a^2}{h^3} + \frac{1}{h}\right) = \frac{4P^2}{E_s B^2} \cdot m \quad (4)$$

For the SBT method of calculation, the value of G_{IC} should be determined from Equation (4). If a standard grade of material is used, the quoted modulus may be used in Equation (4). In the report the value of m, and the range of the crack length a for which this value of m is within ±3%, should be quoted. (Values of G_{IC}, calculated where the value of m is outside of the range ±3%, should be considered suspect.)

Method (5): Corrected Beam Theory (CBT)

The simple beam theory expression for G_{IC} described in Method (4) above will incorrectly estimate the compliance of the specimen since (i) the positions of the loading pins, with their surrounding material, are not taken into account in deriving equation (4), and (ii) as for the DCB specimen, the specimen does not behave as a perfectly built-in cantilever beam. These corrections (Reference 6) lead to equation (10):

$$\frac{dC}{da} = \frac{8m}{E_s B} \cdot \left[1 + 0.43\left(\frac{3}{m}\right)^{\frac{1}{3}} \cdot a^{-\frac{1}{3}}\right] \quad (10)$$

Hence, combining equations (2) and (10):

$$G_{IC} = \frac{4P^2}{E_s B^2} \cdot m \cdot \left[1 + 0.43\left(\frac{3}{m}\right)^{\frac{1}{3}} \cdot a^{-\frac{1}{3}}\right] \quad (11)$$

(In deriving equation (10), the value of m is approximated to $3a^2/h^3$, i.e. the term 1/h in equation (1) is neglected. The error in the value of G_{IC} that is introduced by this approximation is insignificant and round-robin testing has demonstrated good agreement between the values of G_{IC} deduced via equations (11) and (2) for tapered beams manufactured with aluminium alloy substrates (Reference 7).)

Method (6): Experimental Compliance Method (ECM)

The value of the adhesive fracture energy, G_{IC}, may again be ascertained from:

$$G_{IC} = \frac{P^2}{2B}\frac{dC}{da} \qquad (2)$$

For the TDCB geometry, when the values of C are plotted against the crack length a, the resulting graph should be linear. The value of dC/da is given by the slope of the straight line and is used to determine G_{IC} in Equation (2). The value of dC/da and the correlation coefficient, r^2, of the regression analysis should be noted on the results sheet. In the calculation of dC/da, only the propagation values should be included in the regression analysis, i.e. all initiation values should be excluded from this linear fit.

11. Test Report

The recommended format of test reports for the DCB and TDCB geometries are shown in Figures 5 (a)- (b). The test report should contain the following information:

11.1 Test report for the DCB test

(1) Equation (4) (i.e. G_{IC} from SBT, Method 1).

(2) Equation (5) (i.e. G_{IC} from CBT, Method 2).

(3) Equation (9) (i.e. G_{IC} from ECM, Method 3).

(4) Equation (8) (i.e. the value of the modulus, E_f).

Using these equations, the parameters listed below should be calculated:

(1) The initiation points of G_{IC} (NL, VIS, 5% or MAX, see Figure 1) obtained from both the insert (starter film) and from the Mode I precrack. (In the calculation of these values of G_{IC} the corresponding measured value of the crack length a should be used in the Equations

i.e. a_o or a_p). The values determined from the insert (starter film) and from the Mode I precrack should be entered on the same test results sheet (Figure 5a).

(2) The propagation values of G_{IC} (PROP in Figure 2b) determined from the Mode I precrack as a function of crack length a.

(3) The results from both the insert and the Mode I precrack are then used to draw a resistance-curve (R-curve), i.e. G_{IC} versus crack length a (Figure 4. All initiation and propagation values shall be shown on the R-curve. The minimum number of propagation points recorded should be fifteen, if fewer points are used, this should be noted in the report and the results considered suspect (Reference 5).

(4) The flexural modulus E_f of the substrate should be calculated as a function of the crack length a. The flexural or tensile modulus E_s of a substrate arm should also be independently measured, or quoted if a known Standard Grade of material is employed, and the value obtained should be recorded in the report.

(5) After testing, the joints should be broken open to enable the locus of joint failure to be visually assessed. Record whether it is: (i) cohesive-in-the adhesive, (ii) apparently interfacial along the adhesive/substrate interface or (iii) cohesive-in-the-substrate. If a mixture of such failure paths are seen, estimate and record the percentage of each type. (ISO 10365: 1992).

11.2 Test report for the TDCB test

(1) Equation (4) (i.e. G_{IC} from the SBT, Method 4).

(2) Equation (11) (i.e. G_{IC} from the CBT, Method 5).

(3) Equation (2) (i.e. G_{IC} from the ECM, Method 6).

Using these equations, the parameters listed below should be calculated:

(1) The initiation points of G_{IC} (NL, VIS, 5% or MAX, see Figure 1) obtained from <u>both</u> the insert (starter film) <u>and</u> from the Mode I precrack. (In the calculation of these values of G_{IC}

the corresponding measured value of the crack length a should be used.) The values determined from the insert (starter film) and those from the Mode I precrack shall be entered on the same test results sheet (Figure 5b).

(2) The propagation values of G_{IC} (PROP in Figure 2b) determined from the Mode I precrack as a function of crack length a.

(3) The results from both the insert and the Mode I precrack are then used to draw a resistance-curve (R-curve), i.e. G_{IC} versus crack length a (Figure 4). All initiation and propagation values shall be shown on the R-curve. The minimum number of propagation points recorded should be fifteen, if fewer points are used, this should be noted in the report and the results considered suspect (Reference 5).

(4) Also, from the graph of C versus a, the value of the slope, dC/da, and the correlation coefficient, r^2, of the data should be quoted.

(5) After testing, the joints should be broken open to enable the locus of joint failure to be visually assessed. Record whether it is: (i) cohesive-in-the adhesive, (ii) apparently interfacial along the adhesive/substrate interface or (iii) cohesive-in-the-substrate. If a mixture of such failure paths are seen estimate and record the percentage of each type (ISO 10365: 1992).

Annex A: Normative

A.1 Procedure to follow for measuring the compliance of the testing system

Special note: Please ensure that this procedure is carried out by experienced personnel, otherwise damage to equipment may occur when loading the calibration specimen. *It has been observed from round-robin testing that this correction procedure can have a significant effect on the shape of the R-curves and on the values of the back-calculated modulus in the DCB tests.*

1. Set up the tensile loading system in exactly the manner which was used for the fracture testing. It is recommended that this measurement be performed after the fracture tests, because the maximum load during fracture testing will then be known. This load will now be referred to as the calibration load, P_{cal}.

2. A rigid calibration specimen of known compliance, C_{cs}, is required along with a means of connecting it to the loading system. If pins of circular cross section have been used to load the fracture specimens, these should also be used to load the calibration specimen. (*Note: a calibration specimen made from mild steel with a cross-sectional area of 20mm by 25mm and a distance between loading hole centres of 25mm has been found to work satisfactorily, and will possess a compliance which is usually negligible when compared to the system compliance, C_{sy}.*)

3. With the calibration specimen attached, start to load the specimen at a very slow rate e.g. 0.05mm/min up to the calibration load value, P_{cal}. If using a chart recorder to monitor displacement, run this at 100 times the rate of the cross-head. When the load reaches the value of P_{cal}, stop the cross-head and unload the sample. ***The load will rise rapidly during this procedure, so care should be taken not to overload the load-cell!***

4. From the load-displacement trace obtained, draw the best straight line through the second 50% of the data, thus ignoring the initial non-linearity due to take up of play, as shown in Figure A.1. (This take up of play is also ignored in the fracture tests). From this straight line, deduce the total compliance, C_{tot}, of the combined system and calibration specimen in (mm/N), as shown in Figure A.1.

5. Calculate the value of the system compliance, C_{sy}, in (mm/N) from:

$$\mathbf{C_{sy} = C_{total} - C_{cs}} \tag{A.1}$$

6. All displacement values measured during the fracture tests should then be corrected by:

$$\delta_{cor} = \delta - PC_{sy} \tag{A.2}$$

where δ_{cor} is the corrected value of the displacement in (mm) to be used in equations in section 10, δ is the value of the displacement in (mm) measured in the fracture tests and P is the corresponding load. This correction should be made to all displacement values, i.e. at each value of the crack length that was recorded.

Annex A.2: Procedure to detect the occurrence of plastic deformation during a DCB or TDCB adhesive joint test.

A schematic load displacement trace obtained from testing a bonded tapered double cantilever beam specimen is shown in Figure A2. The complete load, propagation and unload cycle obtained during the test from the Mode I precrack is shown. For both the loading and unloading parts of the trace, the best straight lines should be drawn through the data, ignoring any initial non-linearity due to the take up of play in the system. These lines should be extrapolated back to zero load. The distance between the intercepts of these two lines with the displacement axis is termed δ_{offset}. The maximum value of the displacement attained during the test is termed δ_{max}. The values of δ_{offset} and δ_{max} should be measured from the test trace. It is normal for the term δ_{offset} to be non-zero. The value of (δ_{offset}/ δ_{max}) should be calculated for each test and noted on the results sheet.

The occurrence of plastic deformation in the adherends during a fracture test may be observed visually when the amount of deformation is large. If the joint is carefully broken open after the complete test cycle is finished, then plastic deformation of the substrate arms will have occurred if they remain bent on separation. In the case of the tapered beams, this may be seen if the substrates are held back together as they were before separation. The value of (δ_{offset}/ δ_{max}) and the results of a visual check on the straightness of the beams after breaking open should be noted in the report. Experience has shown that plastic deformation of the substrates can be suspected if $\delta_{offset}/\delta_{max} > 0.05$, where δ_{max} is the displacement required to extend the crack by the distance recommended in section 9.3.

Annex B Informative

B.1 Procedure to follow when unstable or 'stick-slip' crack growth is observed during the fracture test.

It is not uncommon for adhesive joints to exhibit unstable or 'stick-slip' crack propagation during a DCB or TDCB fracture test. A schematic example of a load-displacement trace obtained from a TDCB joint exhibiting stick-slip crack growth is shown in Figure B1. The crack grows in shorts bursts separated by periods of crack arrest during this type of propagation. Sometimes the propagation may be partly stable and partly unstable. The reasons for this type of behaviour are not fully understood.

When stick-slip crack propagation is observed during a DCB or TDCB test, it will not be possible to monitor the crack propagation as required by this protocol. The first initiation value of the crack length will be known however, and the crack lengths at subsequent arrest points may be observed using a travelling microscope. Between one arrest point and the next initiation point, the crack will obviously remain stationary. There may then be some stable crack growth before further instability, or the crack may jump directly from the previous arrested value of the crack length. After the crack has propagated sufficiently down the specimen, the joint should be fully unloaded and the unloading trace recorded in the same way as for a stable test. Breaking the joint open may reveal arrest lines on the adhesive that will allow more accurate crack length measurements to be made.

Thus the load, displacement and crack length data will be available at series of initiation and arrest points. As the number of data points will be insufficient to employ the linear regression analysis, needed for the ECM approach and the CBT approach (i.e. for the DCB test), then only the simple beam theory (SBT) method can be used to calculate the values of G_{IC}. Values of G_{IC}(initiation) and G_{IC}(arrest) may be computed using the simple beam theory. However, it should be noted clearly on the results sheet that stick-slip crack propagation was observed, and the type of point, i.e. initiation or arrest, should be clearly stated.

Annex C: Informative

Bibliography

(1) S. Mostovoy, P.B. Crosley, E.J. Ripling: " Use of Crack-Line-Loaded specimens for Measuring Plane-Strain Fracture Toughness", J. of Materials, **2,** 661-681 (1967).

(2) A.J. Kinloch: "Adhesion and Adhesives: Science and Technology", Chapman and Hall, London, 264-296 (1987).

(3) S. Hashemi, A.J. Kinloch, J.G. Williams: "Corrections Needed in Double Cantilever beam tests for Assessing the Interlaminar Failure of Fibre-composites", Journal of Materials Science Letters, 8, 125-129 (1989).

(4) B.R.K. Blackman, J.P. Dear, A.J. Kinloch, S. Osiyemi: "The Calculation of Adhesive Fracture Energies from DCB test Specimens", Journal of Materials Science Letters, **10**, 253-256 (1991).

(5) A.J. Brunner, S. Tanner, P. Davies, H. Wittich: "Interlaminar Fracture testing of Unidirectional Fibre-Reinforced Composites: Results from ESIS-Round Robins" in: Composites Testing and Standardisation ECCM-CTS 2, (P.J. Hogg, K. Schulte, H. Wittich eds.), Woodhead Publishing, 523-532 (1994).

(6) B.R.K. Blackman, H. Hadavinia, A.J. Kinloch, M. Paraschi, J.G. Williams, "The calculation of adhesive fracture energies using the double cantilever beam and tapered double cantilever beam specimens." To be published 2001.

(7) B.R.K. Blackman and A.J. Kinloch. "Fracture tests on structural adhesive joints, in fracture mechanics testing methods for polymers, adhesives and composites." To be published by Elsevier Science, 2001.

(a)

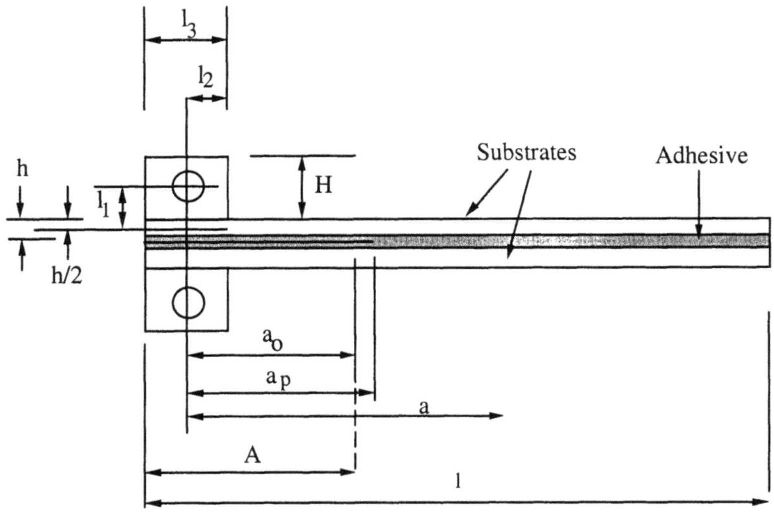
l_3
l_2
Substrates
Adhesive
h
l_1
H
h/2
a_o
a_p
a
A
l

(b)

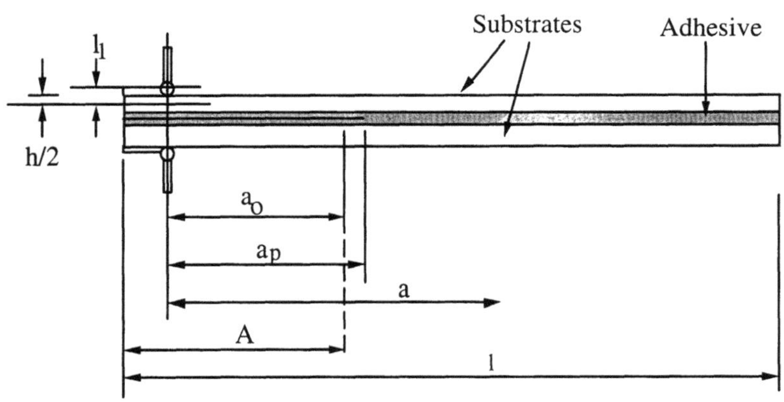
Substrates
Adhesive
l_1
h/2
a_o
a_p
a
A
l

(c)

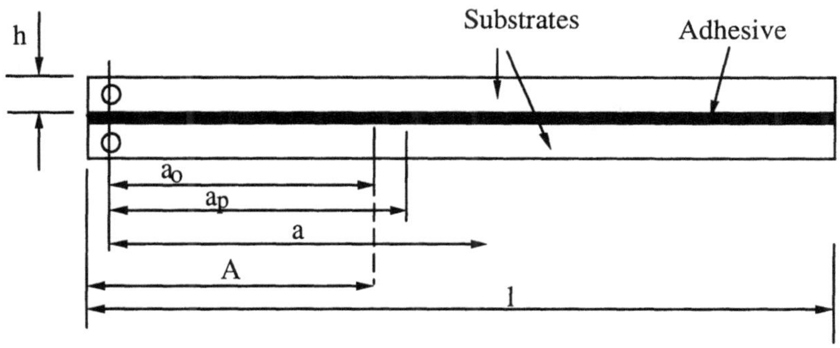
h
Substrates
Adhesive
a_o
a_p
a
A
l

(d)

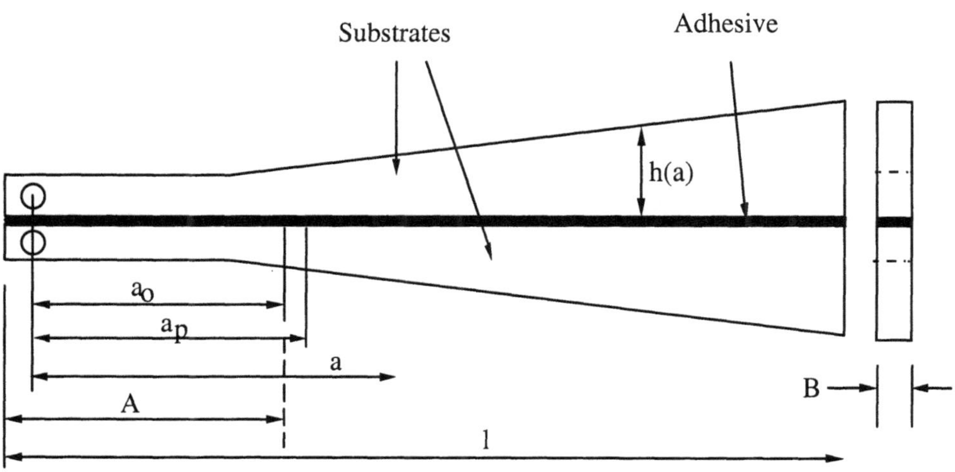
Substrates
Adhesive
h(a)
a_o
a_p
a
A
l
B

Figure 1: Geometry for the adhesive joint specimens.

(a) DCB Specimen with load-blocks.

(b) DCB Specimen with piano hinges (alternative loading arrangement).

(c) DCB specimen with metallic substrates where loading holes may be drilled through the arms of the substrate (alternative loading arrangement).

(d) TDCB specimen.

[The crack length *a* is the distance between the load-line (intersection of the plane through pin-hole centres or the hinge axes and plane of crack) and the tip of the precrack or crack on the edge of the specimen. The value of *h* is the thickness of a substrate arm. Obviously, for the TDCB specimen, the value of *h* is a function of the crack length *a*.]

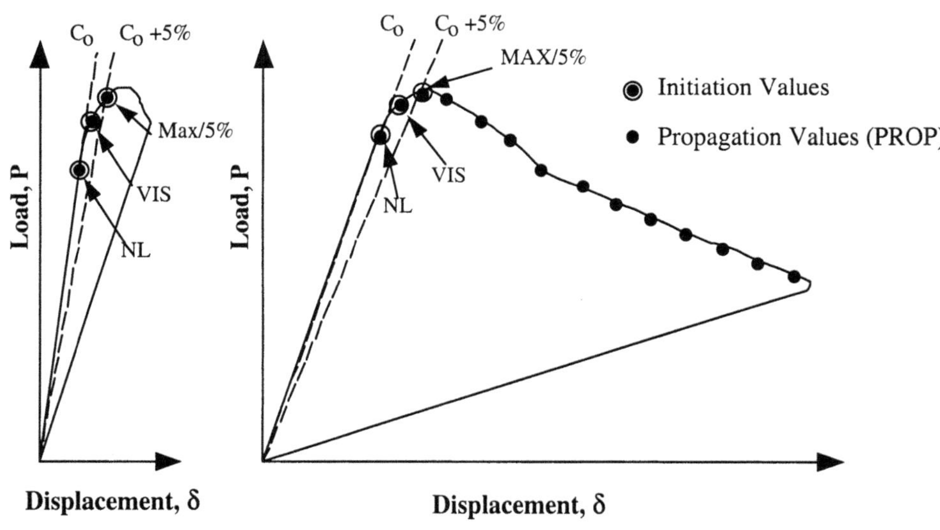

Figure 2: Schematic load-displacement curve for the DCB test.

(a) Testing from the insert with initiation points NL, VIS and MAX/5%.

(b) Testing from the Mode I precrack with initiation points NL, VIS, Max/5%, and propagation points (PROP).

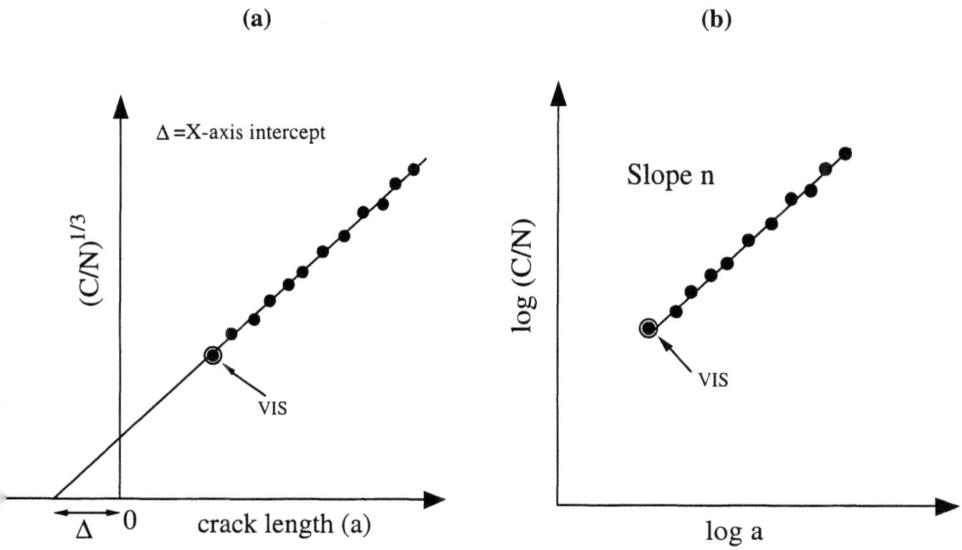

Figure 3: Linear fits used to determine (a) the correction for the Corrected Beam Theory (CBT) Method, and (b) for the slope *n* for the Experimental Compliance Method (ECM) Method. (For the DCB test specimen - and note that the visual point is excluded from the linear regression analysis.)

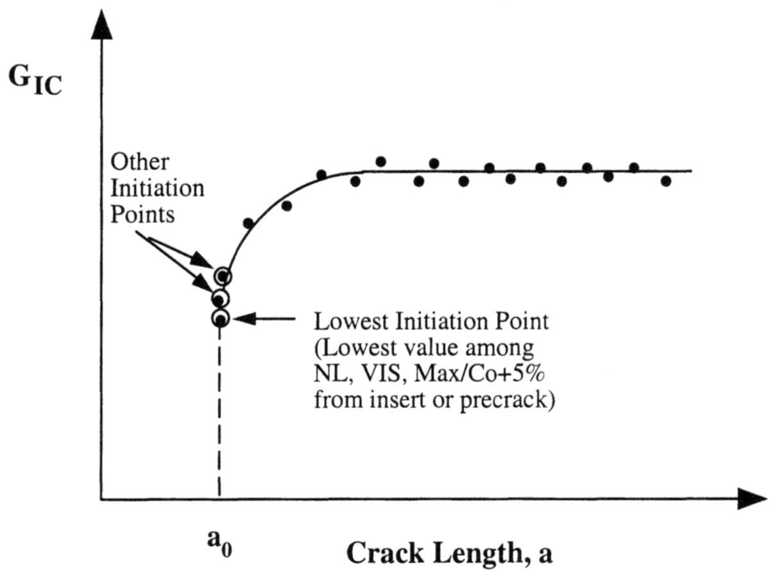

Figure 4. Schematic resistance-curve (R-curve) with G_{IC} value for initiation (i.e. the lowest value among NL, VIS, or MAX/5%) and for propagation (PROP) versus observed crack length a. (For either DCB or TDCB specimens, the specimen type should be stated.)

Figure 5(a) Recommended Test Report Sheet for DCB test

DCB TEST REPORT: PAGE 1 OF 3

Laboratory	
Personnel	
Test date	
Test number / code	

Specimen data

Adhesive	
Substrate	
Surface treatment details	
Specimen length, l (mm)	
Substrate thickness, h, (mm)	
Specimen width, B, (mm)	
Insert film material	
Insert film thickness (μm)	
Insert film total length, A, (mm)	
Insert length from load-line, a_o, (mm)	
Precrack length from load line, a_p, (mm)	
Flexural modulus of substrate, E_S, (GPa)	
Adhesive layer thickness, h_a (mm)	

Joint manufacture and test parameters

Adhesive cure temperature (°C)	
Adhesive cure duration (mins)	
Post cure drying cycle details(°C & hours)	
Fracture test temperature (°C)	
Fracture test relative humidity (%)	
Cross-head loading rate (mm/min)	
Cross-head unloading rate (mm/min)	
End-block dimension, l_1, (mm)	
End block dimension, l_2, (mm)	
Crack growth observations (e.g. stick-slip?)	
Locus of failure (visually assessed)	

Value from unloading line (substrate plasticity check)

δ_{offset} [mm]	δ_{max} [mm]	$\delta_{offset}/\delta_{max}$	Substrate bent?*

(*) After breaking open the joint after testing, any permanent deformation seen?)

Measurement of the system compliance (see Annex A.1)

C_{total} [mm/N]	C_{CS} [mm/N]	C_{sy} [mm/N] Eqn [A.1]

DCB TEST REPORT: PAGE 2 OF 3

Experimentally measured values				Calculated values				
Text	a [mm]	P [N]	δ [mm]	δ_{COR} [mm] Eqn [A.2]	G_{IC}[J/m^2] (*SBT*) Eqn. [4]	G_{IC}[J/m^2] (*CBT*) Eqn. [5]	G_{IC}[J/m^2] (*ECM*) Eqn. [9]	E_f [GPa] Eqn. [8]
NL$^{(insert)}$								
VIS$^{(insert)}$								
MAX/5%$^{(insert)}$								
NL$^{(precrack)}$								
VIS$^{(precrack)}$								
MAX/5%$^{(precrack)}$								
PROP								
PROP								
PROP								
PROP								
PROP								
PROP								
PROP								
PROP								
PROP								
PROP								
PROP								
PROP								
PROP								
PROP								
PROP (*)								
PROP								
PROP								
PROP								
PROP								
PROP								

(*) Minimum number of propagation points required

Mean and standard deviations of propagation values

	G_{IC} [J/m^2] (*SBT*)	G_{IC} [J/m^2] (*CBT*)	G_{IC}[J/m^2] (*ECM*)	E_f [GPa]
Mean value				
Standard deviation				
Coefficient of variation (%)				

DCB TEST REPORT: PAGE 3 OF 3

Linear regression analysis

$(C/N)^{1/3}$ vs a		Log (C/N) vs log (a)	
Δ (mm)	r^2	n	r^2

Intermediate calculated values

a [mm]	F [-] Eqn. [6]	N [-] Eqn. [7]	C [mm/N]	(C/N)^1/3 $[mm/N]^{1/3}$	Log (C/N) [mm/N]	Log (a) [mm]	m [1/mm] Eqn. [1]

Figure 5(b). Recommended Test Report Sheet for the TDCB test

TDCB TEST REPORT:PAGE 1 OF 3

Laboratory	
Personnel	
Test date	
Test number / code	

Specimen data

Adhesive	
Substrate	
Surface treatment details	
Specimen length, l, (mm)	
Specimen geometry factor, m, (mm^{-1})	
Specimen width, B, (mm)	
Insert film material	
Insert film thickness (μm)	
Insert film total length, A, (mm)	
Insert length from load-line, a_o, (mm)	
Precrack length from load line, a_p, (mm)	
Flexural modulus of substrate, E_S, (GPa)	
Adhesive layer thickness, h_a, (mm)	

Joint manufacture and test parameters

Adhesive cure temperature (°C)	
Adhesive cure duration (mins)	
Post cure drying cycle details (°C &hours)	
Fracture test temperature (°C)	
Fracture test relative humidity (%)	
Cross-head loading rate (mm/min)	
Cross-head unloading rate (mm/min)	
Is *m* within +/- 3%?	
Crack growth observations (e.g. stick-slip?)	
Locus of failure (visually assessed)	

Value from unloading line (substrate plasticity check)

δ_{offset} [mm]	δ_{max} [mm]	$\delta_{offset}/\delta_{max}$	Substrate bent?*

(*)After breaking open the joint after testing, any permanent deformation observed?

Measurement of the system compliance (see Annex A.1)

C_{total} [mm/N]	C_{CS} [mm/N]	C_{sy} [mm/N] Eqn [A.1]

TDCB TEST REPORT:PAGE 2 OF 3

Experimentally measured values

Text	a [mm]	P [N]	δ [mm]
NL$^{(insert)}$			
VIS$^{(insert)}$			
MAX/5%$^{(insert)}$			
NL$^{(precrack)}$			
VIS$^{(precrack)}$			
MAX/5%$^{(precrack)}$			
PROP			
PROP			
PROP			
PROP			
PROP			
PROP			
PROP			
PROP			
PROP			
PROP			
PROP			
PROP			
PROP			
PROP			
PROP (*)			
PROP			
PROP			
PROP			
PROP			
PROP			
PROP			
PROP			
PROP			
PROP			
PROP			
PROP			
PROP			
PROP			

Calculated values

δ_{COR} [mm] Eqn. [A.2]	C [mm/N]	G_{IC} [J/m^2] *(SBT)* Eqn. [4]	G_{IC} [J/m^2] *(CBT)* Eqn. [11]	G_{IC} [J/m^2] *(ECM)* Eqn. [2]

(*) Minium number of propagation points.

TDCB TEST REPORT:PAGE 3 OF 3

Linear regression analysis:(C versus a)

dC/da [1/N]	r^2 of regression

Mean and standard deviations of propagation values

	G_{IC} [J/m^2] (*SBT*)	G_{IC} [J/m^2] (*CBT*)	G_{IC} [J/m^2] (*ECM*)
Mean value			
Standard deviation			
Coefficient of variation (%)			

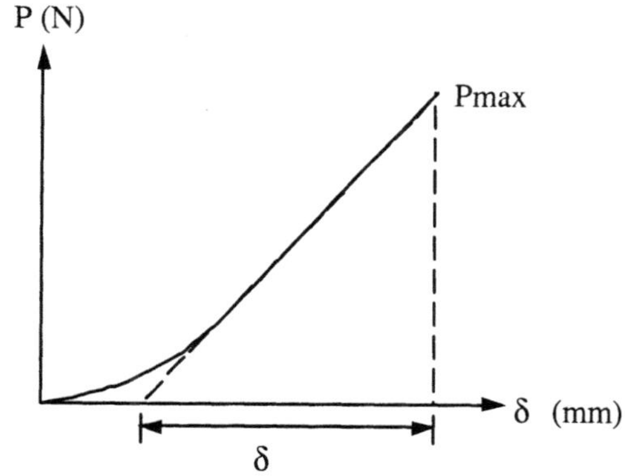

Figure A1. A schematic load-displacement trace obtained during the system compliance measurement. ($C_{total}=\delta/P_{max}$).

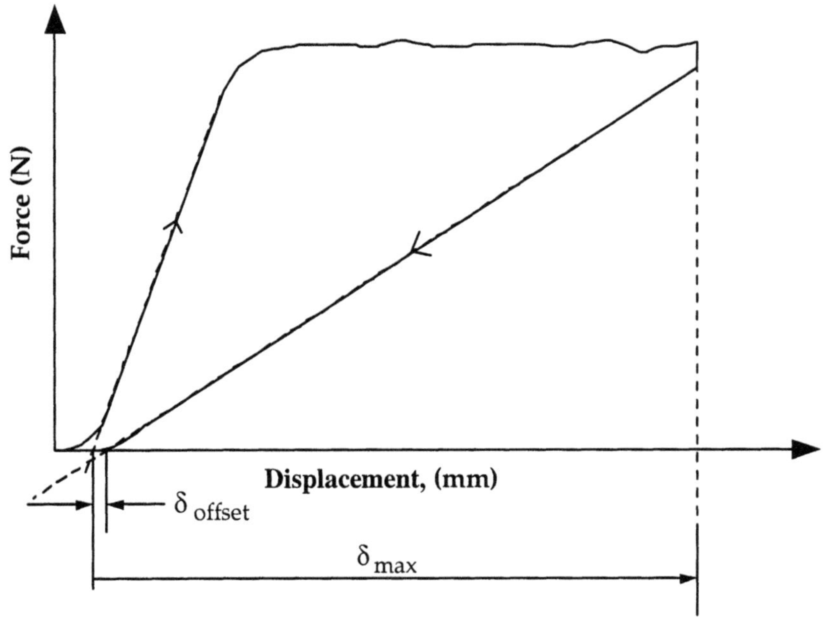

Figure A2. Typical force-displacement trace for a tapered double cantilever beam specimen, showing loading and unloading lines and the displacement offset.

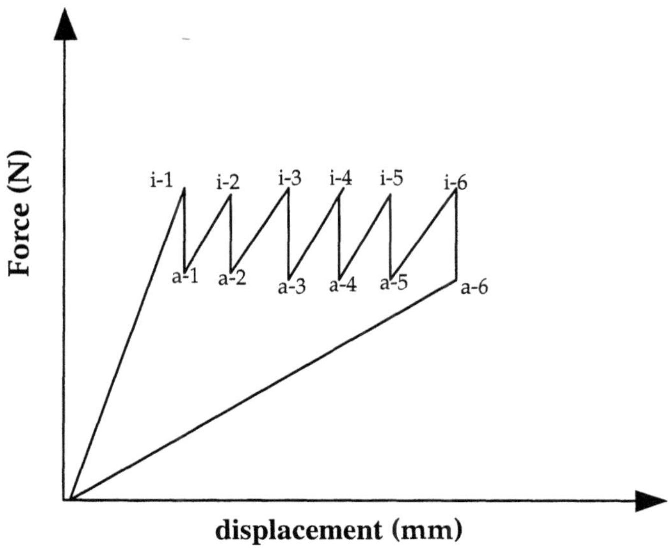

Figure B1. Schematic force-displacement trace for a tapered double cantilever beam specimen, exhibiting unstable 'stick-slip' crack growth behaviour. ('i' indicates initiation points, 'a' indicates arrest points.)

CHAPTER 4

Delamination Fracture Mechanics

INTRODUCTION TO DELAMINATION FRACTURE OF CONTINUOUS FIBRE COMPOSITES

P. DAVIES

1. INTRODUCTION

Continuous fibre reinforced composites are used extensively in applications where low weight and good durability are at a premium. The laminated nature of many of these materials, produced by stacking layers of reinforcement (either pre-impregnated with resin or wetted out by liquid resin during moulding), can result in a tendency for delamination during service. This may be caused by direct out-of-plane loading or at features such as cut-outs or ply-drops which induce high through-thickness stresses. Extensive material developments over the last twenty years have been seen to improve this situation but reliable tests are needed to quantify these improvements. This has led to a major international effort to develop tests to measure the resistance of composites to interlaminar crack initiation and propagation. The work performed within ESIS TC4 in this area over nearly 15 years has been a major part of this effort. An iterative process has been used, involving the drafting of test protocols, running round robin comparisons of results using these protocols, improving them according to the experience of those running the tests, and then validating them by further tests on a range of materials. Since the mid 1980's this has enabled test procedures for mode I (tension), mode II (in-plane shear) and mixed mode (I/II) loading to be evaluated and refined. The subsequent papers will present these efforts in more detail but it is necessary to place this work in context, if the orientations are to be understood.

2. HISTORICAL BACKGROUND

Fracture testing of anisotropic materials has a long history. The composites industry has borrowed methods which were used to study cleavage of other materials, for example mica in the 1930's [1], and metal crystals in the 1950's [2]. The origins of the most popular composite shear delamination tests may be found in studies of timber performed in the 1970's [3]. There was also considerable experience available from testing of adhesively bonded joints [4,5]. In the 60's and 70's several researchers were already publishing results from mode I delamination tests on composites (e.g. [6-9]), and the need to develop standard tests specifically for fibre reinforced composites was recognised by the aerospace industry from the early 1980's. A NASA document issued in 1982 was the basis for much of the early discussion [10]. This document included a mode I delamination test using the DCB (Double Cantilever Beam) specimen, in which measurements of G_{Ic} were made during the propagation of a crack. Resin and composite suppliers started to use this method to characterise new resin systems.

3. MODE I

When the ESIS group started to examine mode I tests in the mid 1980's there did not appear to be any particular problems in taking this method forward to standardisation. However, by the end of the 80's, following several round robin exercises on tough thermoplastic matrix composites, it had become apparent that the values measured during propagation were not always material properties. Significant increases in fracture toughness could be obtained by increasing specimen thickness which favoured the appearance of bridging and multiple cracking [11,12], so a new approach was required. ESIS had also been running tests to measure G_{Ic} at initiation. This approach had been influenced by the results of Benzeggagh and de Charentenay, who were using measurements during the first few millimetres of crack advance to study the mechanisms governing fracture [13]. ESIS proposed measuring the complete fracture resistance curve, from initiation at the defect through propagation. Details of the approach are given in 'Mode I Delamination', but this approach re-oriented research towards the choice of the type of defect required to initiate a delamination, and the definition of when initiation occurs. Subsequent activities, and in particular joint round robins from 1991 onwards with ASTM (American Society for Testing and Materials) and JIS (Japanese Industrial Standards) groups focused on these two areas.

In parallel with the development of the practical aspects of the mode I test the data analysis was also investigated in detail [14]. This resulted in a new data reduction scheme which was subsequently included in the ESIS protocols and later adopted in the draft ISO document.

As a result of this work, and thanks also to a close collaboration between the D30 group of ASTM (American Society for Standards and Materials), JIS (Japanese Industrial Standards group) and ESIS, the mode I test protocol is now close to adoption as an ISO test method, (ISO/DIS 15024) with voting taking place at the time of writing.

4. MODE II

The measurement of mode II delamination resistance has proved much more controversial than mode I, with several different test geometries being proposed. These are described in 'Mode II Delamination', but ESIS has concentrated its efforts on two geometries. The first round robins were performed using an ENF (end notched flexure) geometry, as this was the most widely used method in the mid 1980's. The limitations of this method were quickly apparent, notably the unstable crack advance, and ESIS subsequently adopted the ELS (End Loaded Split) geometry. This is a stable configuration, originally proposed by Vanderkley and Bradley [15]. Several series of tests were run, both to compare results from this specimen with those from the ENF geometry and to investigate other parameters such as the influence of starter defect type, specimen thickness, friction between the sliding faces and loading fixture. Over a period of ten years the test protocol was refined and the final version is given in 'Mode II Delamination'. When it was decided, in 1997, to consider the proposal of a mode II test as a New Work Item for ISO Committee TC61 SC13 this was therefore the natural ESIS suggestion. The ASTM group proposed the ENF specimen, while the JIS group preferred a stabilised version of the ENF (SENF), [16].

International collaboration to examine the different specimen configurations was organised by VAMAS (Versailles Agreement for Materials & Standards) and co-ordinated by the author. Results may be found elsewhere [17], but during this exercise a new test configuration was proposed, the four point loaded ENF (4ENF) [18]. This offered the simplicity of the ENF fixture with the stability of the SENF and ELS, and was therefore included in a second round robin on glass and carbon reinforced materials. The laboratories running these tests were generally very positive and at the ISO meeting in September 1999 the 4ENF method was recommended to be put forward as a new work item. Subsequent administrative difficulties within the ISO Sub-committee 13 have delayed further work on mode II tests, but the ASTM D30.06 proposed a 4ENF test method for sub-committee ballot in February 2000. It should be emphasised, however, that the lessons learnt in the ENF and ELS test developments, particularly with regards to starter defects and friction are still very relevant to this new geometry and have proved invaluable in the drafting of the new 4ENF test protocol.

5. MIXED MODE I/II

In order to determine the full fracture envelope for a material it is necessary to develop tests which allow a combination of mode I and mode II loading to be applied. When the ESIS studies started the CLS (cracked lap shear) specimen, popular for adhesives, was being proposed. An ASTM round robin was run to examine the analysis of mode separation in this specimen [19]. This revealed a number of difficulties and the specimen was not studied further. Another option at the time was the edge delamination tension test, involving the use of special stacking sequences designed to delaminate [20]. These were not always easy to analyse either, as different failure modes could interact before failure, and some of the tougher materials tested did not delaminate. As a result alternative configurations were sought, and one which appeared attractive was the asymmetric double cantilever beam (ADCB) a fixed ratio mixed mode specimen. This can be run on a DCB specimen simply by loading one arm instead of both, using the same fixture as the mode II ELS specimen. A protocol was drafted and several round robins were run with this specimen. This is described in more detail in 'Delamination Fracture of Continuous Fibre composites: Mixed-Mode Fracture'. This method remains the simplest way to obtain one point on the fracture envelope involving significant proportions of mode I and mode II. However, at the same time at NASA a new test fixture was being developed which enabled different mixed mode ratios to be obtained easily in a single fixture [21]. The fixture required for this MMB (mixed mode bending) test is more complicated than that required for the ADCB specimen, but the range of data which can be generated has lead to its more widespread acceptance. Some members of the ESIS group have been involved in ASTM round robins to develop the MMB test procedure.

6. MODE III

This third type of loading, out-of-plane shear, which may be of interest in some particular applications such as helicopter rotors, has not been studied by ESIS, although again there have been some members of the group involved in round robin tests on an edge crack torsion (ECT) test [22] currently being evaluated by ASTM.

7. CURRENT STATUS

The current position regarding standards is summarised in Table 1 below. Details of the technical considerations for each ESIS test protocol will be described in the following papers.

Mode	ISO	National	Industry	ESIS Protocols
I	ISO/DIS 15024 Vote 1999	ASTM D5528 JIS K7086	Various,e.g. prEN, AECMA, Airbus, CRAG	DCB, Basis for ISO ***Mode I Delamination***
II	New work item recommended 4ENF, 1999	JIS K7086 ENF/SENF (ASTM 4ENF)	ENF , e.g. pr EN AECMA	ENF*, ELS ***Mode II Delamination***
I/II	None	ASTM voting (MMB)	None	ADCB ***Delamination Fracture...***

* ENF not further developed beyond 1993 version

Table 1. Standard delamination resistance test methods

8. CONCLUDING REMARKS

The development of tests to measure delamination resistance of fibre reinforced composites has mobilised a very large number of researchers over the last two decades. Much of the work, particularly drafting, checking and correcting procedural documents, has been performed on a voluntary basis, both by members of ESIS and those involved in standards organisations. This is a long and often thankless task, but their efforts are now being rewarded by the appearance of documents with international consensus. This effort is absolutely essential to the safe use of composite materials in critical applications and as more reliable G_c values become available there is an increasing willingness to use fracture mechanics data in design.

There is still much to do. Most of the development work to date has been limited to flat, unidirectional specimens, with a strong emphasis on carbon fibres. The influence of curvature, cracks at interfaces between different fibre orientations, other forms and types of reinforcement, and new material concepts (interlayers, stitching, very ductile resin systems etc.) must be accommodated. This initially requires basic fracture research, but the ability to translate research results into a form which can be integrated into validated test procedures is also vital. Groups such as the ESIS Technical Committees will continue to play an essential role in this latter activity.

The authors of the three papers which follow would like to thank all the members of the Technical Committee 4 (TC4) "Polymers & Composites" of the European Structural Integrity Society (ESIS), formerly known as "European Group on Fracture" (EGF), who have contributed to the development of the ESIS test protocol since the first draft in 1988 (names listed below). The leadership of past and present chairmen of TC4, Prof. J.G. Williams, Prof. A. Pavan, and Prof. H.H. Kausch is gratefully acknowledged. Together with the author, Dr Blackman and Dr Brunner, Dr A.J. Cervenka served several years as session leader for the laminates test protocol development.

M. v. Alberti, S. Andersen, F.J. Belzunce, L. Bertini, B.R.K. Blackman, A.J. Brunner, N. Burgoyne, W.R. Broughton, W.J. Cantwell, D.D.R. Cartié, M.N. Charalambides, A.J. Cervenka, P. Czarnocki, P. Davies, F. Ducept, M. Fischer, P. Flüeler, K. Friedrich, Y. Giraud, B. Goffaux, S. Hashemi, G.E. Hale, M. Hiley, J. Jaussaud, O. Jorgensen, J. Karger-Kocsis, H.H. Kausch, A.J. Kinloch, B. Lauke, R. Lee, D. Martin, G. McGrath, B. Melve, D.R. Moore, D. Neville, I.K. Partridge, A. Pavan, R. Prediger, F. Ramsteiner, C.A. Rebelo, P.E. Reed, I. Robinson, A.C. Roulin-Moloney, N. Roux, T. Schjelderup, S. Seidler, G.D. Sims, G. Steinmetz, A. Torres Marques, N. Trigwell, D. Turmel, I. Verpoest, K. Walls, L. Warnet, J.G. Williams, H. Wittich.

9. REFERENCES

1. Obreimoff JW, Proc. Royal Soc. A127, 1930 p290.
2. Gilman JJ, J. Applied Physics, 31, 1960, p2208.
3. Barrett JD, Foschi RO, Eng. Fract. Mech., 9, 1977, p371.
4. Ripling EJ, Mostovoy S, PatrickRL, Materials Research & Standards, March 1964, p129.
5. Bascom WD, Cottington RL, Timmons CO, J. Appl. Polymer Sci., 32, 1977, p165.
6. Sidey GR, Bradshaw FJ, Proc. 1st Int. Conf. on Carbon Fibres, Plastics & Rubber Inst., 1971 paper 25.
7. McGarry FJ, Mandell JF, SPI 27th Annu. Tech. conf., 1972, section 9A.
8. Phillips DC, Tetelman AS, Composites 3, 1972, p216
9. McKenna GB, Polymer Plast. Technol. Eng., 5 (1), 1975, p23.

10. NASA Reference Publication 1092, Standard tests for toughened resin composites, May 1982, ST5.
11. Davies P, Benzeggagh ML, Chapter 3 in 'Application of Fracture Mechanics to Composite Materials', ed. Friedrich K, Elsevier 1989.
12. Davies P, Kausch HH et al., Comp. Sci & Tech., 43, 1992, p129.
13. de Charentenay FX, Benzeggagh ML, Proc. ICCM3, Vol. 1, 1980, p186.
14. Hashemi S, Kinloch AJ, Williams JG, Proc. Roy. Soc., A427, 1990 p173.
15. Vanderkley PS, MSc thesis Texas A&M University, 1981.
16. Kageyama K, Kikuchi M, Yanagisawa N, ASTM STP 1110, 1991, p210.
17. Davies P, Sims GD et al., Proc 4th Comp. Testing & Standardization conf., Lisbon 1998, p180 and in Plastics, Rubber & Composites 28, 9, 1999 p432.
18. Martin R, Davidson BS, Proc. 4th Int. conf. on Deformation & Fracture of Composites, 1997, p243 and in Plastics, Rubber & Composites 28, 8, p401.
19. Johnson WS, NASA Tech. Memo. 89006, 1986
20. O'Brien TK, ASTM STP 836, 1984, p125.
21. Reeder JR, Crews JH, J. Comp. Tech. & Res. 14, 1992 p12
22. Lee SM, J. Comp. Tech. & Research, 15, 3, 1993, p193.

MODE I DELAMINATION

A.J. BRUNNER, B.R.K. BLACKMAN and P. DAVIES

1. INTRODUCTION

The development of test methods for Mode I delamination testing of continuous fibre-reinforced polymer-matrix composites and test results have been reviewed in several papers that provide detailed references [1-5]. Recent test results for Mode I delamination resistance or fracture toughness mostly from round robin tests are summarised in, e.g., [6-8]. Experimental aspects relating to fracture toughness testing in Mode I and Mode II are discussed, e.g., in [9].

2. HISTORY AND BACKGROUND

The developments leading to the early Mode I draft procedures for fibre-reinforced laminates using the DCB-specimen have been presented in the introductory section. Here, the motivation for the test development and its course up to the present will be summarised.

The main reason for developing test methods for Mode I interlaminar delamination resistance is that delamination is deemed an important failure mode. Besides the tensile or "opening" Mode I loading, other load cases (Mode II shearing [10], Mode III torsional [1], and Mixed Mode I/II [11] loading) have to be considered as well. Mode I loading is certainly of interest if failure envelopes incorporating different modes and mixed mode combinations, respectively, are to be determined [e.g., 12]. These envelopes can be used in the design and dimensioning of parts or structures but the designs often include somewhat arbitrary "knock-down" or "safety factors".

Whether Mode I loading is also of practical interest, is debatable, since most load cases in structural applications of polymer-matrix composites effectively involve mixed mode conditions. It may even be impossible to realise pure Mode I loading in these materials. This becomes obvious if microscopic stress concentrations (e.g., around individual fibres) are considered [1]. However, in the case of Mode I loading, there is evidence that consistent and valid data can be obtained on a macroscopic scale in spite of the complex microscopic stress concentration patterns [1]. From an experimental point of view it can be noted that Mode I loading generally yields the lowest fracture toughness among the different pure or mixed modes [13, 14]. Hence, Mode I data are often regarded as lower (safe) limits for the design and dimensioning of composite structures and parts.

Several groups have contributed to the development of standardised test methods for the fracture toughness measurement of unidirectional fibre-reinforced polymer-matrix composites. The Japanese Standards Association (JSA) was the first to publish a national standard for Mode I and Mode II fracture toughness in 1993 [15], followed about one year later by Subcommittee D30.06 of the American Society for Testing and Materials (ASTM) [16]. In 1995 the Technical Committee 4 "Polymers and Composites" of the European Structural Integrity Society (ESIS) completed their development [17] that had started with a first draft in 1988. Based on an initiative by ASTM, the development since 1995 has become an international effort by these three groups within the International Organisation for Standardisation (ISO), resulting in an ISO Draft International Standard (DIS) [18].

The round robin tests organised within ESIS TC4 are summarised in Table 1. The round robin tests involved between 5 and 20 laboratories, testing typically 5 specimens each. Some of the round robins addressed specific questions, e.g., variation of the insert material type and its thickness [7] or the applicability of the test procedure to glass-fibre-reinforced materials or non-unidirectional lay-ups [e.g., 19-21]. The different versions of the ESIS TC4 Mode I test procedure for DCB-specimens are summarised in Table 2. One of the main contributions of the ESIS TC4 committee was to direct attention to the difference between initiation and propagation values and to emphasise that the whole R-curve is to be determined for an assessment of the behaviour of the laminate.

The limitation of the currently available test protocols to unidirectionally fibre-reinforced laminates can be attributed to several reasons. The first is that early tests on multidirectional laminates yielded multiple cracking and/or crack branching [20], while laminates with woven reinforcements are expected to yield less branching (as shown in a cross-ply round robin [21]). Multiple cracking or branching was not considered suitable for determining a material property. The second is that unidirectional laminates seemed to yield lower, i.e., "conservative" values of the critical energy release rate compared to non-unidirectional laminates [19, 20]. The third is that one of the main applications of the test protocols is quality control at manufacturing plants and relative comparison of different matrix materials for which unidirectional test plates are more easily prepared. However, the second aspect may deserve renewed attention since preliminary results from two types of cross-ply material [21, 29] indicate that unidirectionally reinforced materials may not yield conservative values in all cases.

Table 1: Round robin tests on Mode I DCB organised by ESIS TC4

Date	Material	References	Remarks
1986	CF-Epoxy	-	Based on suggestion from de Charentenay
	CF-PES	22	also Mode II tests
1987	CF-Epoxy	23	Also Mode II tests, exploratory tests
	GF-PA	24	
	GF-PU	-	
1988	CF-Epoxy	-	Also Mode II tests, exploratory tests
	CF-PEEK	-	
1990	CF-Epoxy (IM6/Epoxy)	-	Also Mode II tests, Joint ASTM/EGF/JIS
	CF-PEEK (AS4/APC-2)	7	round robin, NDT-test methods for
	CF-PEEK (IM6/PEEK)		initiation detection [9, 25, 26]
1991	GF-PMMA (GF/Modar)	-	Also Mode II and Mixed Mode I/II tests
1992	GF-Epoxy (hand lay-up)	27	Also Mode II and Mixed Mode I/II tests, including Mode II ENF
1994	CF-Epoxy (IM7/977-1)	28	Insert and precracking Mode I, including
	GF-Epoxy (hand lay-up)		wedge precracking from insert (CF-Epoxy
1997	CF-Epoxy (T300/970)	21	Cross-ply and woven fabric only
1998	CF-Epoxy (T300/970)	21	Cross-ply and unidirectional
1999	CF-Epoxy (T300/977-2)	-	Cross-ply and unidirectional

Beside the efforts described above, research has been aimed at investigating different specimen types, e.g., width tapered [e.g., 30], thickness tapered [e.g., 31, 32], edge modified [e.g., 33], or different loading mechanisms [e.g., 34], or specific effects on the test results, e.g., material modification [e.g., 35-37], multidirectional lay-up [e.g., 38], test parameter variation such as rate-dependence [e.g., 39], and environmental effects such as temperature [e.g., 40]. This list is by no means complete and recent references indicate that fracture toughness testing still offers a wide range of opportunities for research.

3. REVIEW OF TEST METHOD AND RESULTS

The present paper concentrates on the ESIS TC4 draft [17] from 1995 (Version 1995-12-12) with editorial revisions implemented in 1999 (Version 1999-06-03). The document is based on the concepts of Linear Elastic Fracture Mechanics (LEFM). The test uses the so-called Double Cantilever Beam (DCB) specimen that is the most widely used Mode I specimen type. The specimen and the test principle are shown in Figure 1 of the appendix. An opening load produced by a cross-head displacement at constant speed is applied at that end of a rectangular beam that contains a starter crack produced by an insert foil laminated at mid-thickness. For a detailed description the reader is referred to the test protocol in the appendix. Other types of specimens [30-33] never gained the same widespread acceptance as the DCB. The different schemes for data analysis can roughly be classified as either (a) empirical, (b) beam theory, or (c) experimental compliance methods.

Table 2: Synopsis of ESIS TC4 Mode I draft test procedures for unidirectionally fibre-reinforced DCB-specimens

Date	Major Modification	Remarks
June 1987	Guidelines for first RR	One page and list of references
November 1988	First ESIS TC4 draft test procedure	Initiation from thin insert film (NL, 5%, PROP) analysis: CBT, Berry, area method
November 1989	Starter film criterion < 25 μm	Initiation only, analysis: CBT
January 1990	First draft following protocol	Not used in RR
March 1990	Test protocol for ESIS RR and (later) for joint ASTM/EGF/JIS RR	Together with Mode II ENF, starter film criterion < 15 μm, initiation (NL, VIS, MAX/5%), analysis: CBT, Berry, JIS, large displacement and end block corrections
March 1992	Issued as formal report (IFREMER)	Together with Mode II ENF, Mode II ELS and Mixed Mode I/II ADCB
May 1992	Criterion for non-UD materials	Minor corrections added
September 1993	Formal editing (header, date, history of revisions)	Together with Mode II ENF, Mode II ELS and Mixed Mode I/II ADCB
May 1994	First inclusion of spreadsheet SRF and bibliography, adaptation to ISO-style format	
95-12-12	Formal revision, editorial	ESIS draft for ISO CD 15024 *
99-06-03	Editorial revision	

* There were several editorial revisions of ISO CD 15024; the current version is ISO/DIS 15024 dated 99-04-09.

The basis of all methods of data analysis is equation (1) that relates the energy release rate G_{IC} with the change in compliance due to a change in delamination length.

$$G_{IC} = \frac{P^2}{2B}\left(\frac{dC}{da}\right) \tag{1}$$

with critical force P, specimen width B, compliance C (displacement divided by force), and delamination length a. The data analysis methods all use different approaches to evaluate dC/da. An early approach was an empirical line-fitting procedure proposed by Berry [41], plotting log C (compliance) versus log a (delamination length) and determining the slope "n" of a linear fit through the data points which is then used as a factor in an equation for G_{IC} (similar to that for the experimental compliance calibration). The beam theory methods require the measurement of corresponding values of force, displacement and delamination length for evaluating dC/da calculated from simple beam theory (for a perfectly built-in beam). The simple beam theory approach can be improved by adding corrections and the current test procedure uses corrected beam theory (details are given in the appendix). Area methods using the force-displacement curves were initially considered for the data analysis but were dropped because they are only useful for determining propagation values.

The experimental compliance methods are based on the assumption of a certain type of functional dependence of the compliance on the delamination length. By plotting the measured specimen compliance, or a quantity derived from the compliance, sometimes including suitable corrections, versus the delamination length (or, again, a quantity derived from the delamination length) the assumed functional dependence is represented. If plotted in a suitable way, a linear dependence between the respective quantities representing compliance and delamination length is obtained and the slope of a linear fit through the data points yields a proportionality constant. It is usually labelled "m" and an appropriate equation is then used to calculate G_{IC}. In all methods, corrections for (a) large displacements and (b) load-block effects can be applied (details see appendix). In general, the results from the different methods of analysis agree quite well, i.e., within a few percent (e.g., as shown, in Figure 1).

The data analysis yields initiation and propagation values of G_{IC}, the former being defined as either NL-, VIS- or MAX/5%-point, the latter as the plateau values in the R-curve plot (R for resistance, a plot of G_{IC} versus delamination length). Figure 1 shows such an R-curve for one type of CF-epoxy and GF-epoxy. Initiation and propagation values from ESIS round robin tests for this CF-epoxy are listed in Table 3. An experimental fact that has not been fully explained yet is that initiation values of G_{IC} were consistently lower if measured from a precrack than from the insert starter crack. ASTM also noted this in one of their round robins [18]. To date it is not clear whether this is an exception, and the cause or causes responsible for lowering the precrack values are not clear either. A possible explanation could be that delamination propagation from the insert causes microscopic damage ahead of the (macroscopic) delamination tip that leads to an increase in the apparent delamination resistance. The blunting effect of a resin-rich region beyond the insert tip would be similar but can be excluded since a sufficiently thin insert had been used and the insert tip regions were examined in the microscope.

For the ESIS round robin results, it should be noted that delamination growth from the insert film was mostly unstable (in spite of the valid starter film thickness), so the test results would not be valid. Initiation was stable from the Mode I precrack. There is a clear trend that initiation values from the precrack are lower than from the insert film for this material while propagation values are comparable. Data from the joint ASTM/ESIS/JIS round robin for CF-PEEK are listed in the ISO/DIS 15024 [18].

4. PERSPECTIVES AND OPEN PROBLEMS

After more than 15 years of development and discussion, international standardisation of a test method for Mode I interlaminar fracture testing of continuous fibre-reinforced polymer-matrix composites is now within reach. The ISO DIS 15024 "Fibre-reinforced plastic composites – Determination of Mode I interlaminar fracture toughness, G_{Ic}, for unidirectionally reinforced materials (Version 1999-04-09)" is currently being circulated for balloting and has been reviewed at the ISO meeting at Williamsburg (USA) in September 1999. It is now expected that the draft standard will be accepted as an international standard in 2000 after minor editorial changes.

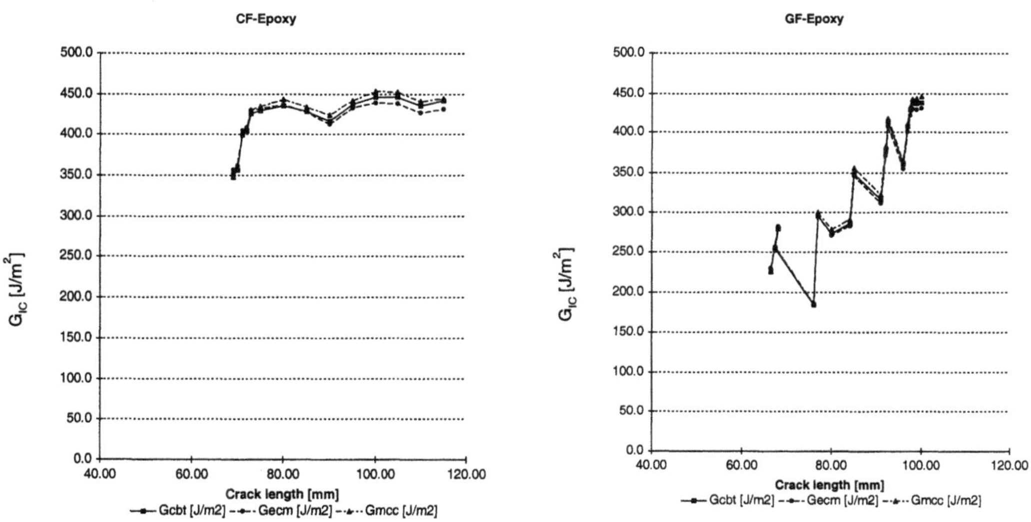

Figure 1: R-curve (G_{IC} versus delamination length) for one type of CF-epoxy (IM7/977-1, from a Mode I precrack, left) and GF-epoxy (hand lay-up, from a 7 μm thick insert, right), respectively. Note the relatively "flat" R-curve for CF-epoxy in comparison with the "steep" R-curve for GF-epoxy. The R-curve on the right shows partly unstable behaviour (values from stable propagation alternating with lower arrest values). The curves have been generated by the spreadsheet described in [26], cbt = corrected beam theory, ecm = experimental compliance calibration, mcc = modified compliance calibration.

Even though the test method developed by the ESIS TC4 served as a basis for the ISO DIS 15024 the ISO document incorporates elements from the previously published JSA (JIS K 7086) and ASTM standards (D 5528). It thus combines the know-how and experience of all three groups involved in the test method development. However, such international standardisation has resulted in several compromises. The most notable differences between the ISO/DIS 15024 [18] and the latest ESIS test protocol [17] are: The ISO/DIS (1) uses the terms critical energy release rate and fracture toughness synonymously, (2) includes the so-called modified compliance calibration method for data analysis (the ESIS document includes an empirical compliance calibration in the form of Berry's method) as well as the corrected beam theory, (3) allows a larger tolerance for thickness variation (± 0.1 mm instead of a maximum thickness variation of 0.1 mm as in the ESIS document), (4) includes prescriptions for test plate fabrication (referring to ISO/DIS 1268-4), (5) includes precision data (from the ASTM round robin [7]), (6) uses a different co-ordinate system for delamination length measurements if a travelling microscope is used, and (7) does not recommend the use of wedge precracking for starting the delamination.

Regardless of which test protocol is used, there remain unsolved problems and open questions. The issue that was probably discussed most during the development is the definition of the "initiation" of delamination propagation. The test procedures finally adopted three different points, the non-linearity in the force-displacement plot (NL), the visual detection of delamination initiation along the edge of the specimen by the operator (or by optical methods; VIS) and the maximum force point or the point on the force-displacement plot that intersected with a straight line through the origin with a slope equal to a 5%-increase in compliance, whichever occurs first (MAX/5%). A large amount of data from round robin tests shows that the NL-point frequently yields the lowest G_{IC}-values (therefore called "conservative") for a laminate. However, there is evidence from non-destructive test methods that delamination propagation may be detected even before the NL-point [9, 25]. Results from another round robin [42] suggest that the determination of the NL-point may be more operator-dependent (typically with a 10% variation) than the MAX/5%-point. The cause for the non-linearity in the force-displacement plot is not clear, and sometimes the NL-point has been observed to coincide with the maximum force.

Table 3: Initiation and propagation values of G_{IC} (Corrected Beam Theory, CBT) for CF-epoxy (IM7/977) from an insert film and from a Mode I precrack from the 1994 ESIS TC4 round robin

Laboratory	G_{IC} (NL) ± Standard deviation [J/m^2]	G_{IC} (VIS) ± Standard deviation [J/m^2]	G_{IC} (MAX*) ± Standard deviation [J/m^2]	G_{IC} (PROP) ± Standard deviation [J/m^2]	Remarks
1	-	(=MAX) **	429 ± 53	397 ± 38	Insert
	356 ± 13	332 ± 57	382 ± 21	417 ± 13	Mode I Precrack
2	397 ± 25	(=MAX) **	435 ± 21	442 ± 36	Insert
	349 ± 21	382 ± 20	389 ± 16	413 ± 22	Mode I Precrack
3	271 ± 108	-	416 ± 83	-	Insert
	314 ± 69	278 ± 78	399 ±22	439 ±33	Mode I Precrack
4	355 ± 39	-	436 ± 32	420 ± 60	Insert
	-	-	-	-	Mode I Precrack
5	-	-	345 ± 34	358 ± 23	Insert
	-	-	-	-	Mode I Precrack
6	503 ± 28 #	(=MAX) **	534 ± 10	452 ± 38	Insert
	-	-	-	-	Mode I Precrack
Average (1-6)	**382**	**466**	**432**	**414**	**Insert**
S.D. ***	96	59	61	38	
COV***	25 %	13 %	14 %	9 %	
Average (1-6)	**340**	**331**	**390**	**423**	**Mode I Precrack**
S.D. ***	22	52	9	14	
C.O.V***	7 %	16 %	2 %	3 %	

* MAX = maximum force or 5% increase in compliance, whichever occurs first
** Included in average with value for MAX
*** S.D. = Standard deviation, C.O.V = Coefficient of variation (Standard deviation divided by average x 100%)
# Only 2 of 5 values have been determined

Another issue that has not been resolved is the occurrence of "unstable" (fast) delamination growth. Data points from unstable growth are currently excluded from the analysis. Unstable growth usually ends in a so-called "arrest" point. Arrest points have been shown to sometimes yield lower G_{IC}-values than delamination initiation [9]. Some laminates tend to show considerable amounts of unstable delamination growth resulting in too few data points for a valid analysis. Even though it could be argued that unstable delamination growth reflected a material property no ways for quantitatively analysing the data have been implemented in the test procedures as yet. In principle, an "arrest" R-curve could be determined analogous to the "real" R-curve (see Figure 1) but its interpretation and use in design and dimensioning is not clear.

Load-rate effects have been investigated in the early round robins but recently received renewed attention [39]. The test procedures now all recommend constant crosshead speeds on the order of a few mm/min. Theoretically, a constant crack tip strain rate would be preferable over a constant load-rate but is difficult to implement experimentally. From a practical point of view, it would be desirable to reduce the test time as much as possible in order to reduce test cost. Besides using higher crosshead speeds a range of crosshead speeds could be used ("faster" during start-up and "slower" during delamination growth). Since rate-effects are material-dependent (glass-fibre reinforced unsaturated polyester is an example of a strongly rate-sensitive material), it would be difficult to investigate all possible material and rate combinations. An empirical approach would be to test at least one specimen at slow speed, if high speeds are used. Fatigue testing rather than quasi-static testing has been investigated by ASTM but not by the ESIS committee and for Mode I fatigue loading there is an ASTM standard [43]. For practical applications, fatigue loading could ultimately be more important than static loading. The determination of G_{IC} under "high" load-rates (typically on the order of 1 m/s or more) is outside the scope of the static test procedures since additional dynamic effects have to be taken into account, but research by members of the ESIS TC4 has detailed the difficulties [44].

Fibre-bridging, i.e., a connection between the two arms of the double beam specimen by fibres or fibre bundles after the delamination tip has passed by, is commonly observed in Mode I DBC tests. This behaviour is considered an "artefact" of the DCB-tests due to the use of unidirectional material. Fibre bridging is considered to be responsible for the increase of the curve of G_{IC} versus delamination length. This curve is usually called the R-curve, R indicating "resistance", from which initiation and propagation values (equal to the "plateau"-region, if observed) can be derived. Mode I tests on cross-ply material [0°/90°] have shown fibre bridging as well, but perpendicular to the direction of delamination propagation instead of parallel to it [45]. This indicates that fibre-bridging may occur even in "engineering" laminates with non-unidirectional lay-up. It has been argued that fibre bridging (and multiple cracking from crack branching) could invalidate the experimental compliance analysis [46]. This was investigated within ESIS TC4, by comparing multi-specimen compliance calibrations (performed on a series of specimens with different insert film length) with single specimen compliance calibrations using data during delamination propagation. For the materials tested the differences were quite small, around 6% in the worst case.

Results based on beam theory and experimental compliance calibration usually agree within a few percent (contrary to some round robin results for Mode II [10]). Any discrepancy between the two types of analysis might indicate the action of additional effects. The validity of such data should then be questioned. However, to date, no general validity criteria have been produced. The question of whether the DCB-test really determines a material property, as tacitly assumed by the test protocols, rather than a system property depending on the specimen geometry and test set-up has been checked experimentally on DCB specimens with different thicknesses (see, e.g., Figure 5 in [6]). There was no thickness effect on initiation values, but a strong effect on propagation values suggesting that the former is a material property whilst the latter is not.

As mentioned above, "engineering" composites do not often use unidirectional lay-ups. The question of the applicability of the test procedures, including data analysis methods, developed for unidirectional materials is currently being explored in ESIS round robins. A further problem is posed by specimens cut from structures or structural parts that do not contain laminated insert starter cracks. The only approach available in these cases is wedge precracking. Preliminary results for one type of glass-fibre reinforced epoxy indicate that wedge precracking can yield G_{IC}-values comparable to those from an insert but with a larger coefficient of variation [26].

Emerging trends in fibre-reinforced composites currently seem to be through-thickness reinforcements, basically providing additional strength in the third dimension. Some specific types of 3D-reinforced DCB-specimens have been tested using the existing procedures but it is too early to decide whether the data analysis methods have to be revised or not [47, 48].

5. SUMMARY

The current status can be summarised as follows:

(1) The ISO/DIS 15024 based on the ESIS test protocol provides a procedure for unidirectional fibre-reinforced laminates that is widely accepted and backed by a large number of round robin results both on CFRP and GFRP materials.

(2) This procedure yields in-laboratory and inter-laboratory variations that are deemed acceptable, i.e., less than about 10% in most cases.

(3) It is deemed worthwhile to use methods of data analysis based on both beam theory and experimental compliance approaches. Differences between the results from the two methods exceeding a few percent might indicate invalid test conditions (e.g., multiple branching or excessive fibre bridging).

(4) During the course of development several issues have been identified that are critical if the test is to be performed correctly. The most important relate to the starter defect and include an upper thickness limit for the insert film used as a crack starter (< 13 μm), the proper choice of the insert film material (non-sticking, non-crimping, temperature resistant beyond the curing temperature of the matrix), and the use of an appropriate release agent coating on the insert films.

(5) Corrections for large displacements and for the effects of load-blocks have been developed.

(6) Factors that have been shown to contribute to larger variations in the test results are the number of data points used for the analysis (≥ 15 are recommended), operator-dependent detection of delamination initiation (both using NL from the force-displacement plot and visual observation along the edges of the specimens), inclusion of arrest points from partly unstable delamination growth (arrest points have to be excluded), and the omission or incorrect application of the correction factors.

(7) Unsolved problems and questions include, e.g., the determination of G_{IC} on engineering laminates with non-unidirectional lay-ups, on composites with matrix systems showing unstable behaviour, on specimens taken from structures or structural parts (without laminated insert films as starter cracks), and on specimens with 3D-reinforcements (are LEFM-methods still valid for the analysis?). The question of the

economic efficiency of the test procedure, including test optimisation, has not been addressed in detail as yet.

6. REFERENCES

[1] P. Davies, B.R.K. Blackman, A.J. Brunner "Standard Test Methods for Delamination Resistance of Composite Materials: Current Status", Applied Composite Materials, 5, No. 6, 345-364 (1998).
[2] T.K. O'Brien "Interlaminar fracture toughness: the long and winding road to standardisation" Composites Part B, 29B, No. 1, 57-62 (1998).
[3] J.W. Gillespie Jr., L.A. Carlsson "Interlaminar Fracture of Laminated Composite Materials" in Delaware Encyclopedia of Composite Materials, Vol. 6 (R.B. Pipes, R.A. Blake, J.W. Gillespie, L.A. Carlsson, eds.) Technomic Publishing, Lancaster, 111-160 (1990).
[4] P. Davies, M.L. Benzeggagh "Interlaminar Mode-I Fracture Testing", in: Application of Fracture Mechanics to Composite Materials (K. Friedrich, ed.), Elsevier, pp. 81-112 (1989).
[5] N. Sela, O. Ishai "Interlaminar fracture toughness and toughening of laminated composites: A review", Composites 20, No. 5, 423-435 (1989).
[6] P. Davies, H.H. Kausch, J.G. Williams, A.J. Kinloch, M.N. Charalambides, A. Pavan, D.R. Moore, R. Prediger, I. Robinson, N. Burgoyne, K. Friedrich, H. Wittich, C.A. Rebelo, A. Torres Marques, F. Ramsteiner, B. Melve, M. Fischer, N. Roux, D. Martin, P. Czarnocki, D. Neville, I. Verpoest, B. Goffaux, R. Lee, K. Walls, N. Trigwell, I.K. Partridge, J. Jaussaud, S. Andersen, Y. Giraud, G. Hale, G. McGrath "Round-robin interlaminar fracture testing of carbon-fibre-reinforced epoxy and PEEK composites", Comp. Sci. & Tech. 43, 129-136 (1992).
[7] T.K. O'Brien, R.H. Martin "Round Robin Testing for Mode I Interlaminar Fracture Toughness of Composite Materials", J. Comp. Technol. & Res. 15, No. 4, 269-281 (1993).
[8] M. Hojo, K. Kageyama, K. Tanaka "Prestandardization study on mode I interlaminar fracture toughness test for CFRP in Japan" Composites 26, No. 4, 243-255 (1995).
[9] A.J. Brunner "Experimental aspects of Mode I and Mode II fracture toughness testing of fibre-reinforced polymer-matrix composites" Comp. Methods Appl. Mech. Engrg., 185, No. 2-4, 161-172 (2000).
[10] P. Davies, B.R.K. Blackman, A.J. Brunner "Introduction to Delamination Fracture of Continuous Fibre Composites: Mode II Delamination", this book.
[11] B.R.K. Blackman, P. Davies, J.G. Williams, A.J. Brunner "Introduction to Delamination Fracture of Continuous Fibre Composites: Mixed Mode I/II Delamination", this book.
[12] M. Cvitkovich, R.W. Lang "Polymer matrix effects on interlaminar crack growth in advanced composites under mixed-mode conditions" Proceedings European Conference on Composite Materials, Composites Testing and Standardisation, CTS-2, Woodhead Publ., 543-551 (1994).
[13] T.K. O'Brien "Interlaminar shear fracture toughness, G_{IIc}: shear measurement or sheer myth?", NASA Technical Memorandum TM110280/Army Research Laboratory Technical Report ARL-TR 1312 (1997).
[14] Compare data from ESIS round robins on CF-epoxy (IM7/977-1) presented in this paper and in the following papers [10, 11].

[15] Japanese Industrial Standard JIS K 7086 – 1993 "Testing methods for interlaminar fracture toughness of carbon fibre reinforced plastics", Japanese Standards Association (1993).

[16] "Standard Test Method for Mode I Interlaminar Fracture Toughness of Unidirectional Fiber-Reinforced Polymer-Matrix Composites", ASTM D 5528-94a, American Society for Testing and Materials, West Conshohocken, USA (1994).

[17] "Determination of the Mode I Delamination Resistance (Critical Energy Release Rate or Fracture Toughness G_{IC}) of Unidirectional Fiber-Reinforced Polymer Laminates Using the Double Cantilever Beam Specimen (DCB)", ESIS TC4 (1995).

[18] "Fibre-Reinforced Plastics Composites: Determination of Mode I Interlaminar Fracture Toughness, G_{IC}, For Unidirectionally Reinforced Materials", ISO/DIS 15024, International Organisation for Standardisation (1999).

[19] A.J. Russell, K.N. Street, "Factors affecting the interlaminar fracture energy of graphite/Epoxy laminates", Proceedings ICCM-IV, (Eds. T. Hayashi, K. Kawata, S. Umekawa), 279-286 (1982).

[20] D.J. Nicholls, J.P. Gallagher "Determination of G_{IC} in Angle Ply Composites Using a Cantilever Beam Test Method", J. Reinforced Plastics and Comp. 2, 2-17 (1983).

[21] B.R.K. Blackman, A.J. Brunner "Mode I fracture toughness testing of fibre-reinforced polymer composites: unidirectional versus cross-ply lay-up", Proceedings 12th European Conference on Fracture, ECF-12: Fracture from Defects, Vol. III (Eds. M.W. Brown, E.R. de los Rios, K.J. Miller), EMAS Publishing, 1471-1476 (1998).

[22] A.C. Roulin Moloney, P. Davies "Interlaminar Fracture of Composite Materials", Proceedings 7th European Conference on Fracture, ECF7, EMAS Publisher, 416-426 (1988).

[23] P. Davies, C. Moulin, H.H. Kausch, M. Fischer "Measurement of G_{IC} and G_{IIC} in Carbon/Epoxy Composites", Comp. Sci. & Technol., 39, 193-205 (1990).

[24] P. Davies, D.R. Moore "Glass/Nylon 66 composites: Delamination resistance testing" Comp. Sci. & Tech. 38, 211-227 (1990).

[25] P. Flüeler, A.J. Brunner "Crack Propagation in Fiber-reinforced Composite Materials Analysed with In-situ Microfocal X-Ray Radiography and Simultaneous Acoustic Emission Monitoring", Proceedings European Conference on Composites Testing and Standardisation, ECCM-CTS (Eds. P.J. Hogg, G.D. Sims, F.L. Matthews, A.R. Bunsell, A. Massiah), European Association for Composite Materials, EACM, 395-404 (1992).

[26] J. Bohse, T. Krietsch, J. Chen, A.J. Brunner "Acoustic Emission Analysis and Micromechanical Interpretation of Mode I Fracture Toughness Tests on Composite Materials", Talk at 2nd ESIS conference on Fracture of Polymers, Composites and Adhesives, Les Diablerets (Switzerland), September 13-15, 1999.

[27] A.J. Brunner, S. Tanner, P. Davies, H. Wittich "Interlaminar Fracture Testing of Unidirectional Fibre-Reinforced Composites: Results from ESIS Round Robins", Proceedings 2nd Conference on Composites Testing and Standardisation, CTS-2, Woodhead Publishing, 523-532 (1994).

[28] A.J. Brunner, P. Flüeler, P. Davies, B.R.K. Blackman, J.G. Williams "Determination of the delamination resistance of fibre-reinforced composites: current scope of test protocols and future potential", Proceedings 7th European Conference on Composite Materials, ECCM-7, Vol. 2, 3-8, Woodhead Publishing (1996).

[29] M.J. Hiley "Delamination between multidirectional interfaces in carbon-epoxy composites under static and fatigue loading", Talk at 2nd ESIS conference on Fracture of Polymers, Composites and Adhesives, Les Diablerets (Switzerland), September 13-15, 1999.

[30] W.D. Bascom, J.L. Bitner, R.J. Moulton, A.R. Siebert, "The interlaminar fracture of organic-matrix woven reinforcement composites", Composites 11, No. 1, 9-18 (1980).

[31] S. Mostovoy, P.B. Crosley, E.J. Ripling "Use of Crack-Line-Loaded Specimens for Measuring Plane-Strain Fracture Toughness", J. Materials 2, 661-681 (1967).

[32] D.C. Phillips "The fracture mechanics of carbon fibre laminates", J. Comp. Mat. 8, 130-141 (1974).

[33] P Robinson, D.Q. Song "A Modified DCB Specimen for Mode I Testing of Multidirectional Laminates" J. Comp. Mat. 26, No. 11, 1554-1577 (1992).

[34] A.L. Glessner, M.T. Takemori, M.A. Vallance, S.K. Gifford "Mode I Interlaminar Fracture Toughness of Unidirectional Carbon Fiber Composites Using a Novel Wedge-Driven Delamination Design" ASTM STP 1012 "Composite Materials: Fatigue and Fracture", (Ed. P.A. Lagace), 181-200 (1989).

[35] S. Matsuda, M. Hojo, S. Ochiai, A. Murakami, H. Akimoto, M. Ando "Effect of ionomer thickness on mode I interlaminar fracture toughness for ionomer toughened CFRP", Composites Part A 30A, 1311-1319 (1999).

[36] J.H. Chen, E. Schulz, J. Bohse, G. Hinrichsen "Effect of fibre content on the interlaminar fracture toughness of unidirectional glass-fibre/polyamide composite" Composites Part A, 30A, 747-755 (1999).

[37] R. Rikards, A. Korjakin, F.G. Buchholz, H. Wang, A.K. Bledzki, G. Wacker "Interlaminar Fracture Toughness of GFRP Influenced by Fiber Surface Treatment" J. Comp. Mat. 32, No. 17, 1528-1559 (1998).

[38] N.S. Choi, A.J. Kinloch, J.G. Williams "Delamination Fracture of Multidirectional Carbon-Fiber/Epoxy Composites under Mode I, Mode II and Mixed-Mode I/II Loading" J. Comp. Mat. 33, No.1, 73-100 (1999).

[39] T. Kusaka, M. Hojo, Y.-W. Mai, T. Kurokawa, T. Nojima, S. Ochiai "Rate dependence of Mode I fracture behaviour in Carbon-Fibre/Epoxy composite laminates" Compos. Sci. & Technol. 58, 591-602 (1998).

[40] W.X. Wang, Y. Takao, F.G. Yuan, B.D. Potter, R.H. Pater "The Interlaminar Mode I Fracture of IM7/LaRC-RP46 Composites at High Temperatures" J. Comp. Mat. 32, No. 16, 1508-1526 (1998).

[41] J.P. Berry "Determination of Fracture Energies by the Cleavage Technique", J. Appl. Phys. 34, No. 1, 62-68 (1963).

[42] P. Davies "Round Robin Analysis of G_{IC} Interlaminar Fracture Test", Appl. Comp. Mat. 3, 135-140 (1996).

[43] ASTM D6115 Standard Test Method for Mode I Fatigue Delamination Growth Onset of Unidirectional Fiber-Reinforced Polymer Matrix Composites (1997).

[44] B.R.K Blackman, J.G. Williams "Impact and High Rate Testing of Composites", NATO Advanced Study Institute "Mechanics of Composite Materials and Structures", NATO Science Series E: Applied Science, Vol. 361 (Eds. C.A. Mota Soares, C.M. Mota Soares, M.J.M. Freitas), Kluwer Academic, 225-234 (1999).

[45] Unpublished results from ESIS TC4 round robin with a cross-ply carbon-fibre epoxy material (T300/970).

[46] X.-Z. Hu, Y.-W. Mai "Mode I delamination and fibre bridging in carbon fibre epoxy composites with and without PVAL coating", Comp. Sci. & Technol. 46, No. 2, 147-156 (1993).

[47] O. Ishai "Interlaminar Fracture Toughness Characterization of Selectively Stitched Thick Fabric Composite Laminates", Proceedings European Conference on Composite

Materials, Composites Testing and Standardisation, CTS-4, IOM Communications, 127-137 (1998).

[48] D.D.R. Cartié, "Delamination behaviour of Z-pinned laminates", Talk at 2nd ESIS conference on Fracture of Polymers, Composites and Adhesives, Les Diablerets (Switzerland), September 13-15, 1999.

Version 99-06-03

Determination of the Mode I Delamination Resistance of Unidirectional Fiber-Reinforced Polymer Laminates Using the Double Cantilever Beam Specimen (DCB)

Détermination de la résistance au délaminage en mode I (taux de dégagement de l'énergie critique G_{IC}), éprouvette double poutre encastrée (DCB), de matériaux composites à matrice polymère renforcés de fibres unidirectionelles

Descriptors: delamination resistance, determination, double cantilever beam, laminate, Mode I, polymer-matrix, energy release rate, test result sheet, unidirectional fiber-reinforced

1 Scope

This standard specifies a method for the determination of the delamination resistance (G_{IC}, critical energy release rate or fracture toughness are sometimes used as equivalent terms) of unidirectional fiber-reinforced polymer laminates under Mode I opening load using the Double Cantilever Beam specimen (DCB). The resistance to the initiation and propagation of a delamination is to be determined from a non-adhesive starter film (insert) and from a Mode I (opening) precrack created by initial loading from the starter film or by wedge opening. The critical energy release rate for Mode I loading can be calculated and a resistance-curve (R-curve, i.e. a plot of the critical energy release rate versus delamination length) be determined.

The method is applicable to unidirectional carbon-fiber and glass-fiber reinforced laminates. The scope is not necessarily limited to these fibers and lay-ups, but for laminates with other types of fibers or lay-ups, no recommendations for specimen dimensions and fiber volume content are given. The procedure can be used as a guideline for testing materials that do not strictly satisfy the requirements, provided that (a) the data can be validated using an independent method, or (b) the results are considered to be order of magnitude estimates only and are quoted as such with the property or properties outside specification clearly indicated, or (c) the procedure is used for the purpose of relative comparison between materials only.

2 Normative References

The following standards contain provisions which through reference in this text constitute provisions of this International Standard. At the time of publication the editions indicated were valid. All standards are subject to revision, and parties to agreements based on this standard are encouraged to investigate the possibility of applying the most recent editions of the standards listed below. Members of IEC and ISO maintain registers of currently valid International Standards.

ISO 291:1997 *Plastics; standard atmospheres for conditioning and testing*
ISO 4588:1995 *Adhesives; preparation of metal surfaces for adhesive bonding*
ISO 5893:1993 *Rubber and plastics test equipment; tensile, flexural and compression types (constant rate of traverse); description*

3 Definitions

For a list of the definitions of symbols and conventions used in this protocol, refer to the central list of symbols in this book.

4 Principle

This standard uses the Double Cantilever Beam (DCB) specimen shown in Figure 1 for the determination of the delamination resistance (critical energy release rate) of unidirectional fiber-reinforced laminates. Opening loads (Mode I) are applied through load-blocks or piano hinges under displacement control at a constant rate. The onset of stable delamination growth and the subsequent delamination propagation from a non-adhesive starter film (insert), if available, and from a Mode I precrack, both at the laminate midplane are monitored. Delamination initiation and propagation readings are recorded on the force-displacement curves. Data reduction yields the critical energy release rates G_{IC} for initiation and propagation of a Mode I delamination that are presented in the form of R-curves (critical energy release rate G_{IC} versus delamination length a).

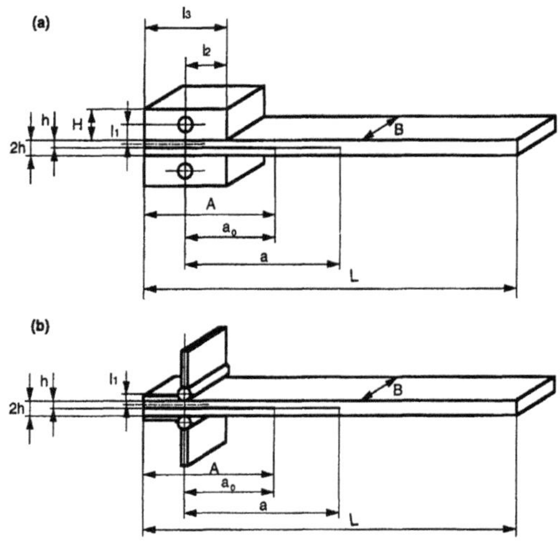

Figure 1: Geometry for the Double Cantilever Beam (DCB) specimen with a starter delamination; the starter film can be extended by a Mode I precrack prepared by DCB-testing or wedge opening. Alternative loading arrangements are (a) load-blocks, and (b) piano hinges. The fiber orientation is parallel to the length L. The delamination length a is the distance between the load-line (intersection of the plane through the pin-hole centers of the load blocks or centers of the hinge axes and plane of delamination) and the tip of the precrack or delamination.

This standard has been developed for unidirectional laminates where the plate stiffness components satisfy the condition $(D_{12})^2/(D_{11}D_{22}) << 1$. Lay-ups not satisfying this condition will result in the formation of a severely curved delamination front invalidating the initiation values.

It has to be noted that using a Mode I precrack for starting the delamination may yield values for the critical energy release rate G_{IC} differing from those obtained from the starter film (insert). In particular for materials with a "steep" R-curve, i.e. with a strong increase in the critical energy release rate G_{IC} versus delamination length a, precracking may yield larger, i.e. non-conservative, values but lower values have been observed as well. Therefore, in order to determine conservative values, whenever possible, both approaches (starter film and Mode I precrack) have to be used. It has also to be noted that different methods for precracking in Mode I may not yield identical critical energy release rates. This may in part be due to the fact that the starter film and wedge precracking yield a fairly straight starter delamination front across the specimen, while Mode I precracking using the method described in this standard may yield a slightly curved starter delamination front. Therefore, the method used to prepare the precrack shall be noted in the report.

5 Apparatus

A tensile testing machine in compliance with ISO 5893, capable of producing constant cross-head displacement rates between 1 and 5 mm/min in displacement control should be used. The load-cell should be calibrated and accurate within ± 1% for the chosen load-range (forces are typically expected to be less than 500 N). The testing machine shall be equipped with a fixture to introduce the load to the pins inserted into the load-blocks or with grips to hold the piano hinges that in each case allow rotation of the specimen end. The testing machine shall be equipped with means for recording the complete force-displacement curves (loading and unloading) that allow a determination of the loads and the corresponding displacements with an accuracy of ± 1 %.

6 Specimens

6.1 Preparation of Specimens

The recommended specimen width B and length L are 20 mm and 125 mm, respectively. The recommended specimen thickness is 3 mm for 60% by volume carbon fiber-reinforced and 5 mm for 60% by volume glass fiber-reinforced composites. Other specimen dimensions may be used, but the specimen width should be between 15 and 30 mm. Increasing the length of the specimen is not critical, shortening will reduce the maximum delamination length that can be investigated and thus yield too few data points for the analysis (see clause 8.2). If possible, the thickness shall be chosen such that the condition

$$2h > 8.28 \sqrt[3]{\frac{G_{IC} a_0^2}{E_{11}}} \tag{1}$$

with 2h the thickness of the specimen, G_{IC} the critical energy release rate, a_0 the initial delamination length ($a_0 = A - l_2$ for load-blocks), and E_{11} the tensile modulus of the specimen along the direction of the fibers, is satisfied (Reference 1). Application of this criterion requires prior knowledge of values of G_{IC} for the material to be tested, and, if available, test results or data from published literature shall be used.

Two types of initial defect (starter defect) are considered, (a) a laminated starter film (insert), and (b) a Mode I precrack obtained from the insert, either by a Mode I DCB test or by wedge opening.

If a starter film (insert) is used, a nonadhesive film should be placed at laminate mid-thickness during lay-up. The film thickness should not exceed 13 μm and be as thin as possible to minimise the disturbance of the laminate. The starter film should extend at least 50 mm beyond the load-line so that the influence of the load-blocks or piano hinges can be neglected. A polymer film is recommended as starter film to avoid problems with folding or crimping that have been observed with aluminium films (Reference 2). For epoxy matrix composites cured at temperatures below 180° C, a thin film of polytetrafluorethylene (PTFE) is recommended. For composites that are manufactured at temperatures above 180° C (polyimide, bismaleimide, thermoplasts) a thin polyimide film is recommended. If a polyimide film is used, the film shall be painted or sprayed with a mold release agent before insertion into the laminate. For materials outside the scope of this test method, different film materials may be required. Mold release agents containing silicone may contaminate the laminate by migration through the individual layers. It is often helpful to coat the film at least once and then bake it before placing the film on the composite. This will help to prevent silicone migration within the composite.

For producing the Mode I precrack from the starter film (insert) the test procedure for testing from the starter film including data recording in accordance with clause 8.1 or the following procedure for wedge opening (Reference 3) may be used. For wedge opening, the specimen shall be clamped at 5 mm beyond the tip of the starter film. The width of the wedge that is driven into the specimen shall be at least the same as that of the specimen and the opening angle shall be as small as possible without the wedge actually touching the tip of the delamination. The wedge shall be driven into the specimen until the tip of the wedge reaches the clamp. The wedge may be driven by hand (tapping on the side) or by using a suitable fixture and a testing machine. Experience has shown that it may be difficult to produce a suitable precrack by wedge opening, the precrack will not always lie in the midplane of the specimen. Deviations of the precrack from the midplane will invalidate the test results and should be noted in the report. The wedge precrack will usually extend a few mm into the clamp but should be short enough to allow a delamination length increment of at least 50 mm beyond the tip of the precrack.

If specimens are cut from a plate, the location of each specimen on the plate should be recorded and specimens should each be identifiable. Measure and record the length L of each specimen to the nearest mm. Measure the width B and the thickness 2h of each specimen to the nearest 0.02 mm, at five points (quarter, center, and three-quarter length and 10 mm from each end) along the specimen. The variation in thickness and average values of the width and the thickness should be recorded for each specimen. The variation in thickness, i.e. the maximum difference between the thickness measurements, should not exceed 0.1 mm for each specimen. Measure the starter delamination length, i.e. the total length of the starter film (insert) on both edges of the specimen. The average value should be recorded but if the starter film length measurements differ by more than 1 mm this should be noted in the report.

Load-blocks or piano hinges (Figure 1) can be used as load-introductions, they should be at least as wide as the specimen. The load-blocks or the piano hinges and the specimen should first be lightly abraded, use of either sandpaper or grit blasting should be sufficient, as the forces required to delaminate the specimens used in these tests are quite low. The load-introduction and the specimen should then be cleaned with a solvent. If a bond failure occurs it may be necessary to consult ISO 4588 for a more sophisticated procedure. Bonding of the load-introductions should be done immediately after the surface preparation. In most cases a cyanoacrylate ("Super glue") adhesive has been found adequate for previous tests on similar specimens. Alternatively, a tough, room-temperature cure adhesive may be used. The surface preparation and the type of adhesive used should be noted in the report. The load-introductions should be well aligned with the specimen, and with each other, and held in position with clamps while the adhesive sets. Specimen edges should be smoothed prior to determining the dimensions. Adding a thin layer of typewriter correction fluid ("white ink") on the edges after conditioning will facilitate the detection of the delamination growth. It should be noted that some typewriter correction fluids contain solvents that may be harmful to the laminate matrix material. For the measurement of the delamination lengths, marks should be drawn at 5 mm intervals along the edge of the specimen extending at least 55 mm ahead of the tip of the starter film and of the precrack, respectively. Additionally, the first 10 and last 10 mm increments shall be marked at 1 mm intervals.

6.2 Number of Specimens

A minimum number of five specimens shall be tested.

7 Conditioning

Moisture conditioning is required for obtaining baseline data in order to test specimens with a uniform moisture content. The drying conditions (temperature and duration) shall be chosen according to the recommendations of the resin supplier. Conditioning should be performed after bonding of the load-blocks or piano hinges. Before testing, the specimens may be stored in a dessicator for at most one day after conditioning. Other conditioning procedures may be applied for the investigation of specific conditioning effects.

8 Test Procedure

8.1 Test Set-up and Data Recording

The test shall be performed under normal conditions in accordance with ISO 291 (23° ± 2° C, 50% ± 5% relative humidity). After mounting the specimen in the fixture of the testing machine, the end of the specimen may have to be supported in order to keep the beam orthogonal to the direction of the applied force. The force and the displacement signals of the testing machine shall be recorded, either on a paper chart or electronically throughout the test, including the unloading cycle. The delamination length may be measured by eye on the specimen edge, or by using a travelling microscope. If used, the travelling microscope or equivalent magnifying device shall have a magnification no greater than 70x and be set in a position to observe the motion of the delamination front on the edge of the specimen. Whether observing by eye or by a magnifying device the position of the delamination shall be pinpointed with an accuracy of at least ± 0.5 mm on the edge of the specimen. In transparent laminates the delamination length may be followed inside the specimen by marking the specimen surface rather than the edge. If unstable delamination growth followed by arrest ("stick-slip") is observed during any stage of the test, it should be noted in the report. The data evaluated according to clause 8.2 may not be valid in this case. Any permanent deformation of the specimen after unloading should be noted in the report. Deviations of the delamination from the midplane of the laminate will invalidate the test results and should be noted in the report.

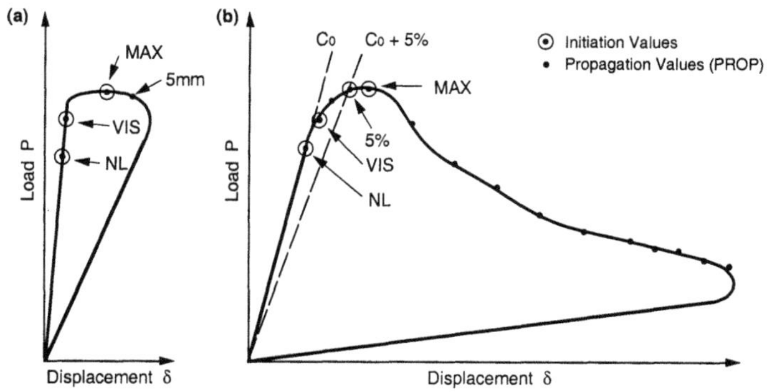

Figure 2: Schematic force-displacement curve (a) testing from the starter film and (b) testing from the Mode I precrack with initiation points NL, VIS, 5%, MAX, and propagation points (PROP). In (a) the MAX occurs before the 5%, in (b) the reverse situation is shown.

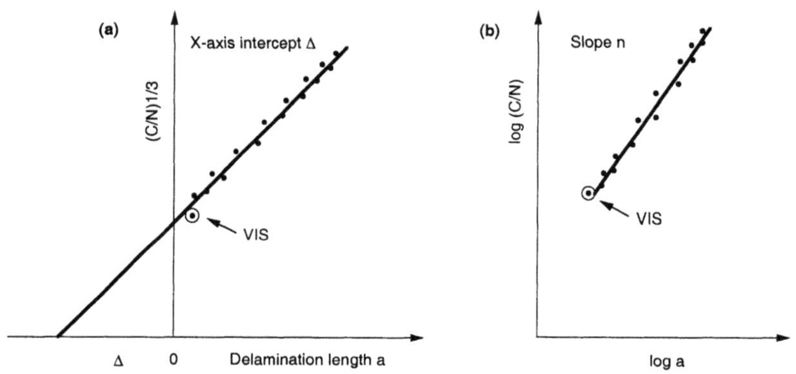

Figure 3: Linear fits used to determine (a) the correction Δ for the Corrected Beam Theory Method, and (b) the slope n for the Experimental Compliance Calibration Method. The VIS point may be excluded from the linear fit (see clause 8.2).

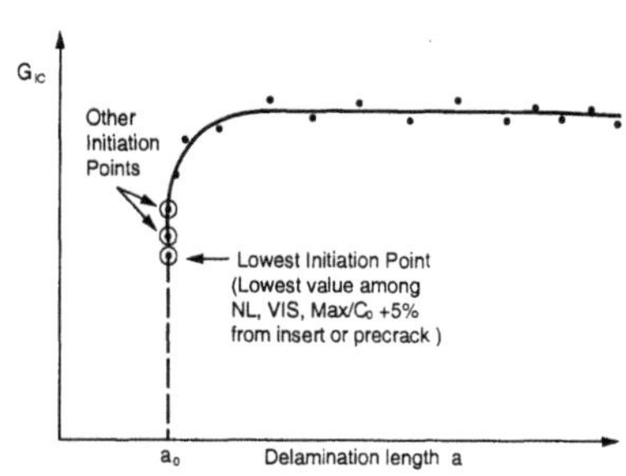

Figure 4: Schematic resistance-curve (R-curve) with G_{IC}-value for initiation (lowest value among NL, VIS, 5% or MAX) and for propagation (PROP) versus observed delamination length a.

(1) Testing from the Starter Film (Insert)

For testing from the starter film (insert), the specimen should be loaded at a constant cross-head speed between 1 - 5 mm/min. The point on the force-displacement curve at which the onset of delamination movement from the starter film is observed on the edge of the specimen should be recorded on the force-displacement curve or in the sequence of force-displacement signals (VIS, Figure 2a). If the start of the delamination growth is difficult to observe, a change in illumination conditions or a cross-head speed from the lower end of the range is recommended. The loading should be stopped at a delamination length increment of 3 - 5 mm. If unstable delamination growth from the starter film is observed, this shall be noted in the report and loading be continued until the delamination length is increased by 3 - 5 mm beyond the arrest point. If the delamination length increment is outside the range of 3 - 5 mm this should be noted in the report. Then the specimen should be completely unloaded at a constant cross-head rate, unloading may be performed at up to 25 mm/min. The position of the tip of the precrack should be marked on both edges of the specimen after unloading, if the specimen is removed from the fixture, this should be noted in the report. If the precrack lengths a_p on the edges of the specimen, i.e. the distance between the load-line and the tip of the precrack, differ by more than 2 mm this may be an indication of asymmetrical loading (Reference 4) and the results should be considered suspect and this be noted in the report.

(2) Testing from the Mode I Precrack

For testing from the Mode I precrack prepared by loading from the starter film (insert) or from wedge opening, the marks on the specimen edge should be checked before testing and adjusted, if necessary, according to clause 6.1. The specimen should be loaded at a constant cross-head rate between 1 - 5 mm/min without stopping or unloading until the final delamination length increment (see below) has been reached. The point on the force-displacement curve at which the onset of delamination movement from the Mode I precrack is observed on the edge of the specimen should be recorded on the plot or in the sequence of force-displacement signals (VIS, Figure 2b). If the start of the delamination growth is difficult to observe, a change in illumination conditions or a cross-head speed from the lower end of the range is recommended. After this, as many delamination length increments as possible should be noted in the first 5 mm on the corresponding force-displacement curves, ideally every 1 mm. Subsequently, delamination lengths are noted every 5 mm, until the delamination has propagated at least 45 mm from the tip of the Mode I precrack, and again every 1 mm for the last 5 mm of delamination propagation, i.e. up to a delamination length increment of 50 mm beyond the tip of the precrack (Figure 2b). After this, the specimen should be unloaded at a constant cross-head rate, unloading may be performed at up to 25 mm/min. The position of the tip of the delamination after unloading should be marked on both edges of the specimen. If the precrack lengths a_p on the edges of the specimen, i.e. the distance between the load-line and the tip of the delamination, differ by more than 2 mm this may be an indication of asymmetrical loading (Reference 4) and the results should be considered suspect and this be noted in the report.

8.2 Data Analysis

The data required for the analysis are the initial delamination length a_0, the delamination lengths a (a = a_0 or a_p + measured delamination length increments), and the corresponding forces P and displacements δ and the width B of the specimen. Several values may be determined from the force-displacement curve, if possible, the following values, shown in Figure 2, should be determined for testing from the starter film and from the Mode I precrack for each specimen:

(1) NL, i.e. deviation from linearity: A region of non-linear behaviour usually precedes the maximum force, even if the unloading curve is linear. The point of deviation from linearity (NL in Figure 2), is determined by drawing a straight line from the origin but ignoring any initial deviations due to take-up of play in the loading system. Experience has shown that it may be difficult to reproducibly determine the position of NL on the force-displacement curve, variations of up to 10% are not uncommon (Reference 5). Performing a linear fit on the force-displacement curve starting at a finite force to avoid nonlinearity due to play and using a consistent criterion for deviation from linearity (e.g. the half-thickness of the plotter trace) is recommended. This is supported by physical evidence from X-ray imaging that shows that the onset of the delamination from the starter film in the interior of the specimen occurs close to the NL point (Reference 6), however, more refined detection methods may indicate an even earlier onset (Reference 4). Experience has shown that the NL point will frequently yield the lowest, i.e. most conservative values of the critical energy release rate (Reference 2).

(2) VIS, i.e. Visual observation: This corresponds to the onset of the delamination, i.e. to the first point at which the delamination is observed to move from the tip of the starter film or of the Mode I precrack on the edge of the specimen (VIS in Figure 2).

(3) 5% or MAX, i.e. 5% increase of compliance or maximum force point: The 5% value corresponds to the point on the force-displacement curve at which the compliance has increased by 5% of its initial value C_0. A best straight line is drawn to determine the initial compliance C_0, ignoring any initial deviation due to take-up of play in the loading system. A new line is then drawn with a compliance equal to C_0 +5% whose intersection with the force-displacement curve yields the force and displacement to be used for the calculation, unless the intersection is at a larger displacement than the maximum force in which case the maximum force and the corresponding displacement have to be used.

Besides the NL, VIS, 5% or MAX points obtained from the starter film and from the Mode I precrack, propagation values (PROP in Figure 2b) can be determined for each delamination length measured during propagation from the Mode I precrack. The values from the Mode I precrack (procedure (2) above) shall be analysed first, and then the values determined from the starter film (procedure (1) above). This is because procedure (1) does not yield a sufficient number of data points for the linear fits used in the data analysis. For the data points from procedure (1) the value of Δ and n, respectively determined from the data points obtained in procedure (2) shall be used (see below). If possible, a single test result sheet (Figure 5) shall be used to report the data for testing from the insert (NL, VIS, MAX/5%) and from the Mode I precrack (NL, VIS, MAX/5% and PROP) for each specimen. Either one of the two methods described below can be used for the analysis, the method chosen should be noted in the report.

Method (1): Corrected Beam Theory (CBT)

The simple beam theory expression for the compliance of a perfectly built-in DCB specimen will underestimate the compliance as the beam is not perfectly built in. A means of correcting for this effect is to treat the beam as containing a slightly longer delamination length a + |Δ|, and |Δ| may be found experimentally by plotting the cube root of the compliance $C^{1/3}$, or the cube-root of the normalised compliance $(C/N)^{1/3}$, if load-blocks are being used (the load-block correction N is described below), as a function of delamination length a (Figure 3). The extrapolation of a linear fit through the data in the plot yields Δ as the x-intercept (Reference 7). If the Δ-value from the fit is positive, a value of Δ = 0 shall be used and this be noted in the report. The VIS and the PROP values are used for the linear fit, but not the NL or 5%/MAX values. If it has not been determined or is considered questionable, the VIS point may be excluded from the linear fit, but this should be noted in the report. The critical energy release rate G_{IC} is given by

$$G_{IC} = \frac{3P\delta}{2B(a+|\Delta|)}F \quad \text{or} \quad G_{IC} = \frac{3P\delta}{2B(a+|\Delta|)}\frac{F}{N} \tag{2}$$

with P the force, δ the displacement, a the delamination length, and B the width of the specimen. F is the large displacement and N the load-block correction. All initiation and propagation values should be calculated. The delamination length for the VIS, NL, 5% or MAX values is the initial delamination length a_0 and a_p (distance between the load-line and the tip of the starter film and precrack, respectively, Figure 1) if no delamination growth has been observed up to those points. Else the effective, observed delamination length should be used for the analysis. The load-block correction N is applied if load-blocks are being used, the large displacement correction F shall be applied for all specimens (F will contribute significantly, if the ratio of displacement δ and delamination length a becomes larger than 0.4). The large displacement correction F and the load-block correction N are calculated as follows (for piano hinges N = 1)

$$F = 1 - \frac{3}{10}\left(\frac{\delta}{a}\right)^2 - \frac{3}{2}\left(\frac{\delta l_1}{a^2}\right) \tag{3}$$

$$N = 1 - \left(\frac{l_2}{a}\right)^3 - \frac{9}{8}\left[1 - \left(\frac{l_2}{a}\right)^2\right]\frac{\delta l_1}{a^2} - \frac{9}{35}\left(\frac{\delta}{a}\right)^2 \tag{4}$$

with l_1 the distance from the center of the loading pin or of the piano hinge axis to the midplane of the specimen beam and l_2 the distance from the loading pin center to its edge (Figure 1). If large displacement corrections F < 0.9 are found this shall be noted in the report. This approach allows the flexural modulus E_f to be calculated by using

$$E_f = \frac{8(a+|\Delta|)^3}{CBh^3} \quad \text{or} \quad E_f = \frac{8(a+|\Delta|)^3}{\left(\frac{C}{N}\right)Bh^3} \tag{5}$$

with a the delamination length, Δ the delamination length correction used in the corrected beam theory, C the compliance and C/N the normalised compliance, respectively (the load-block correction N is defined above), B the width, and h half the thickness of the specimen. This calculation is a useful check on the procedure, as a value independent of delamination length should be obtained, however, no quantitative limits on the variation of the calculated modulus with delamination length can be given. Experience has shown that the calculated value of E_f is frequently larger than the modulus determined from a flexural test and the calculated value of E_f shall not be quoted as modulus value. This effect is attributed mostly to fiber bridging and if this is observed it should be noted in the report.

Method (2): Experimental Compliance Method (ECM) or Berry's Method

An alternative approach is to plot the logarithm of the compliance C or of the normalised compliance C/N, if load-blocks are being used, versus the logarithm of the delamination length a. Only the VIS and the PROP values are used for the linear fits, but not the NL or 5%/MAX values. If it has not been determined or is considered questionable, the VIS point may be excluded from the linear fit, but this should be noted in the report. The slope of this plot, n, can then be used to give G_{IC} as follows

$$G_{IC} = \frac{nP\delta}{2Ba}F \qquad \text{or} \qquad G_{IC} = \frac{nP\delta}{2Ba}\frac{F}{N} \tag{6}$$

with P the force, δ the displacement, a the delamination length, and B the width of the specimen. F is the large displacement and N the load-block correction. The large-displacement correction F and load-block correction N, if applicable, are the same as for the corrected beam theory method (see equations (3) and (4) above). All initiation and propagation values should be calculated. The delamination length for the VIS, NL, 5% or MAX values is the initial delamination length a_0 and a_p (distance between the load-line and the tip of the starter film and precrack, respectively, Figure 1) if no delamination growth has been observed up to those points. Else the effective, observed delamination length should be used for the analysis.

8.3 Data Presentation

All results (NL, VIS, 5% or MAX values from both the starter film and the Mode I precrack and PROP values from the Mode I precrack) are used to draw a resistance-curve (R-curve), i.e. G_{IC} versus delamination length a (Figure 4) for each specimen. The initiation value quoted should be the lowest among the VIS, NL, 5% or MAX values from both the starter film and the Mode I precrack, indicating the type of point in the report. The minimum number of propagation points (PROP) should be 15, if fewer points are used, this shall be noted in the report. In this case, the values of Δ or n determined from the linear fits may be influenced by statistical effects (Reference 8) and it may be difficult to assess whether a constant plateau has been reached in the R-curve.

When quoting characteristic material values from testing several identical specimens (5 specimens are required in clause 6.1), the results shall be averaged as follows: Calculate the arithmetic average and standard deviation of each VIS, NL, MAX, and 5% values separately. The initiation value quoted should be the lowest among the average VIS, NL, 5% or MAX values from both the starter film and the Mode I precrack, indicating the type of point in the report. Then calculate the arithmetic average and the standard deviation of the last ten PROP values or of the last 50% of all PROP values, whichever contains the larger number of data points. The average values and standard deviations should be noted in the report. If the standard deviation of the PROP values exceeds 10% of the average value, a constant plateau value for the propagation may not have been reached and the R-curve plots should be checked. If the R-curve plots do not show a plateau, the average PROP value should be considered suspect and this be noted in the report.

9 Test Report

The test report shall include the following information:

(a) a reference to this test protocol and to the referring standards

(b) a complete identification of the material (e.g. laminate manufacturer, fiber-material, polymer material, maximum cure temperature T_{mc}, duration of curing t_c, location of specimen on plate)

(c) test date, test laboratory, test personnel identification

(d) number and label of specimens tested and type of method used for the analysis

(e) average thickness, average width, maximum thickness variation along the length, and length of each specimen, starter film (insert) material and thickness, length of the starter film; note if starter film lengths measurements differ by more than 1 mm on both edges

(f) conditioning temperature T_d and conditioning duration t_d and temperature T and relative humidity r.h. during the test

(g) type and dimensions of load-introduction, surface preparation, if applicable, and adhesive

(h) type of precracking used (e.g. Mode I test or wedge opening) and, if applicable, whether specimen has been removed from the fixture after precracking

(i) load-rate for loading and unloading, for testing from the starter film and from the Mode I precrack

(j) delamination length a on both edges of the specimen after testing (unloading) from the starter film and from the Mode I precrack; note, if the delamination lengths measurements differ by more than 2 mm on both edges.

(k) x-axis intercept Δ of the linear fit of the cube-root of the compliance $C^{1/3}$ or the normalised compliance $(C/N)^{1/3}$, if applicable, versus the delamination length a, if method (1) is used for the data analysis, and the correlation coefficient r^2 of the linear fit. Note, if the VIS-value has been excluded from the fit.

(l) slope n of plot of the logarithm of the compliance log C or the normalised compliance log (C/N), if applicable versus the logarithm of the delamination length log a, if method (2) is used for the data analysis, and the correlation coefficient r^2 of the linear fit. Note, if the VIS-value has been excluded from the fit.

(m) the calculated flexural modulus E_f of the specimen as a function of the delamination length a, if method (1) is used for data analysis. Note, if the calculated flexural modulus differs from measured values, if available.

(n) $(C_{max/5\%} - C_0)/C_0$, i.e. the percent change in compliance between the initial compliance C_0 and the compliance at the MAX or 5% point, whichever is applicable

(o) copy of the force-displacement curve for each specimen

(p) table of G_{IC} and plot of G_{IC} with all values as defined in clause 8.3 versus delamination length a (R-curve) for each specimen including large displacement corrections and load-block corrections, if applicable. Note, if the large-displacement correction F is lower than 0.9.

(q) average values and standard deviation for each VIS, NL, MAX, and 5% and average value and standard deviation of the last 10 propagation values (PROP) or of the last 50% of the propagation values, whichever contains the larger number of data points, from all specimens tested. Note, if less than 15 propagation values have been recorded. If a specimen is excluded from averaging, the reason for this should be noted in the report.

(r) observations from testing (e.g. deviation of the precrack or the delamination from the midplane, stick-slip, occurrence of fiber-bridging, permanent deformation after unload ing, sticking of starter film, no plateau in the R-curve) that may have affected the test procedure or the results

(s) any deviation from the prescriptions of this protocol (e.g., dimensions of specimens, fiber orientation)

(t) results from additional specimen or material characterisation (e.g., fiber volume fraction, void content), if available

A recommended test result sheet is shown in Figure 5.

10 References

[1] R.A. Naik, J.H. Crews Jr., K.N. Shivakumar: "Effects of T-Tabs and Large Deflections in DCB Specimen Tests" in: Composite Materials; Fatigue and Fracture (T.K. O'Brien ed.), ASTM STP 1110, American Society for Testing and Materials, 169-186 (1991).

[2] T.K. O'Brien, R.H. Martin: "Results of ASTM Round Robin Testing for Mode I Interlaminar Fracture Toughness of Composites Materials", ASTM Journal of Composites Technology and Research, 15, Nr. 4, 269-281 (1993).

[3] M. Hojo, K. Kageyama, K. Tanaka: "Prestandardization study on mode I interlaminar fracture toughness test for CFRP in Japan" Composites 26, Nr. 4, 243-255 (1995).

[4] P. Flüeler, A.J. Brunner: "Crack Propagation in Fiber-Reinforced Composite Materials Analysed with In-situ Microfocal X-ray Radiography and Simultaneous Acoustic Emission Monitoring" in: Composites Testing and Standardisation ECCM-CTS, (P.J. Hogg, G.D. Sims, F.L. Matthews, A.R. Bunsell, A. Massiah eds.) European Association for Composite Materials, 385-394 (1992).

[5] P. Davies: "Round Robin Analysis of G_{IC} Interlaminar Fracture Test", Applied Composite Materials, 3, 135-140 (1996).

[6] T. de Kalbermatten, R. Jäggi, P. Flüeler, H.H. Kausch, P. Davies: "Microfocus Radio graphy Studies During Mode I Interlaminar Fracture Toughness Tests on Composites", Journal of Materials Science Letters, 11, 543-546 (1992).

[7] S. Hashemi, A.J. Kinloch, J.G. Williams: "Corrections Needed in Double Cantilever Beam Tests for Assessing the Interlaminar Failure of Fiber-composites", Journal of Materials Science Letters, 8, 125-129 (1989).

[8] A.J. Brunner, S. Tanner, P. Davies, H. Wittich: "Interlaminar Fracture Testing of Unidirectional Fibre-Reinforced Composites: Results from ESIS-Round Robins" in: Composites Testing and Standardisation ECCM-CTS 2, (P.J. Hogg, K. Schulte, H. Wittich eds.), Woodhead Publishing, 523-532 (1994).

Figure 5: Recommended Test Result Sheet

Recommended Test Result Sheet DCB Mode I, Page 1 of 2			Fiber :	Resin :	
Specimen label	....	# of	Specimens		
Laboratory	....	Test personnel	....	Test date	....
Laminate manufacturer	....	Resin manufacturer	....	Fiber manufacturer	....

Specimen length	L [mm]	....	Load introduction		....	Average NL ±σ
Average thickness	2h [mm]	....	Block thickness	H [mm]	....	
Max. thickness variation	Δ2h [mm]	....	Block length	l_3 [mm]	....	CV NL
Specimen width	B [mm]	....	Block/hinge width	[mm]	....	
Insert material	....		Surface preparation	....		Averge VIS ±σ
Insert thickness	[μm]	....	Adhesive	....		
Insert length	A [mm]	....	Conditioning temperature	T_d [°]	....	CV VIS
Mold release agent	....		Conditioning duration	t_d [h]	....	
Precrack length	a_p [mm]	....	Test temperature	T [°]	....	Av. MAX/5 % ±σ
Max. cure temperature	T_{mc} [°]	....	Relative humidity	r.h. [%]	....	
Cure duration	t_c [h]	....	Loading/unloading rate	[mm/min]	....	CV MAX/5 %
Max. delamination length	A_{max} [mm]	....	Distances l_1/l_2	[mm]	....	Minimum initiation value :
% Change compliance	[%]	....	Correction □ (CBT-Method)	[mm]	....	

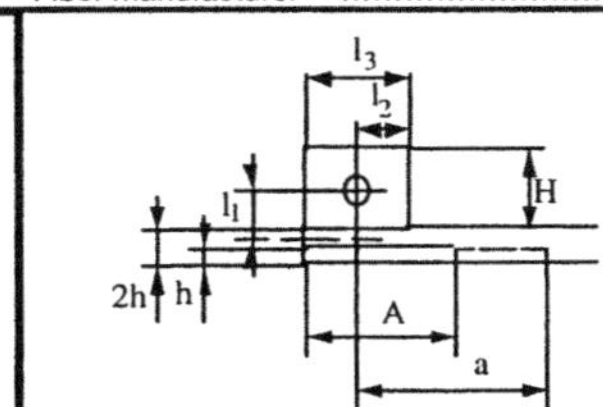

Text	a [mm]	Force [N]	δ [mm]	Specimen and Insert Location on Plate	Comment	G_{IC} CBT [J/m²]	G_{IC} ECM [J/m²]	E-Mod. [GPa]
NL (insert)				Length .... mm				
MAX/5 % (insert)								
VIS (insert)								
NL (precrack)								
MAX/5 % (precrack)								
VIS (precrack)								
PROP								
PROP								
PROP								
PROP								
PROP								
PROP								
PROP								
PROP								
PROP								
PROP								
PROP								
PROP								
PROP								
PROP								
PROP				Width .... mm				

Recommended Test Result Sheet DCB Mode I, Page 2 of 2 **Fiber :** **Resin :**

Specimen label	....	# of	Specimens		
Laboratory	....	Test personnel	....	Test date	....
Laminate manufacturer	....	Resin manufacturer	....	Fiber manufacturer	....

CBT	
slope of fit	
Y-intercept of fit	
Δ [mm]	
correlation coefficient	
or	
ECM	
slope of fit	
Y-Intercept of fit	
m	
correlation coefficient	

Difference insert length ≤ 1 mm ?	yes / no
Difference precrack length ≤ 2 mm ?	yes / no
Precrack tip 3-5 mm from insert ?	yes / no
Difference total delamination ≤ 2 mm ?	yes / no
Specimen removed after precracking ?	yes / no
VIS included in linear fit ?	yes / no
Stable delamination growth through test ?	yes / no
All correction factors F ≥ 0.9 ?	yes / no
Displacement / a < 0.4 ?	yes / no
Reference to ISO 291 ?	yes / no
Reference to ISO 4588 ?	yes / no
Reference to ISO 5893 ?	yes / no

Observations and Comments

Text	F	N	F / N	C / N	$(C/N)^{2/3}$	(BC / N)	a / 2h	δ / a	Checks ok ?
NL (insert)									
MAX/5 % (insert)									
VIS (insert)									
NL (precrack)									
MAX/5 % (precrack)									
VIS (precrack)									
PROP									
PROP									
PROP									
PROP									
PROP									
PROP									
PROP									
PROP									
PROP									
PROP									
PROP									
PROP									
PROP									
PROP									
PROP									

MODE II DELAMINATION

P. DAVIES, B.R.K. BLACKMAN and A.J. BRUNNER

1. INTRODUCTION

Introducing shear loads into composite materials has presented major problems to the composite industry. Tests to determine shear moduli and strengths abound (45° tensile, off-axis tensile, rail shear, Iosipescu) and all are open to criticism. Shear loading of cracked specimens involves many of the difficulties associated with these shear tests, but adds others such as friction between sliding crack faces, instabilities in some of the specimen geometries and non-linear behaviour. Indeed, there has been some discussion recently over whether it is theoretically possible to produce a pure mode II test [1] as the local failure mode is generally observed to be tensile. However, the local stresses round reinforcements in many two-phase materials can be very complex but this does not stop useful measurements of global properties being made. ESIS TC4 has therefore spent considerable time and effort in examining mode II tests, and produced a protocol for mode II testing.

Applying a pure shear loading to a crack has inspired several researchers. Carlsson and Gillespie presented a summary of mode II testing in 1989 [2], but over the ten years since then significant advances have been made. Flexure specimens have generally been preferred, although other geometries such as tubes in torsion [3] and cracked panels [4,5] have also been used. Figure 1 shows the most popular solutions.

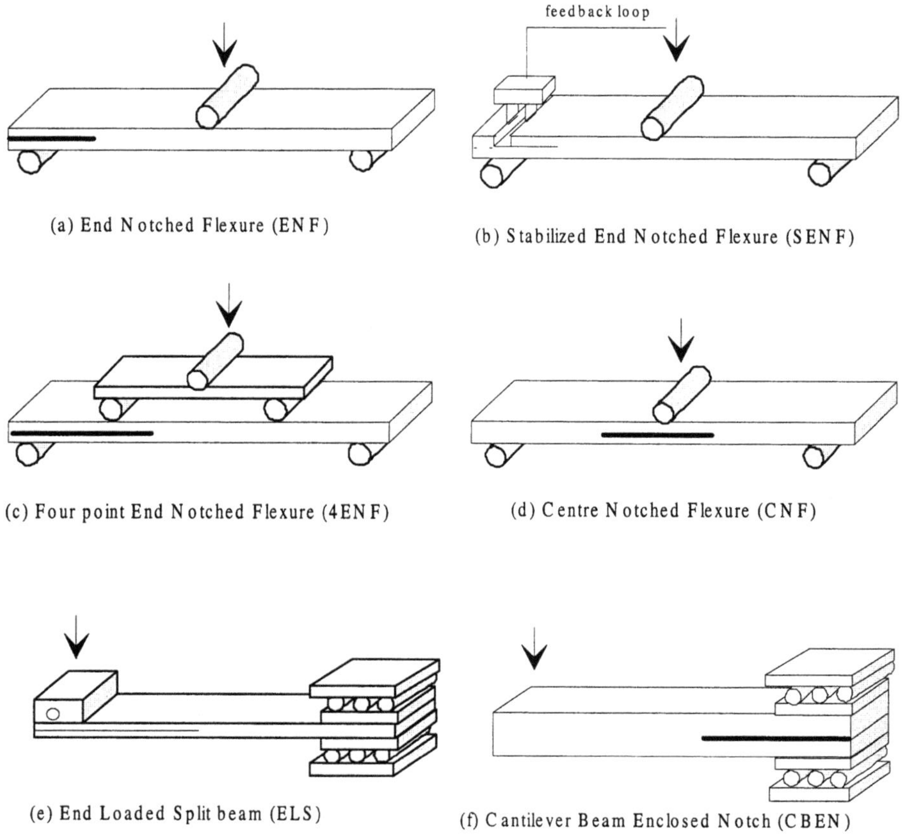

Figure 1. Schematic diagram of different mode II specimens.

2. HISTORICAL BACKGROUND

The most widely-used test for mode II delamination resistance, using the ENF specimen (Figure 1a), was originally developed for testing wood in the 1970's [6]. Russell & Street applied it to composites [7] and for over ten years this was the most popular mode II test for fibre reinforced materials. The ELS specimen was first used at Texas A&M University by Bradley and co-workers, particularly Vanderkley [8]. The other specimens shown above were developed in the 80's and 90's. The SENF and 4ENF were two developments aimed at stabilising the ENF crack propagation. The former was developed in Japan and uses a feedback loop to maintain a constant crack shear displacement rate [9], while the latter, proposed by Martin and Davidson, is based on the intrinsic stability of the four point loading fixture [10]. The CNF geometry, a symmetrical mode II configuration, was developed at the University of Delaware [11] and has been applied to impact studies [13], while the CBEN was originally used in studies of fatigue crack propagation in thick leaf springs at the University of Compiègne [12].

ESIS started mode II studies using both the ENF and ELS test specimens. An ENF test protocol was drafted in 1988, which was examined in a joint round robin with ASTM, in 1990. In parallel an ELS protocol was also developed as this specimen was considered to be better suited to measurement of G_{IIc} values on account of its improved crack propagation stability. As the last version of the ESIS ENF procedure was identical to the ASTM draft procedure it is not given here. Only the ELS development will be discussed in detail, although most of the problems encountered apply to all mode II testing. Table 1 shows the mode II tests run involving ESIS participants.

Date	Specimens	Material	Description	Ref.
1986	ENF, ELS	C/Epoxy, C/PES	First tests, mode I and II	14
1987	ENF, ELS	C/Epoxy, G/PA, G/PU	First tests on tough materials	15
1988	ENF, ELS	C/Epoxy, C/PEEK	First draft ENF protocol	16
1989	ELS		First draft ELS protocol	
1990	ENF		Joint ASTM/JIS/ESIS protocol	
1991	ENF, ELS	G/Acrylic	Transparent, observe initiation	17
1992	ENF, ELS	G/Epoxy	Mixed mode envelope	18
1993	ELS, (ENF)	C/Epoxy	Film and mode II precracks	
1995	ELS, ENF	G/Epoxy (hand lay-up)	Initiation study	19
1996	ENF, ELS, SENF, (4ENF)	C/Epoxy	VAMAS specimen comparison, ESIS/ASTM/JIS	20
1998	4ENF	C/Epoxy, G/Epoxy	VAMAS specimen comparison continued, ESIS/ASTM/JIS	21

C: Carbon, G: Glass

Table 1. ESIS mode II development.

3. PROBLEMS TO BE ADDRESSED IN MODE II TESTING

There are several difficulties to be overcome in developing a mode II test. Many of the points to be verified are applicable to mode I and mixed mode tests, but some additional aspects are also important. Thus points to be checked, all of which have been examined during the ESIS development of the mode II protocols (for unidirectional composites), are:

1. Specimen geometry independence
2. Starter defect influence
3. Stability
4. Data analysis
5. Influence of friction

Specimen geometry independence

Verifying that G_{IIc} values are independent of specimen geometry includes not only evaluating different types of specimen (i.e. ENF, ELS) and the influence of loading fixtures, but also then examining the influence of width, length and thickness to ensure the domain of applicability of the analyses proposed. ESIS studies have concentrated on the ELS configuration, but in order to convince potential users of the advantages offered by the ELS specimen it was necessary to provide comparisons, so most series of tests were run using both ENF and ELS specimens. In the early round robins ELS specimens gave values which were consistently about 20% higher than those measured by ENF specimens, for materials of very different toughness [16]. The comparisons are between unstable ENF values and stable ELS values and may be attributed to the latter including an R-curve contribution. A similar difference was noted for glass reinforced materials [18]. Another comparison exercise was performed more recently on a brittle carbon/epoxy composite, co-ordinated by ESIS within the VAMAS (Versailles Agreement on Materials and Standards) framework. In this case there were small differences which depended strongly on the starter defect considered [20]. However the stable propagation values measured by ELS, SENF and 4ENF were very similar. An important point to consider in preparing the ELS test is the horizontal load component which may be introduced if a suitable slider mechanism is not used. The protocol (see Appendix) shows suitable designs.

Starter defect influence

The importance of the starter defect is primordial in mode II tests for two reasons. First, as most of the published mode II data has been measured using the ENF specimen with an a/L ratio of 0.5, these values correspond to an unstable crack jump. There is no way to check after the test whether this jump was caused by a resin rich region causing blunting or by a sticking insert. This problem has been overcome by the development of stable mode II tests. The second problem is that starter films tend to give higher values than precracks, as shown in Figure 2 for brittle epoxy and tough thermoplastic materials.

This is the inverse situation to mode I, where fibre bridging causes higher values from precracks, and makes it very difficult to establish a coherent approach for mode I and mode II. The way round this problem has been to multiply the number of tests. A first test from the insert is followed by a second from the mode II precrack created. A third test may also be needed, from a mode I precrack as some results suggest mode I precracks give lower values than mode II precracks [24]. An additional difficulty with mode II precracks is that it is hard to determine their exact length. Mode I precracks are usually easy to distinguish on fracture surfaces, but mode II precrack lengths are usually determined by a compliance calibration performed before the test. The ESIS protocol at the end of this paper includes tests from the insert, and from mode I and mode II precracks. Some examples of results from a recent round robin are given in Table 2 below. The mode I and mixed mode values measured on the same material can be found in 'Mode I Delamination' and 'Delamination Fracture of Continuous Fibre Composites: Mixed Mode Fracture'.

(a) Carbon/epoxy

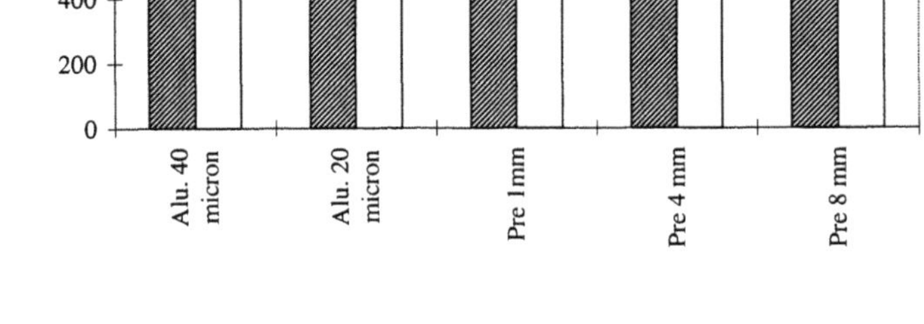

(b) Carbon/PEEK

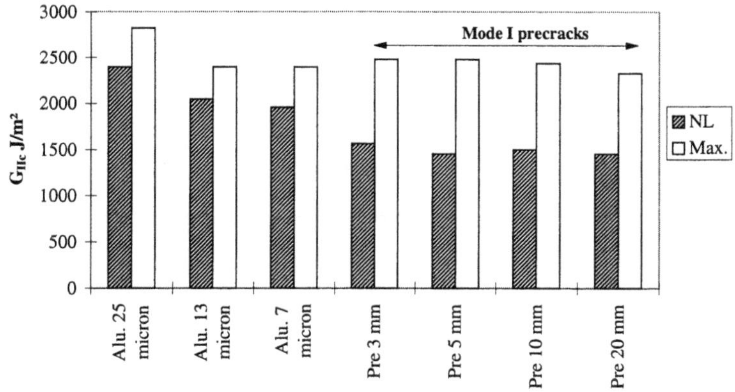

Figure 2. Mode II (ENF) initiation, influence of starter defect type [22,23]

Group	G_{IIc} J/m² NL	G_{IIc} J/m² VIS	G_{IIc} J/m² max/5%	G_{IIc} J/m² Propagation	Defect type
1	-	2188 ±185	2121 ±185	Unstable	Insert
	784 ±46	948 ±119	1029 ±159	1180 ±192	Mode I precrack
	801 ±157	1060 ±155	1142 ±133	1353 ±77	Mode II precrack
2	2339 ±431	2414 ±341	2419 ±304	Unstable	Insert
	977 ±182	1185 ±129	1337 ±133	1444 ±68	Mode I precrack
	1167 ±112	1308 ±1388	1388 ±91	1346 ±68	Mode II precrack
3	2553 ±652	-	2688 ±708	Unstable	Insert
	1205 ±128	-	1397 ±167	1488 ±285	Mode I precrack
Mean	**2458**	**2301**	**2464**	**Unstable**	**Insert**
Std. dev^n.	501	285	474		
C.V.	20%	12%	19%		
Mean	**988**	**1067**	**1255**	**1371**	**Mode I precrack**
Std. dev^n.	211	127	173	168	
C.V.	21%	12%	14%	12%	
Mean	**984**	**1184**	**1265**	**1349**	**Mode II precrack**
Std. dev^n.	232	177	169	69	
C.V.	24%	15%	13%	5%	

Table 2. Values of G_{IIc} measured using ELS specimens, from different types of starter defect, IM7/977. Note unstable initiation from insert. ± one standard deviations in brackets. Corrected beam theory analysis, E_f = 136 GPa.

Stability

The crack propagation stability can be determined by differentiation of the strain energy release rate G with respect to crack length a at constant displacement. When dG/da is negative crack growth will be stable. Stability ranges of the different test specimens are shown in Table 3 below. It is possible to obtain stable propagation with the ENF specimen but the crack length has to be so long initially that very little valid crack advance is possible before the crack meets the compression zone under the central load point.

Specimen	Stability	Usual stable propagation	Reference
ENF	a/L > 0.7	0	25
ELS	a/L > 0.55	30 mm	8
SENF	always stable	20 mm	9
4ENF	always stable	35 mm	10
CNF	a/L > 0.87	0	11

Table 3. Crack propagation stability ranges of common mode II specimens

Data Analysis

For both the mode II tests used by ESIS the data analyses have been thoroughly examined. In general two approaches are used to obtain the compliance dependence on crack length necessary for the Irwin-Kies equation $G_c = \frac{P^2}{2B}\frac{dC}{da}$. These are beam theory and experimental compliance calibration. The beam theory expressions may be based either on critical load and displacement measurements or, when the modulus value is known, only load measurements may be required. Generally the crack length dependence of compliance in mode II is weak. Also, extensive micro-cracking frequently precedes macroscopic crack advance, making it difficult to pinpoint the crack tip. For these reasons beam theory methods have been preferred, and a considerable amount of finite element work has been carried out to check the beam theory solutions. The results initially diverged for the ENF specimen, but there now appears to be reasonably good agreement with modified beam theory expressions for ENF and ELS specimens following the introduction of a number of corrections [2,26,27]. Figure 3 shows an example of this correlation. The beam theory values of G are slightly higher than experimental compliance (about 10% on average), but the former depend very strongly on the value of flexural modulus used. Correlation coefficients for the experimental compliance calibrations were generally very good (above 0.99).

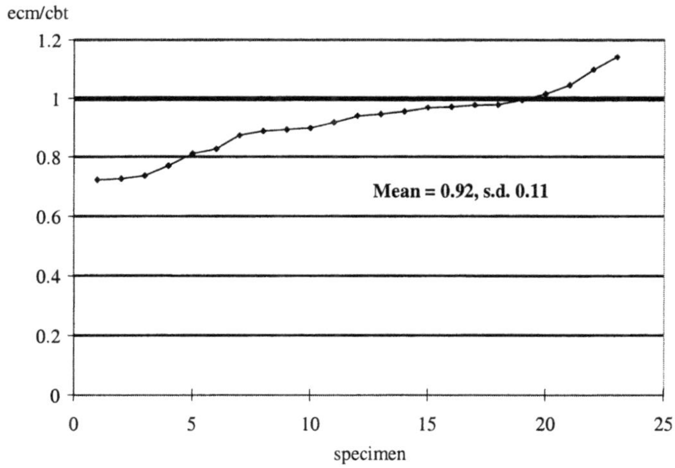

Figure 3. Example of correlation between corrected beam theory (cbt) and compliance calibration (ecm) G_{IIc} values for twenty-three ELS specimens, results from round robin on IM7/977. Data sorted in increasing order.

The latest version of the ELS protocol, Appendix 1, includes the two methods, modified beam theory analysis using modulus and load values, and an experimental compliance method. The specimen is dimensioned so that crack propagation is stable and can be followed visually to produce the compliance calibration. An option was introduced in early versions of the protocol, in which the specimen was inverted in the fixture, so the cracked end was clamped (this was called 'inverse ELS'), and then subjected to a low load. This allowed an uncracked modulus to be determined for the specimen, which could be introduced into a beam theory expression. However, this approach was not pursued as it proved impossible to reproduce exactly the same clamping conditions in this test and the subsequent crack propagation when the specimen was replaced in the fixture and loaded with the uncracked end clamped. Corrections were needed, requiring several measurements at different free lengths, and if the correction was not applied, the inverse ELS- and 3-point-bending moduli on identical specimens differed by up to 20%. A standard three point flexure test is now advocated to determine flexural modulus.
It is advisable to perform both methods of data analysis, beam theory and compliance calibration, as differences between the two can give an indication of the presence of fracture mechanisms which may invalidate some of the test data. Large differences between propagation values for the two methods may reflect multiple crack propagation.

Influence of friction

It is clear that when a crack propagates by the relative sliding of two crack faces the friction between these faces can have a significant effect on the measured fracture toughness. Finite element and shear deformation beam analysis were performed by Carlsson et al. [2]. Their results, assuming that all friction forces act at the load point, indicated that the sliding friction contribution to the ENF could be estimated as:

$$g(\mu) = \frac{4}{3}\mu\frac{h}{a}$$

where $g(\mu)$ is a non-dimensional energy release rate parameter indicating the reduction in available strain energy when friction is accounted for in the analysis, h is the specimen half thickness, a is the crack length. Thus for typical geometry's this friction contribution was estimated as 2 to 5%. Experimental results on specimens from the carbon/PEEK panels used in the 1988 round robin tests, showed effects rather larger than predicted by this expression, up to 20 % [28]. Tests were run on specimens with and without a PTFE film between the crack faces, Figure 4. If the film was not used a strong apparent specimen geometry dependence was noted, while for specimens with the film inserted the results were independent of specimen geometry.

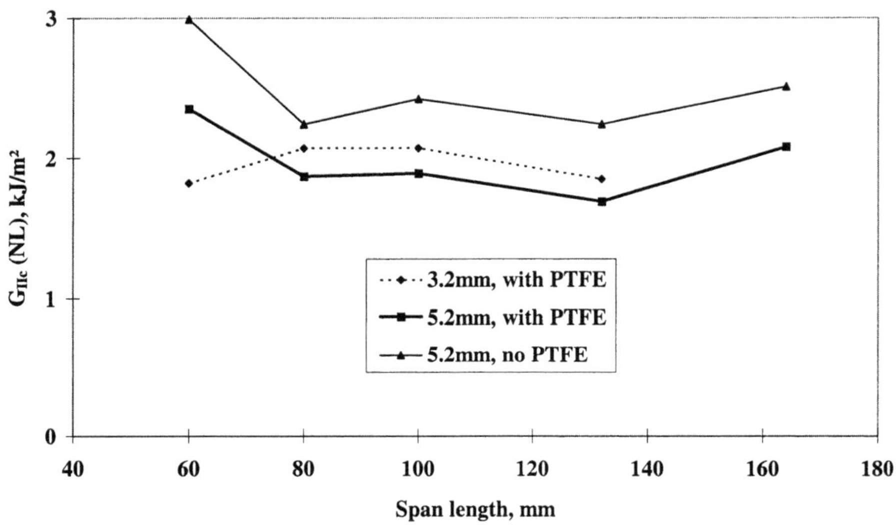

Figure 4. Friction effect, carbon/PEEK, replotted from [28].

An analysis was presented recently showing that both beam theory and experimental compliance methods will overestimate the value of G_{IIc}, and it may be possible to include a correction for friction in the test protocol [29]. Japanese work on friction effects in ENF specimens also showed the importance of spacer films [30], and more recently Kageyama and co-workers have examined friction effects in the 4ENF specimen [31].

The recommendation in the current ESIS protocol is to include either a thin film of PTFE or a pencil lead between the crack faces to minimise friction.

4. CURRENT STANDARDS SITUATION

There is no European nor ASTM mode II standard at present, but the JIS group has a mode II test procedure based on the ENF specimen and an option was added to allow stabilising the test [32]. There are also aerospace industry documents based on the ENF test from a mode I precrack (e.g. AECMA, European Association of Aerospace Industries, [33]). In order to try to obtain an international consensus on which of the three tests then being used might be proposed to ISO (International Standards Organisation) as a New Work Item, it was decided at the ISO meeting in London in 1995 that a series of tests would be performed on the same material by participants of ASTM, ESIS and the JIS. The three specimen configurations proposed at the time, (ENF, SENF and ELS) would be used. A set of mutually-agreed test procedures was prepared [34] and over 160 carbon/epoxy specimens were distributed and tested. A second series of tests was conducted later on a further 60 carbon and glass/epoxy 4ENF specimens, as this test had been proposed in the meantime. This second series was completed in the summer of 1999. The details of these tests are available elsewhere [20,21] but the three stable configurations produced similar R-curves. In February 2000 ASTM proposed the latest 4ENF test procedure for sub-committee ballot.

5. APPLICATION TO MULTI-DIRECTIONAL COMPOSITES

Virtually all the standards development work has been performed on unidirectional specimens but there has been increasing interest in recent years in the application of these tests to multi-directional laminates. For example, Davidson et al have performed theoretical studies of the influence of stacking sequence on ENF behaviour [35], using plate theory to optimise specimen design, while Ozdil et al have presented theory and experimental results for ENF tests on angle ply glass reinforced composites [36]. For the ELS specimen, Choi et al tested carbon/epoxy laminates with 45° plies at the mid-plane and noted high initiation values and R-curve effects which had not been seen in unidirectional specimens [37]. While crack propagation paths will clearly be dependent on stacking sequence, the extension of the protocol based on beam theory equations to measure initiation values of G_{IIc} does not appear to present any major difficulties.

6. CONCLUDING REMARKS

The last ten years have seen an important research effort being applied to the development of stable mode II tests. Two new methods have been proposed, using the SENF and 4ENF specimens, and a third, the ELS, has now been extensively tested. The comparison of the results from these three stable propagation specimens on a fairly brittle unidirectional carbon/epoxy composite suggests that the tests give similar values when tests are run from the same type of starter defect. This is very promising and indicates that material properties are being measured, in spite of reserves which have been voiced over the validity of mode II testing. Further experience with a range of materials is now necessary to evaluate the range of application of the test procedure.

7. REFERENCES

[1]. O'Brien TK, NASA Tech. Memo. 110280, February 1987.
[2]. Carlsson LA, Gillespie JW Jr., Chapter 4 in 'Application of Fracture Mechanics to Composite Materials', ed Friedrich K, 1989, Elsevier Science Publishers.
[3]. Giare GS, Campbell D, Eng. Fract. Mech., 27, 1987, p683.
[4]. Sidey GR, Bradshaw FJ, Proc. 1st Int. conf. on Carbon Fibres, Plastics & Rubber Inst., London 1971, paper 25.
[5]. Lakshminarayana HV, J. Comp. Mats, 18, 1984, p227.
[6]. Barrett JD, Foschi RO, Eng. Fract. Mech., 1977, 9, p371.
[7]. Russell AJ, Street KN, Proc. ICCM4, Tokyo, 1982 p279.
[8]. Vanderkley PS, MSc thesis, Texas A&M University, December 1981.
[9]. Kageyama K, Kikuchi M, Yanagisawa N, ASTM STP 1110, 1991, p210
[10]. Martin RH, Davidson BS, Plastics Rubber & Composites, 28, 8, 1999 to appear.
[11]. Maikuma H, Gillespie JW Jr., Whitney JM, J. Comp. Mats, 23, August 1989, p756.

[12]. Prel YJ, Davies P, Benzeggagh ML, de Charentenay FX, ASTM STP 1012, 1989, p251
[13]. Maikuma H, Gillespie JW Jr, Wilkins DJ, J. Comp. Mats., 24, Feb. 1990, p124.
[14] Roulin-Moloney AC, Davies P, Proc ECF7, Budapest 1988, EMAS publishers, p416.
[15]. Davies P, Moore DR, Comp. Sci. & Tech., 38, 1990, p211.
[16]. Davies P, Kausch HH, Williams JG et al. Comp. Sci & Tech., 43, 1992, p129.
[17]. Davies P, Proc. ECCM-CTS1, September 1992, p405.
[18]. Brunner AJ, Tanner S, Davies P, Wittich H, Proc. ECCM CTS2, Hamburg 1994, Woodhead Publishing, p523.
[19]. Davies P, Ducept F, Brunner AJ, Blackman BRK, de Morais AB, Proc ECCM-CTS3, Woodhead Publishing, 1996, p9.
[20]. Davies P, Sims GD, Blackman BRK, Brunner AJ, Kageyama K, Hojo M, Tanaka K, Murri G, Rousseau C, Gieseke B, Martin RH, Plastics, Rubber & Composites, 1999, Vol 28, 9 to appear.
[21] Davies P, Summary of results from second VAMAS mode II round robin exercise using the 4ENF specimen, IFREMER internal report reference TMSI/RED/MS 99.82, July 1999.
[22] Davies P, Cantwell WJ, Kausch HH, J. Materials Sci. Letters, 9, 1990, p1349.
[23] Davies P, Moulin C, Kausch HH, Fischer M, Comp. Sci & Tech., 39, 1990, p193.
[24] Turmel DJP, Szpicak JA, Singh S, Partridge IK, Proc. 3rd Int. Conf. on Deforation & Fracture of Composites, March 1995, Inst. of Materials.
[25] Carlsson LA, Gillespie JW Jr, Pipes RB, J. Comp. Mats., 20, 1986, p594
[26] Williams JG, Int. Journal of Fracture, 36, 1988, p101
[27] Wang Y, Williams JG, Comp. Sci & Tech., 1992, 43, p251.
[28] Davies P, J. Thermoplastic Composites, 10, July 1997, p353
[29] Blackman BRK, Williams JG, Proc. ECF12, Sheffield, Sept. 1998.
[30] Tanaka K, Kageyama K, Hojo M, Composites, 26, 4, 1995 p257.
[31] Kageyama K, Kimpara I, Suzuki T, Ohsawa I, Kanai M, Tsuno H, Proc ICCM12, July 1999.
[32] JIS 7086, Testing methods for interlaminar fracture toughness of carbon fiber reinforced plastics, 1993.
[33] AECMA /C7, Determination of the interlaminar fracture toughness energy Mode II-G_{IIc}, 12/1995, prEN 6034.
[34] IFREMER Mode II test procedures for a VAMAS International round robin, Report DITI-GO-MM 96-11, April 1996, available from P. Davies.
[35] Davidson BD, Kruger R, Konig M, J. Comp. Mats, 29, 16, 1995, p2108.
[36] Ozdil F, Carlsson LA, Davies P, Comp. Sci & Tech., 58, 1998, p1929
[37] Choi NS, Kinloch AJ, Williams JG, J. Comp. Mats., 33, 1, 1999 p73.

Appendix 1. Mode II ELS protocol

Version 99-12-03

Determination of the Mode II Delamination Resistance of Unidirectional Fiber-Reinforced Polymer Laminates Using the End Loaded Split Specimen (ELS)

Détermination de la résistance au délaminage en mode II, éprouvette ELS (End Loaded Split), de matériaux composites à matrice polymère renforcés de fibres unidirectionelles

Descriptors: delamination resistance, determination, end loaded split, laminate, Mode II, polymer-matrix, energy release rate, test result sheet, unidirectional fiber-reinforced

1 Scope

This standard specifies a method for the determination of the delamination resistance of unidirectional fiber-reinforced polymer laminates under Mode II shear load using the End Loaded Split specimen (ELS). The resistance to the initiation and propagation of a delamination is to be determined from a non-adhesive insert and from a Mode I (opening) or a Mode II (shear) precrack. The critical energy release rate for Mode II loading can be calculated and a resistance-curve (R-curve, i.e. a plot of the critical energy release rate versus delamination length) be determined.

The method is applicable to unidirectional carbon-fiber and glass-fiber reinforced laminates. The scope is not necessarily limited to these fibers and lay-ups, but for laminates with other types of fibers or lay-ups, no recommendations for specimen dimensions and fiber volume content are given. The procedure can be used as a guideline for testing materials that do not strictly satisfy the requirements, provided that (a) the data can be validated using an independent method, or (b) the results are considered to be order of magnitude estimates only and are quoted as such with the property or properties outside specification clearly indicated, or (c) the procedure is used for the purpose of comparison between nominally equivalent materials only.

2 Normative References

The following standard contains provisions which through reference in this text constitute provisions of this standard. At the time of publication the editions indicated were valid. All standards are subject to revision, and parties to agreements based on this standard are encouraged to investigate the possibility of applying the most recent editions of the standards listed below. Members of IEC and ISO maintain registers of currently valid International Standards.

ISO 291:1997	*Plastics; standard atmospheres for conditioning and testing*
ISO 4588:1995	*Adhesives; preparation of metal surfaces for adhesive bonding*
ISO 5893:1993	*Rubber and plastics test equipment; tensile, flexural and compression types(constant rate of traverse); description*
ISO 14125:1998	*Fibre-reinforced plastic composites - Determination of flexural properties*

3 Definitions

For the purpose of this standard the following symbols and conventions apply

a delamination length, distance between the load-line (intersection of the plane through the pin-hole center of the load-block normal to the specimen width and the plane of delamination) and the tip of the precrack or delamination on the edge of the specimen (Figure 1)

A starter delamination (insert) length, distance between end of specimen on which the load-block is mounted and tip of the insert (Figure 1)

B width of the specimen

C compliance δ/P of the specimen

C_0 initial compliance of the specimen neglecting start-up effects, e.g. due to play in the specimen fixture

C_{max} compliance of the specimen at maximum load

$C_{5\%}$ initial compliance C_0 of the specimen increased by 5%

d_{max} maximum horizontal displacement of the clamping arrangement and of the load-point, respectively

δ displacement of the cross-head of the testing machine

Δ_I delamination length correction (determined from Mode I test)

Δ_{II} correction for rotation at the delamination tip calculated as $\Delta_{II} = 0.42\ \Delta_I$

E elastic modulus determined from "three-point bending" flexural test

G_{IIC} critical energy release rate for Mode II shear load

$2h$ total thickness of the specimen (thickness of each specimen beam is h)

H thickness of the load-block

l total length of the specimen

l_1 distance from the center of the loading pin to the midplane of the specimen beam to which the load-block is attached (Figure 1)

l_2 distance between the center of the pin-hole of the load-block and its edge, measured towards the tip of the insert (starter film) or the tip of the Mode I or Mode II precrack (Figure 1)

l_3 total length of the load-block (Figure 1)

L free length of the specimen between load-line and clamp

m slope of a plot of C versus a^3

MAX maximum load on the load-displacement curve (Figure 2)

NL onset of non-linearity on the load-displacement curve (Figure 2)

P load measured by the load-cell of the testing machine

PROP increments of the delamination length during stable delamination growth (propagation) that are marked on the load-displacement curve (Figure 2)

r^2 correlation coefficient of linear fit

r.h. relative humidity during test

t_c duration of curing

t_d duration of conditioning (drying)

T test temperature

T_{mc} maximum cure temperature

T_d conditioning (drying) temperature

θ_1, θ_2 correction factors used in expression for large displacement and load block corrections

VIS onset of visually recognisable delamination growth on the edge of the specimen that is marked on the load-displacement curve (Figure 2)

5% point of intersection of a straight line with the load-displacement curve, with the slope of the straight line corresponding to $C_{5\%}$

4 Principle

This standard is using the End Loaded Split (ELS) specimen shown in Figure 1 for the determination of the delamination resistance of unidirectional fiber-reinforced laminates. Shear loads (Mode II) are applied through a load-block under displacement control at a constant rate. Stable delamination growth (propagation) from a non-adhesive insert (starter film) and from a Mode I or a Mode II precrack all at the laminate midplane is monitored, and delamination initiation and propagation readings (both from insert and precrack) are recorded on the load-displacement curves. Data reduction yields the critical energy release rates G_{IIC} for initiation and propagation of a Mode II delamination that are presented in the form of R-curves (critical energy release rate G_{IIC} versus delamination length a). The advantage of the geometry shown in Figure 1 is that crack propagation is stable for ratios of crack length a to free length L > 0.55.

It has to be noted that using a Mode I or a Mode II precrack for starting the delamination may yield values for the critical energy release rate G_{IIC} differing from those obtained from the insert. Therefore, in order to determine conservative values, both approaches (insert and precrack) have to be used. It has also to be noted that different methods for precracking (in Mode I or Mode II) and precracks obtained from different modes (Mode I or Mode II) may not yield identical critical energy release rates. Using other procedures for precracking in Mode I or Mode II than described in this standard is not recommended but may be performed for the purpose of comparison. Those procedures should comply with the prescriptions of the standards or test protocols for the respective Modes and be described and documented in the report.

5 Apparatus

A tensile testing machine in compliance with ISO 5893, capable of producing a constant load-rate between 1 and 5 mm/min in displacement control should be used. The load-cell should be calibrated and accurate within ± 1% for the chosen load-range (loads are typically expected to be in the range of 100 - 200 N). The testing machine shall be equipped with a fixture to introduce the load to the pin inserted into the load-block that allows rotation of the specimen end. The recommended loading jig requires either a clamping arrangement to freely slide in bearings in the horizontal direction (side-ways) with a fixed load point (Figure 1b) or a fixed clamping arrangement with a loading fixture that allows free horizontal movement (side-ways) of the load-point (Figure 1c). The load shall be applied vertically on the load-block, either pushing downward, if the load-block is on the top side of the specimen or pulling upward, if the load-block is on the bottom side of the specimen, provided the clamp is symmetrical with respect to the specimen. The testing machine shall be equipped with means for recording the complete load-displacement curves (loading and unloading) that allow a determination of the loads and the corresponding displacements with an accuracy of ± 1 %.

6 Specimens

6.1 Preparation of Specimens

The recommended specimen width B and length l are 20 mm and 170 mm, respectively. The specimen length shall not be less than the length of the insert or of the starter delamination plus 110 mm. The free length L is typically 100 mm, with initial crack length 60 mm so the ratio a/L is 0.6. The recommended specimen thickness is 3 mm for 60% by volume carbon fiber-reinforced and 5 mm for 60% by volume glass fiber-reinforced composites. Other specimen dimensions may be used, but the specimen width should be between 15 and 30 mm. Increasing the length of the specimen is not critical, shortening will reduce the maximum delamination length that can be investigated and thus yield too few data points for the analysis (see clause 8.4). If specimens are too thin or not sufficiently stiff, delamination growth may not be induced or occur at large displacements only.

Three types of initial defect (starter defect) are considered, (a) a laminated starter film (insert), (b) a Mode I precrack, obtained either by a Mode I test or by wedge opening, and (c) a Mode II precrack, obtained by a Mode II test. At least two types of initial defects have to be used in the tests.

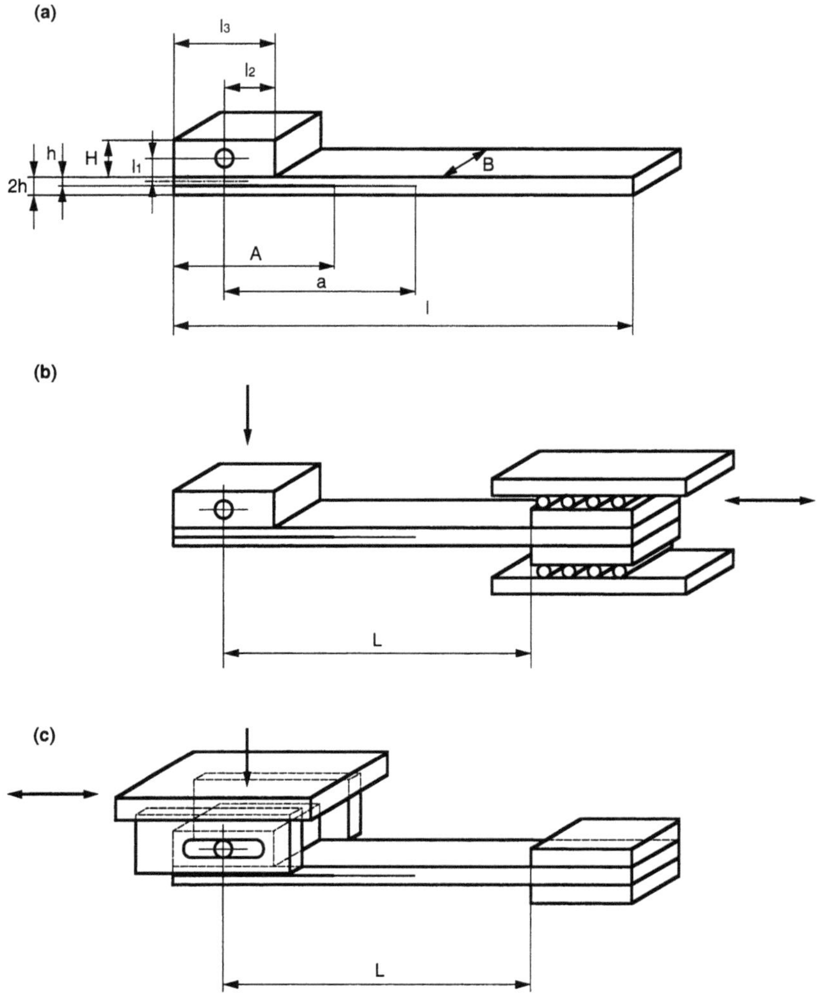

Figure 1: Geometry for the End Loaded Split (ELS) specimen with one load-block. The fiber orientation is parallel to the length l. The delamination length a is the distance between the load-line (intersection of the plane through the pin-hole center of the load-block normal to the specimen width B and the plane of delamination) and the tip of the delamination. a) Specimen with load-block, b) clamping arrangement free to slide with fixed load point, and c) fixed clamping arrangement with load-point free to slide. In both clamping arrangements it is possible to push the load-block downwards, if it is above the specimen, or to pull it upwards, if it is below the specimen, provided the clamp is symmetrical with respect to the specimen.

If a starter film is used, an insert (starter film) should be placed at laminate mid-thickness during molding. This film should be as thin as possible to minimise the disturbance of the laminate. Less than 15 μm thickness is recommended and the starter film should be coated with a release agent. Starter film length should be at least 50 mm from the load-line so that the influence of the load-block can be neglected.

For specimens with an insert (starter film) the test procedure for testing Mode I Double Cantilever Beam (DCB) specimens or wedge opening can be used to prepare the Mode I precrack. If DCB-specimens are used, the second load-block has to be removed before testing from the precrack. For wedge opening, the specimens shall be clamped at 5 mm beyond the tip of the insert. If material or specimens without insert have to be tested, precracking by wedge opening is the only choice for starting the delamination. In this case, the specimen should be clamped at most 60 mm from the end on which the precrack should be formed. Clamping the specimen at shorter distances and repeating the wedging procedure, after moving the clamp, is allowed. The (final) precrack should extend at least 50 mm beyond the load-line so that the influence of the load-block can be neglected. The (final) precrack should, however, be short enough to allow a delamination length increment of at least 30 mm beyond the tip of the precrack, before the delamination arrives within 10 mm of the clamped end. The width of the wedge shall be at least the same as that of the specimen and the opening angle shall be as small as possible. Notching the specimen edge with a razor blade or a diamond saw will provide a firm hold for the wedge. The wedge is driven into the specimen until the tip of the wedge reaches the clamp. The wedge may be driven by hand (tapping on the side) or by using a suitable fixture and a testing machine. Experience has shown that it may be difficult to produce a suitable precrack by wedge opening, frequently the precrack will not lie in the midplane of the specimen. Deviations of the precrack from the midplane will invalidate the test results and should be noted in the report.

The Mode II shear precrack shall be prepared in accordance with clause 8.3 but loading be stopped as soon as the delamination is seen to move and the specimen then be completely unloaded. Experience has shown that Mode II delamination growth from an insert may be unstable and, if observed, this should be noted in the report. In that case, preparing the Mode II precrack from a short Mode I precrack is permitted. The position of the tip of the delamination after precracking should be marked on both edges of the specimen. If the delamination lengths a on the edges of the specimen, i.e. the distance between load-line and the tip of the delamination, differ by more than 2 mm the results should be considered suspect and this be noted in the report.

If specimens are cut from a plate, the location of each specimen on the original plate should be recorded and specimens should each be identifiable. Measure and record the length l of the specimens to the nearest mm. Measure the width B and the thickness 2h of each specimen to the nearest 0.02 mm, at five points (quarter, center, and three-quarter length and 10 mm from each end) along the specimen. The variation in thickness and average values of the width and the thickness should be recorded for each specimen. The variation in thickness, i.e. the maximum difference between the thickness measurements, should not exceed 0.1 mm for each specimen. Measure the starter delamination length, i.e. the total length of the insert (starter film) or of the Mode I or Mode II precrack on both edges of the specimen. The average value should be recorded but if the insert length measurements differ by more than 1 mm the results should be considered suspect and this be noted in the report.

One load-block (Figure 1) is used as load-introduction, it should be of the same width as the specimen. The load-block and the specimen should first be lightly abraded, use of either sandpaper or grit blasting should be sufficient, as the loads required to delaminate the specimens used in these tests are quite low. The load-block and the specimen should then be cleaned with a solvent. If a bond failure occurs it may be necessary to consult ISO 4588 for a more sophisticated procedure. Bonding of the load-block should be done immediately after the surface preparation. In most cases a cyanoacrylate ("Super glue") adhesive has been found adequate for previous tests on similar specimens. Alternatively, a tough, room-temperature cure adhesive may be used. The surface preparation and the type of adhesive used should be noted in the report. The load-block should be well aligned with the specimen and held in position with a clamp while the adhesive sets. Specimen edges should be smoothed prior to determining the dimensions. Adding a thin layer of typewriter correction fluid ("white ink") on the edges after conditioning will facilitate the detection of the delamination growth. It should be noted that some typewriter correction fluids contain solvents that may be harmful to the laminate matrix material. For the measurement of the delamination length, marks should indicate every 1 mm from the tip of the insert and of the Mode I or Mode II precrack for at least the first 5 mm, then, for testing from the Mode I or Mode II precrack, marks should be applied every 5 mm, and every 1 mm should be marked after 25 mm at least up to 30 mm.

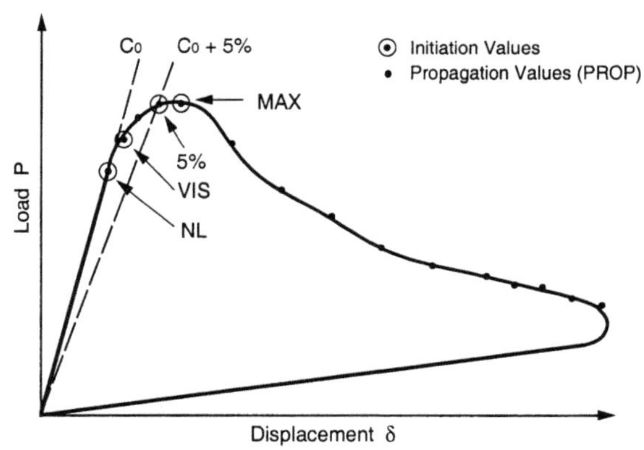

Figure 2: Schematic load-displacement curve for testing from the insert and from the Mode I or Mode II precrack with initiation points NL, VIS, 5%, MAX, and propagation points indicated.

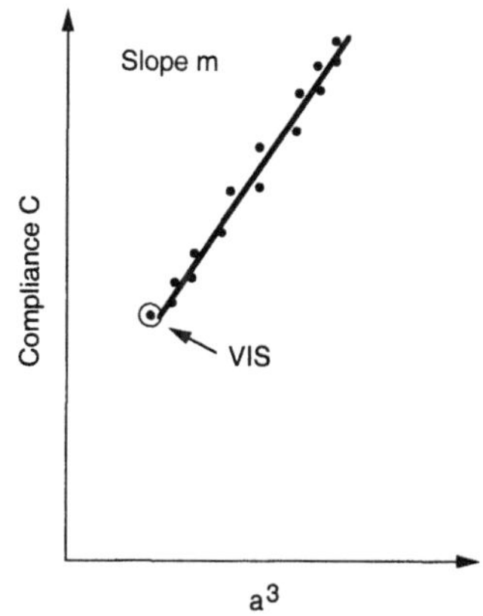

Figure 3: Linear fit used to determine the slope m for the Experimental Compliance Calibration Method.

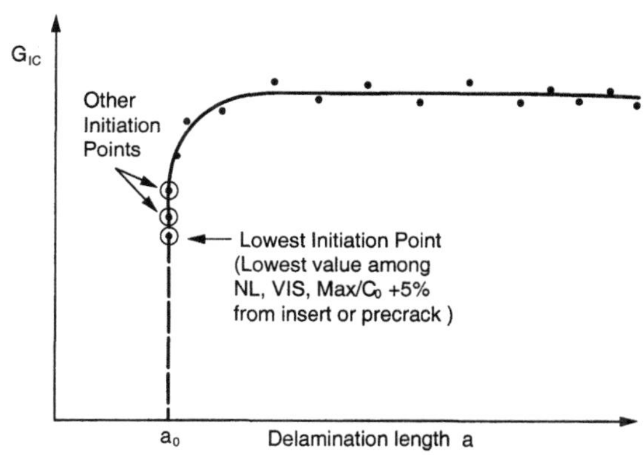

Figure 4: Schematic resistance-curve (R-curve) with G_{IC}-value for initiation (lowest value among NL, VIS, 5% or MAX) and for propagation (PROP) versus observed delamination length a.

6.2 Number of Specimens

A minimum number of five specimens each shall be tested from the insert and from the Mode I or Mode II precrack unless a smaller number is prescribed.

7 Conditioning

Moisture conditioning is required for obtaining baseline data in order to test specimens with a uniform moisture content. The drying conditions (temperature and duration) shall be chosen according to the recommendations of the resin supplier. Conditioning should be performed after bonding of the load-block. Before testing, the specimens may be stored in a dessicator for at most one day after conditioning. Other conditioning procedures may be applied for the investigation of specific conditioning effects.

8 Test Procedure

8.1 A Note on the Application

If this test procedure is used to prepare specimens with Mode II (shear) precracks as starter cracks for testing the delamination resistance in other Modes, it is recommended to consult the appropriate standards or test protocols for additional requirements on specimen characterisation and preparation.

8.2 Test Preparation

A value of the E-modulus from a three-point bending test is required if the beam theory analysis (see clause 8.4) is used to evaluate the data. The experimental compliance method does not require modulus values but the method is not applicable unless additional requirements are fulfilled (see clause 8.4). The modulus value shall be determined before delamination testing on that part of the specimen that does not contain the insert or the Mode I or Mode II precrack. The three-point bending test shall be performed and the flexural modulus E be calculated in accordance with ISO 14'125.

8.3 Test Set-up and Data Recording

The test shall be performed under normal conditions in accordance with ISO 291 (23° ± 2° C, 50% ± 5% relative humidity) unless prescribed otherwise. The load and the displacement signals of the testing machine shall be recorded, either on a paper chart or electronically throughout the test, including the unloading cycle. The reproducibility and tightness of the clamping is crucial in this test. The delamination length may be measured by eye on the specimen edge, or by using a travelling microscope. In transparent laminates the delamination length may be followed inside the specimen by marking the specimen surface rather than the edge. If unstable delamination growth followed by arrest ("stick-slip") is observed during any stage of the test, it should be noted in the report. The data evaluated according to clause 8.4 may not be valid in this case. Any permanent deformation of the specimen after unloading should be noted in the report. Deviations of the delamination from the midplane of the laminate will invalidate the test results and should be noted in the report. The maximum horizontal displacement d_{max} of the sliding fixture (clamp or load-point) shall be recorded by determining the initial (before loading) and final (before unloading) positions of the clamp or of the load-point, respectively, and be noted in the report.

Test parameters and data recording are the same for testing from the insert and testing from the Mode I or Mode II precrack. The free length is generally of the order of 100 mm, so that a reasonable delamination propagation can take place, but a shorter free length may be necessary to promote propagation in some materials.

Testing from the Insert and from the Precrack

For testing from the precrack (Mode I or Mode II), the marks on the specimen edge should be checked before testing and adjusted, if necessary, according to clause 6.1. A thin film of PTFE or a pencil lead should be inserted between the starter delamination faces to open the delamination slightly and to minimise friction between the faces. Measure the initial position of the sliding fixture (clamp or load-point) before the start of loading. For testing from the insert (starter film) and from the Mode I or Mode II precrack, the specimen should be loaded at a constant cross-head rate between 1 and 5 mm/min. For specimens with nominal length, a cross-head speed from the lower, for longer specimens from the upper end of the range is recommended. The point on the load-displacement curve at which the onset of delamination movement from the insert or the tip of the precrack is observed on the edge of the specimen should be recorded on the load-displacement curve or in the sequence of load-displacement signals (VIS, Figure 2). If the start of the delamination growth is difficult to observe, a change in illumination conditions or a cross-head speed from the lower end of the range is recommended. After this, as many delamination length increments as possible should be noted in the first 5 mm on the corresponding load-displacement curves, ideally every 1 mm. Subsequently, delamination lengths are noted every 5 mm, and again every 1 mm for the last 5 mm of delamination propagation (ideally, the total delamination length increment should be at least 30 mm). The loading should be stopped before the delamination arrives within 10 mm of the clamped end. Measure the final position of the sliding fixture before unloading in order to calculate its maximum horizontal displacement d_{max}. After this, the specimen should be unloaded at a constant cross-head rate, unloading may be performed at up to 25 mm/min. The position of the tip of the delamination should be marked on both edges of the specimen. If the delamination lengths a on the edges of the specimen, i.e. the distance between the load-line and the tip of the delamination, differ by more than 2 mm the results should be considered suspect and this be noted in the report.

8.4 Data Analysis

The data required for the analysis are delamination lengths a, and the corresponding loads P and displacements δ. Several values may be determined from the load-displacement curve, if possible, the following initiation values, shown in Figure 2, should be determined for testing from the insert (starter film) and from the Mode I or Mode II precrack for each specimen:

(1) NL, i.e. deviation from linearity: A region of non-linear behaviour usually precedes the maximum load, even if the unloading curve is linear. The point of deviation from linearity (NL in Figure 2), is determined by drawing a straight line from the origin but ignoring any initial deviations due to take-up of play in the loading system. Experience has shown that it is difficult to reproducibly determine the position of NL on the load-displacement curve. Performing a linear fit on the load-displacement curve starting at 5% of the maximum load and using a consistent criterion for deviation from linearity (e.g. the half-thickness of the plotter trace) is recommended. If non-linearity due to large displacements is observed then this value should not be used for the analysis.

(2) VIS, i.e. Visual observation: This corresponds to the onset of the delamination, i.e. to the first point at which the delamination is observed to move from the tip of the insert or of the Mode I precrack on the edge of the specimen (VIS in Figure 2).

(3) 5% or MAX, i.e. 5% increase of compliance or maximum load point: The 5% value corresponds to the point on the load-displacement curve at which the compliance has increased by 5% of its initial value C_0. A best straight line is drawn to determine the initial compliance

C_0, ignoring any initial deviation due to take-up of play in the loading system. A new line is then drawn with a compliance equal to C_0 +5% whose intersection with the load-displacement curve yields the load and displacement to be used for the calculation, unless the intersection is at a larger displacement than the maximum load in which case the maximum load and the corresponding displacement have to be used.

Besides the initiation points (NL, VIS, 5% or MAX), propagation values (PROP in Figure 2) can be determined for each delamination length measured during propagation from the insert (except when preparing a Mode II precrack) and from the Mode I or Mode II precrack. A separate test result sheet shall be used for the values determined from the insert (starter film) and those from the Mode I or Mode II precrack. Either one of the two methods described below can be used for the analysis, the method chosen should be noted in the report.

Method (1): Corrected Beam Theory (CBT)

The simple beam theory is an approximation and a correction for the rotation at the delamination tip has to be included. There are two prerequisites if this method is to be used. First, it has been shown that multiplying the value Δ_I obtained from Mode I tests by 0.42 gives a good approximation to the value of Δ_{II} corresponding to Mode II loading of the ELS specimen (reference 1). If Mode I tests on the same material have not been performed the Δ_I value should be set to $\Delta_I = 0$ and this be noted in the report. Second, a value of E from a three-point bending test before the delamination test is necessary for calculating the beam theory value. The critical energy release rate G_{IIC} is given by

$$G_{IIC} = \frac{9P^2(a + \Delta_{II})^2}{4B^2 E h^3} \tag{1}$$

with P the load, a the delamination length, Δ_{II} the correction (delamination tip rotation) determined from Mode I tests using the Double Cantilever Beam specimen, B the width of the specimen, the value of E the modulus parallel to the fiber direction, and h the half-thickness of the specimen (total specimen thickness 2h). All initiation and propagation values, if applicable, should be calculated, the delamination length for the initiation values is the distance between the load-line and the tip of the insert or precrack (Figure 1). Large displacement and load-block corrections have to be applied to equation (3) as follows

$$G_{IIC(corrected)} = G_{IIC(CBT)}\left[1 - \theta_1\left(\delta/L\right)^2 - \theta_2\left(\delta\, l_1/L^2\right)\right] \tag{2}$$

with δ the displacement, l_1 the distance from the center of the load-block to the midplane of the specimen beam to which the load-block is attached (Figure 1), and L the free length of the specimen. The correction factors θ_1 and θ_2 are calculated as follows:

$$\theta_1 = \frac{3}{20} \frac{\left[15 + 50\left(a/L\right)^2 + 63\left(a/L\right)^4\right]}{\left[1 + 3\left(a/L\right)^3\right]^2} \qquad (3)$$

$$\theta_2 = -3\left(L/a\right) \frac{\left[1 + 3\left(a/L\right)^2\right]}{\left[1 + 3\left(a/L\right)^3\right]} \qquad (4)$$

with a the delamination length and L the free length of the specimen.

Method (2): Experimental Compliance Method (ECM)

If the loading and unloading curves (load-displacement curve) are both linear, an alternative approach is to plot the compliance C versus the cube of the delamination length a^3. Only the VIS and the PROP values are used for the linear fits, but not the NL or 5%/MAX values. If it has not been determined or is considered questionable, the VIS point may be excluded from the linear fit, but this should be noted in the report. The slope of this plot, m, can be used to give G_{IIC} as follows

$$G_{IIC} = \frac{3P^2 m a^2}{2B} \qquad (5)$$

with P the load, m the slope of the plot of the compliance C versus the cube of the delamination length a^3, a the delamination length, and B the width of the specimen. All initiation and propagation values, if applicable, should be calculated, the delamination length for the initiation values is the distance between the load-line and the tip of the insert or precrack (Figure 1). The same large-displacement and load-block corrections θ_1 and θ_2, respectively, are used as for the corrected beam theory method (see above)

$$G_{IIC(corrected)} = G_{IIC(ECM)}\left[1 - \theta_1\left(\delta/L\right)^2 - \theta_2\left(\delta\, l_1/L^2\right)\right] \qquad (6)$$

The results from testing from the insert and from the Mode I or Mode II precrack are separately used to draw a resistance-curve (R-curve), i.e. G_{IIC} versus delamination length a (Figure 4). For each type of initiation value (NL, VIS, MAX/5%) the arithmetic average and the standard deviation of all specimens tested shall be calculated. The minimum number of propagation points recorded for each specimen should be 15, if fewer points are used, this should be noted in the report and the results considered suspect (Reference 2).

If more than one specimen of a material is tested, the results shall be averaged as follows to yield characteristic material values: Calculate the arithmetic average and standard deviation of each type of initiation value (VIS, NL, MAX, and 5%) separately, then calculate the arithmetic average and the standard deviation of the last ten PROP values or of the last 50% of all PROP values, whichever is larger. The average values and standard deviations should be noted in the report. If the calculated standard deviation exceeds 10% of the average value, a constant plateau value for the propagation may not have been reached and the R-curve plots should be checked. If the R-curve plots do not show a plateau, the average PROP value should be considered suspect and this be noted in the report.

9 Test Report

The test report shall include the following information:

(a) a reference to this test protocol and to the referring standards
(b) a complete identification of the material (e.g. laminate manufacturer, fiber-material, polymer material, maximum cure temperature T_{mc}, duration of curing t_c, location of specimen on plate)
(c) test date, test laboratory, test personnel identification
(d) number and label of specimens tested and type of method used for the analysis
(e) average thickness, average width, maximum thickness variation along the length, and length of each specimen, insert (starter film) material and thickness, length of the insert; note if insert lengths measurements differ by more than 1 mm on both edges
(f) conditioning temperature T_d and conditioning duration t_d and temperature T and relative humidity r.h. during the test
(g) dimensions of the load-block, surface preparation, if applicable, and adhesive
(h) type of precracking used (e.g. Mode I test or wedge opening) and, if applicable, whether the specimen has been removed from the fixture after precracking
(i) load-rate for loading and unloading, for testing from the insert and from the Mode I or Mode II precrack
(j) the maximum horizontal displacement d_{max} of the sliding fixture (clamp or load-point)
(k) length of delamination after unloading for testing from the insert and from the Mode I or Mode II precrack; note, if delamination lengths measurements differ by more than 2 mm on both edges
(l) Δ_{II}, i.e. the correction for delamination tip rotation obtained from the delamination length correction Δ_I for Mode I tests using Double Cantilever Beam (DCB) specimens
(m) E-modulus from "three-point bending" test, if method (1) is used for the data analysis (see clause 8.2)
(n) slope m of plot of the compliance C versus the cube of the delamination length a^3, if method (2) is used for the data analysis (see clause 8.2), and the correlation coefficient r^2 of the linear fit. Note, if the VIS-value has been excluded from the fit.
(o) $(C_{max/5\%} - C_0)/C_0$, i.e. the percent change in compliance between the initial compliance C_0 and the compliance at the MAX or 5% point, whichever is applicable
(p) copy of the load-displacement curve for each specimen
(q) table of G_{IIC} (all initiation and propagation values) and plot of G_{IIC} (all initiation and propagation values) versus delamination length a (R-curve) for each specimen including large displacement and load-block corrections

(r) average values and standard deviation for each type of initiation value (VIS, NL, MAX, and 5%) and average value and standard deviation of the last 10 propagation values (PROP) or of the last 50% of the propagation values, whichever contains the larger number of data points, from all specimens tested. Note, if less than 15 propagation values have been recorded. If a specimen is excluded from averaging, the reason for this should be noted in the report.
(s) observations from testing (e.g. deviation of the precrack or the delamination from the midplane, stick-slip, occurrence of fiber-bridging, permanent deformation after unloading, sticking of insert foil, no plateau in the R-curve) that may have affected the test procedure or the results
(t) any deviation from the prescriptions of this protocol (e.g., dimensions of specimens, fiber orientation)
(u) results from additional specimen or material characterisation (e.g. fiber volume fraction, void content), if specified

A recommended test result sheet is shown in Figure 5.

10 References

[1] Y. Wang, J.G. Williams: "Corrections for Mode II Fracture Toughness Specimens of Composites Materials", Composite Science and Technology 43, 251-256 (1992).

[2] A.J. Brunner, S. Tanner, P. Davies, H. Wittich: "Interlaminar Fracture Testing of Unidirectional Fibre-Reinforced Composites: Results from ESIS-Round Robins" in: Composites Testing and Standardisation ECCM-CTS 2, (P.J. Hogg, K. Schulte, H. Wittich eds.), Woodhead Publishing, 523-532 (1994).

Recommended Test Result Sheet ELS Mode II, Page 1 of 2 — Fiber : — Resin :

Specimen label	....	# of Specimens			
Laboratory	....	Test personnel	....	Test date	....
Laminate manufacturer	....	Resin manufacturer	....	Fiber manufacturer	....

Free length	L_F [mm]	....	Load introduction		....
Average thickness	2h [mm]	....	Block thickness	H [mm]	....
Max. thickness variation	Δ 2h [mm]	....	Block length	l_3 [mm]	....
Specimen width	B [mm]	....	Block/hinge width	[mm]	....
Insert material	....	....	Surface preparation		....
Insert thickness	[μm]	....	Adhesive		....
Insert length	A [mm]	....	Conditioning temperature	T_d [°]	....
Mold release agent	....	....	Conditioning duration	t_d [h]	....
Precrack length	a_0 [mm]	....	Test temperature	T [°]	....
Max. cure temperature	T_{mc} [°]	....	Relative humidity	r.h. [%]	....
Cure duration	t_c [h]	....	Loading/unloading rate	[mm/min]	....
Max. delamination length	A_{max} [mm]	....	Distances l_1/l_2	[mm]	....
Corrections	Δ_I, Δ_{II} [mm]	....	Slope m (ECM-Method)	[-]	....
% Change compliance	[%]	....			

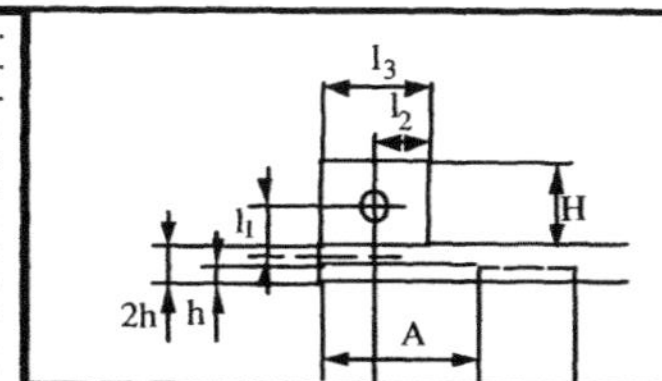

Minimum initiation value

Average NL ±σ	
CV NL	
Averge VIS ±σ	
CV VIS	
Av. MAX/5 % ±σ	
CV MAX/5 %	

Text	a [mm]	Load [N]	δ [mm]	Specimen and Insert Location on Plate	Comment	G_{IIC} CBT [J/m²]	G_{IIC} ECM [J/m²]	E-Mod. [GPa]
NL (insert)								
MAX/5 % (insert)								
VIS (insert)								
NL (precrack)				Length .... mm				
MAX/5 % (precrack)								
VIS (precrack)								
PROP								
PROP								
PROP								
PROP								
PROP								
PROP								
PROP								
PROP								
PROP								
PROP								
PROP								
PROP								
PROP								
PROP								
PROP								

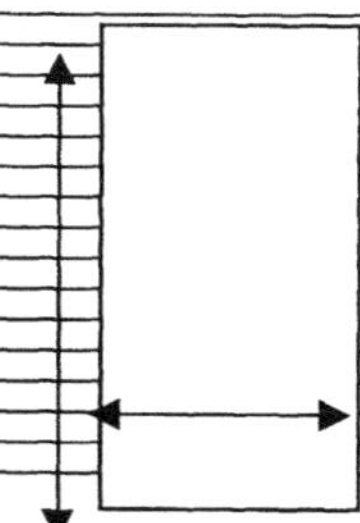

Recommended Test Result Sheet ELS Mode II, Page 2 of 2 **Fiber :** **Resin :**

Specimen label	....	# of	Specimens		
Laboratory	....	Test personnel	....	Test date	....
Laminate manufacturer	....	Resin manufacturer	....	Fiber manufacturer	....

CBT	
E_{11}	
Δ_{11} [mm]	
or	
ECM	
slope of fit	
Y-Intercept of fit	
m	
correlation coefficient	

Difference insert length ≤ 1 mm ?	yes / no
Difference precrack length ≤ 2 mm ?	yes / no
Precrack tip 3-5 mm from insert ?	yes / no
Difference total delamination ≤ 2 mm ?	yes / no
Specimen removed after precracking ?	yes / no
VIS included in linear fit ?	yes / no
Stable delamination growth through test ?	yes / no
Total delamination length increment > 40 mm ?	yes / no
a / L > 0.55 ?	yes / no
Reference to ISO 291 ?	yes / no
Reference to ISO 4588 ?	yes / no
Reference to ISO 5893 ?	yes / no

Observations and Comments

Text	θ_1	θ_2	**Corr. Factor**	**C / a^3**	**a / L**	**a / L > 0.55 ?**	**δ / L**	**δ / L < 0.2 ?**	**Checks ok ?**
NL (insert)									
MAX/5 % (insert)									
VIS (insert)									
NL (precrack)									
MAX/5 % (precrack)									
VIS (precrack)									
PROP									
PROP									
PROP									
PROP									
PROP									
PROP									
PROP									
PROP									
PROP									
PROP									
PROP									
PROP									
PROP									
PROP									
PROP									

DELAMINATION FRACTURE OF CONTINUOUS FIBRE COMPOSITES: MIXED-MODE FRACTURE

B.R.K. BLACKMAN, A.J. BRUNNER and P. DAVIES

1. INTRODUCTION

A delamination in a composite structure is unlikely to experience pure mode I or mode II loading. The loading is more likely to be a combination of modes I, II and III. Of the possible combinations, combined mode I and mode II has received the most attention and is termed mixed-mode I/II. This paper deals only with mixed-mode I/II loading.

2. HISTORY AND BACKGROUND TO THE TEST DEVELOPMENT

Over the past two decades, a number of different test specimens have been developed for mixed-mode I/II testing. A cracked lap shear (CLS) specimen was investigated by ASTM [1,2] and then abandoned. An edge delamination tension (EDT) geometry utilised coupons with special stacking sequences to induce a specific mixed-mode ratio [3,4]. The Arcan fixture has been used to vary the mixed-mode ratio [5] and ESIS has utilised the ADCB geometry initially proposed at Texas A & M [6], which is discussed in this paper. More recently, a mixed-mode bending specimen (MMB) was introduced by Reeder and Crews [7] which was later modified [8] to reduce the non-linearities which existed in the original test.

Recently, the MMB test has received a great deal of attention and has become popular with many workers in the field. The main advantage of the test is that it enables the whole mixed-mode failure envelope, from pure mode I to pure mode II, to be measured using a single apparatus. Indeed, the MMB test is currently being balloted within ASTM as a prospective mixed-mode standard and ESIS has actively participated in the two round robin test validation exercises. However, the MMB test is not without problems and so there are some advantages that the ADCB test method offers. Perhaps the main advantage is that the ADCB test uses the same apparatus and the same specimen type, DCB with one load-block, that is used for mode II ELS test, making its use along side the ELS test a very attractive option. It is convenient to conduct DCB, ELS and ADCB tests together consecutively on a laminate to enable the failure envelope to be drawn.
Table 1 lists the ESIS round robin activities on mixed-mode testing. The first experimental activity was in 1991, following the initial drafting of the ESIS ADCB protocol in June of that year. Ten laboratories participated in a round robin on a glass reinforced modar composite. Mode I, II and mixed-mode I/II tests were performed. Additional round robins were run in 1992, 1993 and 1994 using the ADCB test and revisions were made to the protocol in 1992 and 1993 based upon the round robin results. In 1994 and again in 1998, ESIS participated in the ASTM co-ordinated round robins using the MMB test. A draft standard using this test has been written and is currently under consideration as an ASTM standard. In the next section, the ADCB test protocol will be reviewed, with an emphasis on the remaining problems to be solved.

Table 1. ESIS TC4 Round robin activities on mixed-mode I/II testing

Date	Material	Labs	Remarks
1991	Glass/modar	10	ADCB tests. DCB, ELS also conducted
1992	Glass/epoxy	10	ADCB tests. DCB, ELS also conducted
1993	Glass/epoxy	17	ADCB tests. DCB, ELS also conducted
1994	Carbon/epoxy	11	ADCB tests. DCB, ELS also conducted
1994	*Carbon/epoxy*	*5*	*Part of ASTM MMB Activity* (1st RR)
1998	*Carbon/epoxy*	*4*	*Part of ASTM MMB Activity* (2nd RR)

3. REVIEW OF ADCB PROTOCOL AND DISCUSSION OF PROBLEMS

Position of insert

It is recommended that a PTFE film insert is located at the mid-thickness of the laminate during manufacture to act as a crack starter. The maximum allowable thickness of this film is 13 microns, which is consistent with the DCB [9] and the ELS [10] test protocols. Altering the position of the film to a location not at the mid-thickness has been used by some workers e.g. [11] to intentionally alter the mixed-mode ratio. With the film placed centrally, the ratio of GI/GII =4/3 and is very nearly constant and independent of crack length. The relation between the mixed-mode ratio and the film location is:

$$\frac{G_I}{G_{II}} = \frac{1}{3}\left[\frac{h_1}{h_2}\left(1+\frac{h_1}{h_2}\right)\right]^2$$

Where h_1 is the distance between the plane of the insert film and the top of the beam, and h_2 is the distance from the plane of the insert film to the bottom of the beam.

Pre-cracking

ESIS round robin activities have highlighted the importance of pre-cracking specimens prior to testing [12]. Results from the 7th ESIS round robin conducted in 1994 and presented later in this paper and in the two preceding papers on mode I and mode II testing show the importance of growing a natural crack ahead of the insert film. Work conducted by the group has shown clearly that the most conservative values of initiation toughness (i.e. the lowest) are obtained following a mode I pre-cracking procedure [13]. ESIS came to the view that testing should be based upon two approaches, i.e. testing from both the insert and from a pre-crack. The values of initiation toughness should then be compared and the lower of the insert or pre-crack quoted. The ADCB protocol requires that a pre-crack be generated prior to testing, although considerable flexibility is allowed concerning how this is performed i.e. mode I, II, I/II or wedge pre-cracking to extend the film insert are all permissible.

Stability

It has been shown [14] that stable crack growth will be obtained from the ADCB test provided that the ratio of the crack length, a, to free length, L_F, be ≥ 0.41. The relevant crack length will be that of the extended crack after pre-cracking. It has been observed however, that unstable crack growth can occur when testing directly from an insert even if the condition $(a/L_F)>0.41$ is met. This was another justification for the argument to always pre-crack the specimens before testing and this further confirmed the belief that testing should be conducted both from the insert and from a pre-crack. The free length, L_F, can be adjusted to allow for more crack growth propagation and typical values of L range from 90mm to 120mm, for standard size specimens.

Mode Partitioning

There has been considerable debate e.g. [15, 16] as to how the components of mixed-mode loading should be partitioned. The ADCB protocol uses a global partitioning scheme rather than one based upon a local singular field approach. Charalambides et al [15] argued that the global partitioning scheme was more appropriate to delamination in composites because (i) the requirement for symmetrical deformations at the crack tip was not necessary for a global partitioning scheme and (ii) the size of the damage zone in laminates implies that the singular field will not be dominant and will thus not control the failure mechanism. In the global partitioning scheme the total energy release rate G, often termed $G_{I/II}$, is partitioned into a mode I component, G_I, and a mode II component, G_{II}. Then:

$$G_{I/II} = G_I + G_{II}$$

And a mixed-mode failure criterion may be expressed as:

$$G_{I/IIC} = G_{IC}^{\text{mixed}} + G_{IIC}^{\text{mixed}}$$

In the ADCB test, i.e. for $h_1=h_2$, there is no difference in result between using the local or global partitioning schemes.

Data Analysis

Two methods of data analysis have been implemented in the ADCB mixed-mode protocol. The first uses beam theory [17] to derive an expression for dC/da and hence Gc, the second determines dC/da directly from the experimental data. The beam theory analysis requires various correction factors which were discussed in [11] and [14] and the analysis method is termed the corrected beam theory (CBT) method. An experimental compliance calibration is required for the second analysis method to enable the value of dC/da to be obtained. Such a calibration requires that both the loading and unloading curves are linear and the experimental data is curve fitted according to:

$$C = C_o + ma^3$$

where C is the beam compliance (=δ/P), a the crack length and m and C_o are constants. The relation is generally found to be linear, so linear regression of the data yields the slope, m, necessary for the determination of $G_{I/IIc}$.

Problems with the test and analysis

(a) Only a single mixed-mode ratio of G_I/G_{II} (=4/3) is obtained, and thus considerable extrapolation of the data is required when plotting the mixed-mode failure envelope (see later).
(b) The test also suffers, but to a lesser extent, with the problems associated with friction [20], [21] and potential micro-cracking ahead of the crack tip. These were discussed in the previous paper on mode II. Of course, high mode II MMB tests suffer to an even greater extent from the problems associated with friction.
(c) The use of the corrected beam theory to analyse the data requires a correction term, Δ_I, which is measured in a mode I DCB test. Therefore, this correction term is not available if DCB tests have not previously been performed.
(d) The CBT method also requires a value for the axial E modulus of the laminate. If this is not provided by the manufacturers, then the protocol requires that this value should be measured using a flexural test [18].
(e) If insufficient crack propagation is obtained or if unstable crack growth is observed, the experimental compliance method may be inaccurate.

4. REVIEW AND DISCUSSION OF ROUND-ROBIN RESULTS

The results presented here are from the 7th ESIS Round-robin conducted in 1994 on the carbon-fibre epoxy composite, IM7-977/1. Participants were provided with specimens to perform Mode I, Mode II and Mixed-mode delamination tests. The Mode I and Mode II results have been discussed in the two preceding papers and those data will be used to plot the failure envelope for the laminate by combining the DCB, ADCB and ELS results on to a single graph. Each participating laboratory was supplied with twenty specimens, five of which were to be tested using the then current version of the ADCB test protocol. Participants were asked to initiate the delamination from the insert rather than to pre-crack the specimens.

The results obtained from the different laboratories are summarised in Table 2. All reported unstable initiation from the insert, causing the force to drop rapidly, followed by a period of crack arrest and then re-initiation from effectively a mixed-mode pre-crack. A period of stable crack propagation followed. Due to the instability, the loading was generally linear up to the point of unstable initiation and hence the three initiation values defined in the protocol by (NL), (VIS) and (MAX/5%) coincided. Therefore, for each lab, just one pair of G_{Ic}^{mixed} and G_{IIc}^{mixed} values at initiation are shown in Table 2. All the data have been analysed using the corrected beam theory with an E modulus of 136 GPa. Typically, the correction from mode I, Δ_I, = 7.5mm.

Table 2: Mixed-mode results from the 7th Round robin on the IM7/977-1 composite. Initiation was from the insert, and data analysis was via the CBT method.

Lab	Initiation (J/m²)		Propagation (J/m²)	
	G_{Ic}^{mixed}	G_{IIc}^{mixed}	G_{Ic}^{mixed}	G_{IIc}^{mixed}
1	526±89	350±60	339±74	235±51
2	514±62	342±41	325±48	225±33
3	435±51	312±37	282±26	205±19
4	433±69	302±48	285±21	204±17
5	456±63	357±48	275±18	209±13
Mean (±SD)	473±44	333±24	301±29	216±14
COV (%)	9.3%	7.2%	9.6%	6.5%

Values are the mean and standard deviations from five repeat tests for each lab. The mean, standard deviation and coefficients of variation for all the data are shown at the bottom of the table. Due to unstable initiation, NL, VIS and Max/5% points coincide.

It is clear from these data that the initiation values obtained from the insert were non-conservative, i.e. the propagation values of G_{Ic}^{mixed} and G_{IIc}^{mixed} were all lower than the initiation values. One of the participating labs (Lab 5) tested additional samples to further investigate this effect. Tests were conducted using additional samples which had been pre-cracked in mode I. The data obtained are shown in Table 3.

Table 3. Comparison of ADCB mixed-mode data: Testing from the insert versus testing from a mode I pre-crack.

Crack starter	Initiation (J/m²)		Propagation (J/m²)	
	G_{Ic}^{mixed}	G_{IIc}^{mixed}	G_{Ic}^{mixed}	G_{IIc}^{mixed}
Insert	456+63	351+48	275+18	209+13
Pre-crack	258+28	199+21	309+23	236+18

Data from one lab only. Mean and standard deviations of five repeat tests are shown.

These data confirm that, for initiation, pre-cracking the samples prior to testing yields the more conservative values of G_{Ic}^{mixed} and G_{IIc}^{mixed}. It can be observed that the propagation values obtained from the pre-cracked specimens are higher than for the specimens where initiation had been from the insert. The reason for this was that the propagation values obtained from the specimens tested from the insert were affected by the unstable crack jump.

One laboratory (lab 6) tested the specimens in mixed-mode using the MMB test apparatus, rather than the ADCB test rig. The lever arm of the MMB rig was set so that a ratio of G_I/G_{II} was 1.33 (actual values ranged from 1.315 to 1.345). Values of G_I and G_{II} were obtained from a corrected beam theory analysis for initiation from the insert and from a mode I pre-crack. These data are shown in Table 4.

Table 4. Results from laboratory (lab 6), using the mixed-mode bend (MMB) test apparatus rather than the ADCB apparatus. Tests were on the IM7/977-1 composite with G_I/G_{II} =1.33. All values are for initiation.

Crack Starter	NL (J/m^2) G_{IC}^{mixed}	NL (J/m^2) G_{IIC}^{mixed}	VIS (J/m^2) G_{IC}^{mixed}	VIS (J/m^2) G_{IIC}^{mixed}	MAX (J/m^2) G_{IC}^{mixed}	MAX (J/m^2) G_{IIC}^{mixed}
Insert	197±11%	140±13%	405±35%	289±39%	566±19%	405±19%
PC (I)	159±38%	120±36%	310±11%	231±11%	346±14%	258±14%

Comparison of Initiation from the Insert versus a Mode I Pre-crack.

It is interesting to note that non-linear initiation values are rather lower for the MMB specimen than for the ADCB specimen. This may indicate that some non-linearity comes from the test fixture, although these values are also quite operator dependent. The MMB maximum values are quite similar to ADCB propagation values. A comparison of all the data is shown in Figure 1 below.

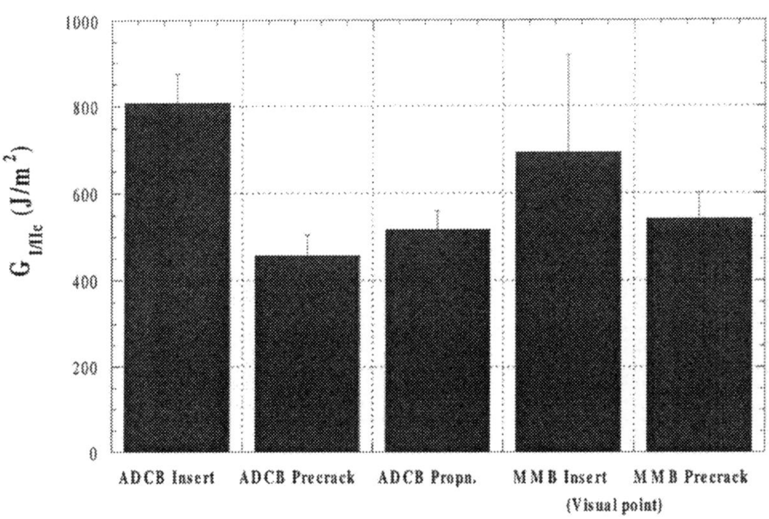

Figure 1. Comparison of $G_{I/IIc}$ values obtained from the 7th Round robin showing the mean and standard deviations of all data. Columns represent: (i) Initiation from the insert for ADCB test (5 labs), (ii) Initiation from a mode I pre-crack for ADCB test (1 lab), (iii) Mean propagation for ADCB test (5 labs), (iv) Visual initiation from the insert for the MMB test (1 lab) and (v) Visual initiation from a mode I pre-crack for MMB test (1 lab).

The results obtained from DCB, ADCB and ELS testing can be combined onto a single failure envelope plot in which the fracture energy, Gc, is plotted against the %G_{II}. The envelope uses the values of G_{IC}, $G_{I/IIC}$ and G_{IIC} which are plotted at 0%, 43% and 100% respectively. Values for the IM7/977-1 composite have been obtained from this paper and the two preceding papers to draw the envelope. Figure 2 shows the failure envelopes obtained when values of Gc are obtained: (i) at visual initiation from the insert, (ii) at visual initiation from a mode I pre-crack and (iii) using the mean propagation values.

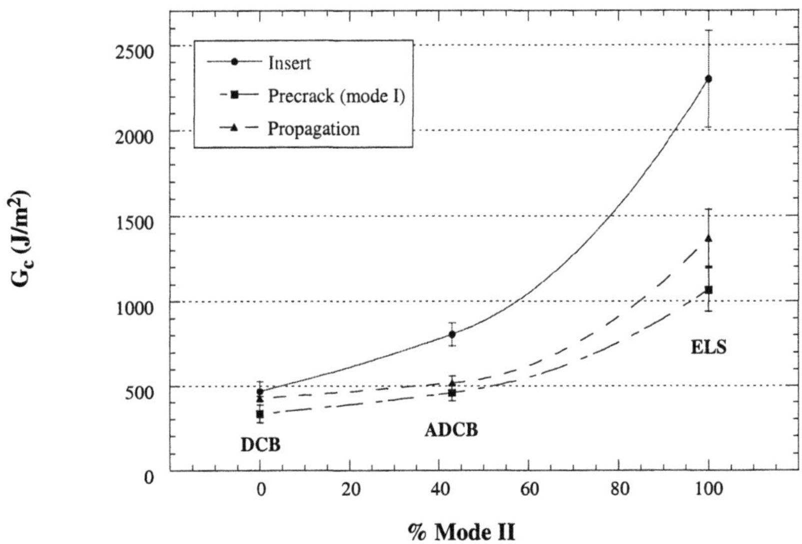

Figure 2. Complete failure envelope for the IM7/977-1 composite obtained using data from the DCB, ADCB and ELS tests from the ESIS 7th Round robin. Values of Gc were obtained: (i) For visual initiation from the insert, (ii) For visual initiation from a mode I pre-crack and (iii) Using the mean propagation values. The envelopes (curves) have simply been fitted to the experimental data. **(Note: For the ADCB tests, the pre-crack was generated in mixed-mode and the pre-crack data was obtained from a single lab.)**

Several noteworthy features can be identified from the failure envelopes shown in Figure 2. For all curves, the value of Gc increases with increasing %mode II, the standard deviation of the data (represented by the error bars) increases with %mode II and the curves becomes very steep towards the 100% mode II point. However, the most significant observation is that the values of Gc obtained from initiation at the insert are higher than equivalent values obtained from pre-cracks at all mixed-mode ratios. The values of initiation from the insert become increasingly less conservative as the % mode II is increased and this demonstrates the importance of the pre-cracking procedure that ESIS proposed for all interlaminar tests.

5. CONCLUDING REMARKS

The ADCB mixed-mode test has received considerable attention and ESIS have run a number of validation exercises using this geometry. However, due to a single value of the mode mix being obtained by this test, most workers now favour the mixed-mode bending (MMB) test where the mode mix can be varied from pure mode I to pure mode II. However, when performed in parallel with DCB and ELS tests, the ADCB method provides useful mixed-mode data with minimal additional effort and cost.

6. REFERENCES

[1] Johnson, W.S., NASA TM 89006, 1986.
[2] Gustafson, C-G., Hojo, M. and Holm, D., J. Comp. Materials 23, 1989, 146.
[3] Johannesson, T. and Blikstad, M., Proc. ISCMS, Beijing, June 1986, p495.
[4] O'Brien, T.K., ASTM STP 836, 1984, p.125.
[5] Arcan M., Hashin, Z. and Voloshin, A., Exp. Mechanics, April 1978, 141.
[6] Venderkley, P.S., M.S. Thesis, Texas A & M University, 1981.
[7] Crews, J.H. and Reeder, J.R., NASA TM 100662, 1988.
[8] Reeder, J.R. and Crews, J.H.J., J. Comp. Tech. & Res. 14, 1992, p12-14.
[9] ESIS TC4 Protocol for DCB testing. Version 1995.
[10] ESIS TC4 Protocol for ELS testing, Version 1999.
[11] Kinloch, A.J., Wang, Y., Williams, J.G., Yayla, P. . Composites Science and Technology **47**, (1993) 225-237.
[12] Brunner, A.J., Tanner, S., Davies, P. Proc. CTS-2 1994, 523-532.
[13] Brunner, A.J., Blackman, B.R.K., and Davies, P., Mode I Delamination, ESIS book.
[14] Hashemi, S., Kinloch, A.J., and Williams, J.G. Proc. R. Soc. Lond. A **427**, 173-199 (1990).
[15] Charalambides, M., Kinloch, A.J., Wang, Y. and Williams, J.G. International Journal of Fracture **54**: 269-291, 1992.
[16] Suo, Z., and Hutchinson, J.W., Materials Science and Engineering A107, 1989, 135-143.
[17] Williams, J.G., International Journal of Fracture 36, 1988, 101-119.
[18] ISO 14125: 1998, Fibre-reinforced plastic composites- Determination of flexural properties.
[19] O'Brien, T.K., NASA TM 110280, TR 1312, 1997.
[20] Davies, P., J. Thermoplastic Composites, 10, July 1997, 353.

7. ACKNOWLEDGEMENTS

The authors wish to express their thanks to the ESIS TC4 laboratories that contributed mixed-mode data to the Round-robin i.e. Politecnico di Milano, Imperial College, Swiss Federal Laboratories for Materials Testing and Research (EMPA), University of Porto, Cranfield University and BASF. For the supply of composite specimens, the authors would like to thank ICI Plc and Cytec Fiberite.

Version 00-05-03

Determination of the Mixed Mode I/II Delamination Resistance of Unidirectional Fibre-Reinforced Polymer Laminates Using the Asymmetric Double Cantilever Beam Specimen (ADCB)

Détermination de la résistance au délaminage en mode mixte I/II, éprouvette double poutre encastrée asymétrique (ADCB), de matériaux composites à matrice polymère renforcés de fibres unidirectionelles

Descriptors: delamination resistance, determination, double cantilever beam, laminate, Mixed Mode I/II, polymer-matrix, energy release rate, test result sheet, unidirectional fibre-reinforced

1 Scope

This standard specifies a method for the determination of the delamination resistance of unidirectional fibre-reinforced polymer laminates under Mixed Mode I/II loading using the Asymmetric Double Cantilever Beam specimen (ADCB). The resistance to the initiation and propagation of a delamination is to be determined from a non-adhesive insert and from a Mode I (opening) or a Mixed Mode I/II pre-crack. The critical energy release rate for Mixed Mode I/II loading and their Mode I and Mode II components, respectively, can be calculated and a resistance-curve (R-curve, i.e. a plot of the critical energy release rate versus delamination length) be determined.

The method is applicable to unidirectional carbon-fibre and glass-fibre reinforced laminates. The scope is not necessarily limited to these fibres and lay-ups, but for laminates with other types of fibres or lay-ups, no recommendations for specimen dimensions and fibre volume content are given. The procedure can be used as a guideline for testing materials that do not strictly satisfy the requirements, provided that (a) the data can be validated using an independent method, or (b) the results are considered to be order of magnitude estimates only and are quoted as such with the property or properties outside specification clearly indicated, or (c) the procedure is used for the purpose of comparison between nominally equivalent materials only.

2 Normative References

The following standard contains provisions which through reference in this text constitute provisions of this standard. At the time of publication the editions indicated were valid. All standards are subject to revision, and parties to agreements based on this standard are encouraged to investigate the possibility of applying the most recent editions of the standards listed below. Members of IEC and ISO maintain registers of currently valid International Standards.

ISO 291:1997	*Plastics; standard atmospheres for conditioning and testing*
ISO 4588:1995	*Adhesives; preparation of metal surfaces for adhesive bonding*
ISO 5893:1993	*Rubber and plastics test equipment; tensile, flexural and compression types (constant rate of traverse); description*
ISO 14125:1998	*Fibre-reinforced plastic composites - Determination of flexural properties*

3 Definitions

For a list of the definitions of symbols and conventions used in this protocol, refer to the central list of symbols in this book.

4 Principle

This standard uses the Asymmetric Double Cantilever Beam (ADCB) specimen shown in Figure 1 for the determination of the delamination resistance of unidirectional fibre-reinforced laminates. Mixed Mode I/II loads with a fixed ratio of Mode I to Mode II component of 4:3 are applied through a load-block under displacement control at a constant rate. Stable delamination growth (propagation) from a non-adhesive insert (starter film) and from a Mode I or a Mixed Mode I/II pre-crack, all at the laminate mid-plane is monitored, and delamination initiation and propagation readings (both from insert and pre-crack) are recorded on the load-displacement curves. Data reduction yields the critical energy release rates $G_{I/IIC}$ for initiation and propagation of a Mixed Mode I/II delamination, and G_{IC}^{mixed} and G_{IIC}^{mixed}, the respective Mode I and Mode II components that are presented in the form of R-curves (critical energy release rate $G_{I/IIC}$ or components G_{IC}^{mixed} and G_{IIC}^{mixed} versus delamination length a).

It has to be noted that using a Mode I or a Mixed Mode I/II pre-crack for starting the delamination may yield values for the critical energy release rate $G_{I/IIC}$ or the respective Mode I and Mode II components differing from those obtained from the insert. Therefore, in order to determine conservative values, both approaches (insert and pre-crack) have to be used. It has also to be noted that different methods for pre-cracking (in Mode I or Mixed Mode I/II) may not yield identical critical energy release rates. Using other procedures for pre-cracking in Mode I or Mixed Mode I/II than described in this standard is not recommended but may be performed for the purpose of comparison. Those procedures should comply with the prescriptions of the standards or test protocols for the respective Modes and be described and documented in the report.

5 Apparatus

A tensile testing machine in compliance with ISO 5893, capable of producing a constant load-rate between 1 and 5 mm/min in displacement control should be used. The load-cell should be calibrated and accurate within ± 1% for the chosen load-range (loads are typically expected to be in the range of 100 - 200 N). The testing machine shall be equipped with a fixture to introduce the load to the pin inserted into the load-block that allows rotation of the specimen end. The recommended loading jig requires either a clamping arrangement to freely slide in bearings in the horizontal direction (side-ways) with a fixed load-point (Figure 1b) or a fixed clamping arrangement with a loading fixture that allows free horizontal movement (side-ways) of the load-point (Figure 1c). The load shall be applied vertically on the load-block, either pulling upward, if the load-block is on the top side of the specimen and pushing downward, if the load-block is on the bottom side of the specimen. The testing machine shall be equipped with means for recording the complete load-displacement curves (loading and unloading) that allow a determination of the loads and the corresponding displacements with an accuracy of ± 1 %.

6 Specimens

6.1 Preparation of Specimens

The recommended specimen width B and length L are 20mm and 170 mm, respectively. The specimen length shall not be less than the length of the insert or of the starter delamination plus 110 mm. The free length L_F is typically 100mm. The recommended specimen thickness is 3 mm for 60% by volume carbon fibre-reinforced and 5 mm for 60% by volume glass fibre-reinforced composites. Other specimen dimensions may be used, but the specimen width should be between 15 and 30 mm. Increasing the length of the specimen is not critical, shortening will reduce the maximum delamination length that can be investigated and thus yield too few data points for the analysis (see clause 8.4). In Figure 1, the fibre orientation is parallel to the length L. The delamination length a is the distance between the load-line (intersection of the plane through the pin-hole centre of the load-block normal to the specimen width B and the plane of delamination) and the tip of the delamination. In both clamping arrangements it is possible to pull the load-block upwards, if it is above the specimen, or to push it downwards, if it is below the specimen, provided the clamp is symmetrical with respect to the specimen.

Three types of initial defect (starter defect) are considered, (a) a laminated starter film (insert), (b) a Mode I pre-crack, obtained by either a Mode I test or by wedge opening, and (c) a Mixed Mode I/II pre-crack obtained from a Mixed Mode I/II test. At least two types of initial defects have to be used in the tests.

If a starter film is used, (as is strongly recommended) an insert (starter film) should be placed at the laminate mid-thickness during moulding. This film should be as thin as possible to minimise the disturbance of the laminate and should be less than 13μm thick. For epoxy matrix composites cured at temperatures below 180°C, a thin film of polytetrafluoroethylene, (PTFE) is recommended. For composites that are manufactured above 180°C (e.g. polyimides, bismaleimides and thermoplastics), a thin film of polyimide is recommended. If a polyimide film is used the film shall be painted or sprayed with a mould release agent before insertion into the laminate. The starter film length should be at least 50 mm from the load-line so that the influence of the load-block can be neglected.

For specimens with an insert (starter film) the test procedure for testing Mode I Double Cantilever Beam (DCB) specimens or wedge opening can be used to prepare the Mode I pre-crack. If DCB-specimens are used, the second load-block has to be removed before testing from the pre-crack. For wedge opening, the specimens shall be clamped at 5 mm beyond the tip of the insert. If material or specimens without insert have to be tested, pre-cracking by wedge opening is the only choice for starting the delamination. In this case, the specimen should be clamped at most 60 mm from the end on which the pre-crack should be formed. Clamping the specimen at shorter distances and repeating the wedging procedure, after moving the clamp, is allowed. The (final) pre-crack should extend at least 50 mm beyond the load-line so that the influence of the load-block can be neglected. The (final) pre-crack should, however, be short enough to allow a delamination length increment of at least 40 mm beyond the tip of the pre-crack, before the delamination arrives within 10 mm of the clamped end. The width of the wedge shall be at least the same as that of the specimen and the opening angle shall be as small as possible. Notching the specimen edge with a razor blade or a diamond saw will provide a firm hold for the wedge. The wedge is driven into the specimen until the tip of the wedge reaches the clamp. The wedge may be driven by hand (tapping on the side) or by using a suitable fixture and a testing machine. Experience has shown that it may be difficult to produce a suitable pre-crack by wedge opening, frequently the pre-crack will not lie in the mid-plane of the specimen. Deviations of the pre-crack from the mid-plane will invalidate the test results and should be noted in the report.

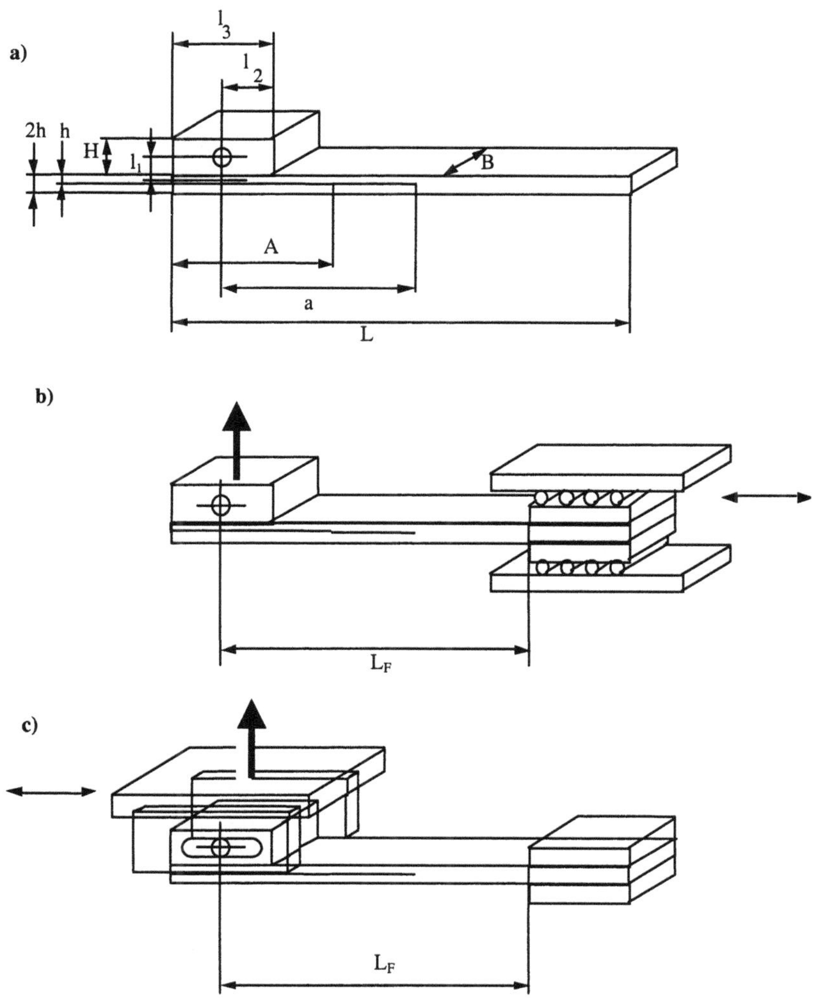

Figure 1: Geometry for the Asymmetric Double Cantilever Beam (ADCB) specimen with one load-block. a) Specimen with load-block, b) clamping arrangement free to slide with fixed load point, and c) fixed clamping arrangement with load-point free to slide.

The Mixed Mode I/II pre-crack shall be prepared in accordance with clause 8.3 but loading be stopped as soon as the delamination is seen to move and the specimen then be completely unloaded. The position of the tip of the delamination after pre-cracking should be marked on both edges of the specimen.

If specimens are cut from a plate, the location of each specimen on the original plate should be recorded and specimens should each be identifiable. Measure and record the length L of the specimens to the nearest mm. Measure the width B and the thickness 2h of each specimen to the nearest 0.02 mm, at five points (quarter, centre, and three-quarter length and 10 mm from each end) along the specimen. The variation in thickness and average values of the width and the thickness should be recorded for each specimen. The variation in thickness, i.e., the maximum difference between thickness measurements, should not exceed 0.1 mm for each specimen. Measure the starter delamination length, i.e. the total length of the insert (starter film) and of the Mode I or Mixed Mode I/II pre-crack on both edges of the specimen. The average values should be recorded but if the starter crack length measurements differ by more than 2 mm, or the pre-crack length measurements by more than 2 mm, the results should be considered suspect and this be noted in the report.

One load-block (Figure 1) is used as load-introduction, it should be of the same width as the specimen. The load-block and the specimen should first be lightly abraded, use of either sandpaper or grit blasting should be sufficient, as the loads required to delaminate the specimens used in these tests are quite low. The load-block and the specimen should then be cleaned with a solvent. If a bond failure occurs it may be necessary to consult ISO 4588 for a more sophisticated procedure. Bonding of the load-block should be done immediately after the surface preparation. In most cases a cyanoacrylate ("Super glue") adhesive has been found adequate for previous tests on similar specimens. Alternatively, a tough, room-temperature cure adhesive may be used. The surface preparation and the type of adhesive used should be noted in the report. The load-introduction should be well aligned with the specimen and held in position with clamps while the adhesive sets. Specimen edges should be smoothed prior to determining the dimensions. Adding a thin layer of typewriter correction fluid ("white ink") on the edges after conditioning will facilitate the detection of the delamination growth. It should be noted that some typewriter correction fluids contain solvents that may be harmful to the laminate matrix material. For the measurement of the delamination length, marks should indicate every 1 mm from the tip of the insert and of the Mode I or Mixed Mode I/II pre-crack for at least the first 5 mm, then, for testing from the Mode I or Mixed Mode I/II pre-crack marks should be applied every 5 mm, and every 1 mm should be marked after 35 mm at least up to 40 mm.

6.2 Number of Specimens

A minimum number of five specimens shall be tested from the insert and from the Mode I or Mixed Mode I/II pre-crack unless a smaller number is prescribed.

7 Conditioning

Moisture conditioning is required for obtaining baseline data in order to test specimens with a uniform moisture content. The drying conditions (temperature and duration) shall be chosen according to the recommendations of the resin supplier. Conditioning should be performed after bonding of the load-block. Before testing, the specimens may be stored in a desiccator for at most one day after conditioning. Other conditioning procedures may be applied for the investigation of specific conditioning effects.

8 Test Procedure

8.1 A Note on the Application

If this test procedure is used to prepare specimens with Mixed Mode I/II pre-cracks as starter cracks for testing the delamination resistance in other Modes, it is recommended to consult the applicable standards or test protocols for additional requirements on specimen characterisation and preparation.

8.2 Test Preparation (Determination of the E modulus)

A value of the E-modulus from a three-point bending test is required if the beam theory analysis (see clause 8.4) is used to evaluate the data. The experimental compliance method does not require modulus values but the method is not applicable unless additional requirements are fulfilled (see clause 8.4). The modulus value shall be determined before delamination testing on that part of the specimen that does not contain the insert or the Mode I or Mixed Mode I/II pre-crack. The three-point bending test shall be performed and the flexural modulus E be calculated in accordance with ISO 14125.

8.3 Test Set-up and Data Recording

The test shall be performed under normal conditions in accordance with ISO 291 (23° ± 2° C, 50% ± 5% relative humidity) unless prescribed otherwise. The load and the displacement signals of the testing machine shall be recorded, either on a paper chart or electronically throughout the test, including the unloading cycle. The delamination length may be measured by eye on the specimen edge, or by using a travelling microscope. In transparent laminates the delamination length may be followed inside the specimen by marking the specimen surface rather than the edge. If unstable delamination growth followed by arrest ("stick-slip") is observed during any stage of the test, it should be noted in the report. The data evaluated according to clause 8.4 may not be valid in this case. Any permanent deformation of the specimen after unloading should be noted in the report. Deviations of the delamination from the mid-plane of the laminate will invalidate the test results and should be noted in the report. The maximum horizontal displacement d_{max} of the sliding fixture shall be recorded by determining the initial (before loading) and final (before unloading) positions of the clamp or of the load-point, respectively, and be noted in the report.

Test parameters and data recording are the same for testing from the insert and testing from the Mode I or Mixed Mode I/II pre-crack. The free length is generally of the order of 100 mm, so that a reasonable crack propagation can take place, but a shorter free length may be necessary to promote propagation in some materials.

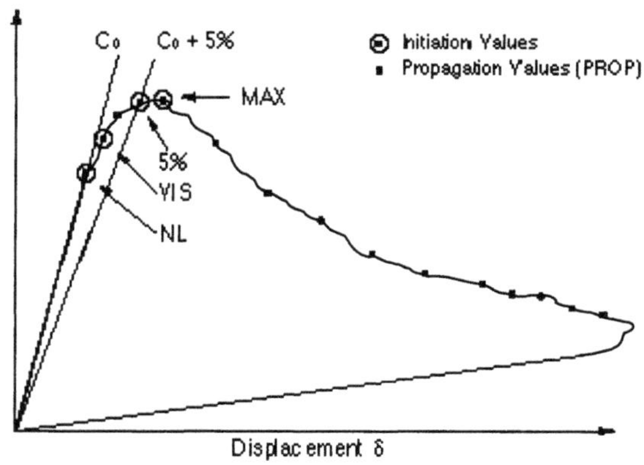

Figure 2: Schematic load-displacement curve for testing from a Mode I pre-crack with initiation points NL, VIS, 5%, MAX, and propagation points (PROP).

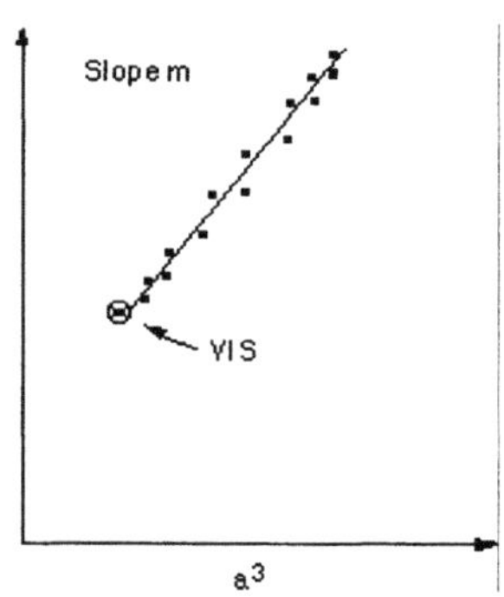

Figure 3: Linear fit used to determine the slope m for the Experimental Compliance Calibration Method.

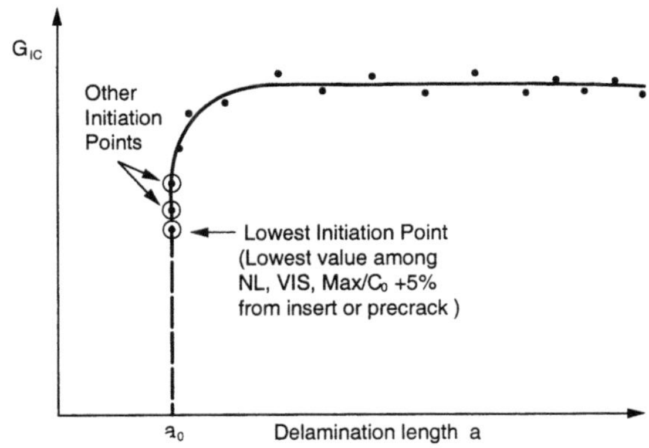

Figure 4: Schematic resistance-curve (R-curve) with values of $G_{I/IIC}$ for initiation (lowest value among NL, VIS, 5% or MAX) and for propagation (PROP) versus observed delamination length a.

8.4 Testing from the Insert and from the Pre-crack

Measure the initial position of the sliding fixture (clamp or load point) before the start of loading. For testing from the insert (starter film) and from the Mode I or Mixed Mode I/II pre-crack, the specimen should be loaded at a constant cross-head rate between 1 and 5 mm/min. For specimens with nominal length, a cross-head speed from the lower end of the range is recommended and for longer specimens a speed from the upper end of the range is recommended. The point on the load-displacement curve at which the onset of delamination movement from the insert or the tip of the pre-crack is observed on the edge of the specimen should be recorded on the load-displacement curve or in the sequence of load-displacement signals (VIS, Figure 2). If the start of the delamination growth is difficult to observe, a change in illumination conditions or a cross-head speed towards the lower end of the range is recommended. After this, as many delamination length increments as possible should be noted in the first 5 mm on the corresponding load-displacement curves, ideally every 1 mm. Subsequently, delamination lengths are noted every 5 mm, and again every 1 mm for the last 5 mm of delamination propagation (ideally, the total delamination length increment should be at least 40 mm). The loading should be stopped before the crack arrives within 10 mm of the clamped end. Measure the final position of the sliding fixture before unloading in order to calculate its maximum horizontal displacement d_{max}. After this, the specimen should be completely unloaded at a constant cross-head rate, unloading may be performed at up to 25 mm/min. The position of the tip of the delamination should be marked on both edges of the specimen. If the delamination lengths a on the edges of the specimen, i.e. the distance between load-line and the tip of the delamination differ by more than 2 mm the results should be considered suspect and this be noted in the report.

8.5 Data Analysis

The data required for the analysis are delamination lengths a, and the corresponding forces P and displacements δ. Several values may be determined from the load-displacement curve, if possible, the following initiation values, shown in Figure 2, should be determined for testing from the insert (starter film) and from the Mode I or Mixed Mode I/II pre-crack for each specimen:

(1) NL, i.e. deviation from linearity: A region of non-linear behaviour usually precedes the maximum load, even if the unloading curve is linear. The point of deviation from linearity (NL in Figure 2), is determined by drawing a straight line from the origin but ignoring any initial deviations due to take-up of play in the loading system. Experience has shown that it is difficult to reproducibly determine the position of NL on the load-displacement curve. Performing a linear fit on the load-displacement curve starting at 5% of the maximum load and using a consistent criterion for deviation from linearity (e.g. the half-thickness of the plotter trace) is recommended. If non-linearity due to large displacements is observed, then this value should not be used for the analysis.

(2) VIS, i.e. visual observation: This corresponds to the onset of the delamination, i.e. to the first point at which the delamination is observed to move from the tip of the insert or of the Mode I pre-crack on the edge of the specimen (VIS in Figure 2).

(3) 5% or MAX, i.e. 5% increase of compliance or maximum load point: The 5% value corresponds to the point on the load-displacement curve at which the compliance has increased by 5% of its initial value C_0. A best straight line is drawn to determine the initial compliance C_0, ignoring any initial deviation due to take-up of play in the loading system. A new line is then drawn with a compliance equal to C_0 + 5% whose intersection with the load-displacement curve yields the load and displacement to be used for the calculation, unless the intersection is at a larger displacement than the maximum load in which case the maximum load and the corresponding displacement have to be used.

Besides the initiation points (NL, VIS, 5% or MAX), propagation values (PROP in Figure 2b) can be determined for each delamination length measured during propagation from the insert (except when preparing a Mixed Mode I/II pre-crack) and from the Mode I or Mixed Mode I/II pre-crack. A separate test result sheet shall be used for the values determined from the insert (starter film) and those from the Mode I or Mixed Mode I/II pre-crack. Either one of the two methods described below can be used for the analysis, the method chosen should be noted in the report.

Method (1): Corrected Beam Theory (CBT)

Analysis is performed using corrected beam theory which corrects for the effects of the beam not being perfectly built-in as is assumed by simple beam theory and also for large displacements in the test. The error in the built-in beam assumption is corrected by replacing the measured crack length, a, with a slightly longer crack, $a+\Delta$. The mode I component of the energy release rate is calculated using the correction $a+\Delta_I$ where Δ_I is obtained from a Mode I

DCB test. The mode II component of the energy release rate is calculated using the correction $a+\Delta_{II}$, where $\Delta_{II}=0.42\Delta_I$ [1]. If Mode I tests on the same material have not been performed, the value of Δ_I should be set to $\Delta_I = 0$ and this be noted in the report. (See the mode I DCB test protocol for a description of how Δ_I is determined). The correction for large displacements arises because of the shortening of the moment arm during the test, and this is corrected for by employing the multiplication factor, F. The factor F also accounts for the rotation of the load-block above the beam during the test. The corrected beam theory requires a value for the axial modulus of the arms of the specimen. This value of E should be obtained from a three-point bending test as described in clause 8.2.

The total mixed-mode critical energy release rate in the ADCB test, $G_{I/IIC}$ can be partitioned into the Mode I and Mode II components, G_{IC}^{mixed} and G_{IIC}^{mixed} respectively such that:

$$G_{I/IIC} = G_{IC}^{mixed} + G_{IIC}^{mixed} \tag{1}$$

Where G_{IC}^{mixed} and G_{IIC}^{mixed} are given by:

$$G_{IC}^{mixed} = \frac{3P^2(a+\Delta_I)^2}{B^2Eh^3} \cdot F \tag{2a}$$

$$G_{IIC}^{mixed} = \frac{9P^2(a+\Delta_{II})^2}{4B^2Eh^3} \cdot F \tag{2b}$$

with P the force, a the delamination length, Δ_I the correction for the Mode I component determined from Mode I tests using the Double Cantilever Beam specimen, $\Delta_{II} = 0.42\ \Delta_I$ the correction for the Mode II component, B the width of the specimen, E the modulus parallel to the fibre direction, h the half-thickness of the specimen (total specimen thickness 2h) and F the large displacement correction factor given by:

$$F = \left[1 - \theta_1\left(\frac{\delta}{L_F}\right)^2 - \theta_2\left(\frac{\delta l_1}{L_F^{\,2}}\right)\right] \tag{3}$$

with δ the displacement, l_1 the distance from the centre of the load-block to the mid-plane of the specimen beam to which the load-block is attached (Figure 1), and L_F the free length of the specimen. The factors θ_1 and θ_2 are calculated as follows:

$$\theta_1 = \frac{3}{20} \cdot \frac{\left[15 + 150\left(\frac{a}{L_F}\right)^2 + 367\left(\frac{a}{L_F}\right)^4\right]}{\left[1 + 7\left(\frac{a}{L_F}\right)^3\right]^2} \quad (4)$$

$$\theta_2 = 3\left(\frac{L_F}{a}\right)\frac{\left[1 + 7\left(\frac{a}{L_F}\right)^2\right]}{\left[1 + 7\left(\frac{a}{L_F}\right)^3\right]} \quad (5)$$

with *a* the delamination length. All initiation and propagation values, if applicable, should be calculated, the delamination length for the initiation values is the distance between the load-line and the tip of the insert or pre-crack (Figure 1).

Method (2): Experimental Compliance Method (ECM) Method

If the loading and unloading curves (load-displacement plot) are both linear, an alternative approach is to plot the compliance C versus the cube of the delamination length, a^3. Only the VIS and the PROP values are used for the linear fits, not the NL or 5%/MAX values. If it has not been determined or is considered questionable, the VIS point may be excluded from the linear fit, but this should be noted in the report. The slope of this plot, m, can then be used to calculate the value of $G_{I/IIC}$ as follows:

$$G_{I/IIC} = \frac{3P^2ma^2}{2B} \cdot F \quad (6)$$

with P the force, m the slope of the plot of the compliance C versus the cube of the delamination length a^3, a the delamination length, and B the width of the specimen and F the correction factor defined in equation (3).

All initiation and propagation values, if applicable, should be calculated, the delamination length for the initiation values is the distance between the load-line and the tip of the insert or pre-crack (Figure 1).

This gives the overall energy release rate, which may then be split into Mode I and Mode II components by assuming the ratio $G_I/G_{II} = 4/3$, so

$$G_{IC}^{\,mixed} = 0.57(G_{I/IIC}) \quad (7)$$

$$G_{IIC}{}^{mixed} = 0.43(G_{I/IIC}) \tag{8}$$

The results from testing from the insert and from the Mode I or Mixed Mode I/II pre-crack are separately used to draw resistance curves (R-curves), i.e. $G_{IC}{}^{mixed}$ and $G_{IIC}{}^{mixed}$ versus delamination length a (Figure 4). The initiation value quoted for each specimen shall be the lowest among the (NL, VIS, MAX/5%) values for both Mode I and Mode II components. The minimum number of propagation points recorded for each specimen should be 15, if fewer points are used, this should be noted in the report and the results considered suspect [2].

As more than one specimen of a material is to be tested, the results shall be averaged as follows to yield characteristic material values: Calculate the arithmetic average and standard deviation of each type of initiation value (VIS, NL, MAX, and 5%) separately for both Mode I and Mode II components, then calculate the arithmetic average and the standard deviation of the last 10 PROP values or of the last 50% of all PROP values, whichever contains the larger number of data points for both Mode I and Mode II components. The average values and standard deviations should be noted in the report. If the calculated standard deviation exceeds 10% of the average value, a constant plateau value for the propagation may not have been reached and the R-curve plots should be checked. If the R-curve plots do not show a plateau, the average PROP value should be considered suspect and this be noted in the report.

9 Test Report

The test report shall include the following information:

(a) a reference to this standard and to the referring standards

(b) a complete identification of the material (e.g. laminate manufacturer, fibre-material, polymer material maximum cure temperature T_{mc}, duration of curing t_c, location of specimen on plate)

(c) test date, test laboratory, test personnel identification

(d) number and label of specimens tested and type of method used for the analysis

(e) average thickness, average width, maximum thickness variation along the length, and the length of each specimen, insert (starter film) material and thickness, length of the insert; note if insert length measurements differ by more than 1 mm on both edges

(f) conditioning temperature T_d and conditioning duration t_d and temperature T and relative humidity r.h. during the test

(g) dimensions of the load-block, surface preparation, if applicable, and adhesive

(h) type of pre-cracking used (e.g. Mode I test or wedge opening) and, if applicable, whether the specimen has been removed from the fixture after pre-cracking

(i) load-rate for loading and unloading, for testing from the insert and from the Mode I or Mixed Mode I/II pre-crack

(j) the maximum horizontal displacement d_{max} of sliding fixture (clamp or load-point)

(k) length of delamination after unloading for testing from insert and from the Mode I or Mixed Mode I/II pre-crack; note, if delamination length measurements differ by more than 2 mm on both edges

(l) Δ_I and Δ_{II}, i.e. the corrections for crack tip rotation obtained from the delamination length correction Δ_I for Mode I tests using Double Cantilever Beam (DCB) specimens, if method (1) is used for the data analysis (see clause 8.5)

(m) E-modulus from "three point bending" test, if method (1) is used for the data analysis (see clause 8.5)

(n) slope m of plot of the compliance C versus the cube of the delamination length a^3, if method (2) is used for the data analysis, and the correlation coefficient r^2 of the linear fit (see clause 8.5)

(o) $(C_{max/5\%} - C_0)/C_0$, i.e. the percent change in compliance between the initial compliance C_0 and the compliance at the MAX or 5% point, whichever is applicable

(p) copy of the load-displacement curve for each specimen

(q) table of G_{IC}^{mixed} and G_{IC}^{mixed} (all initiation and propagation values) and plot of G_{IC}^{mixed} and G_{IC}^{mixed} (all initiation and propagation values) versus delamination length a (R-curve) for each specimen including large displacement and load-block corrections

(r) average values and standard deviation for each type of initiation value (VIS, NL, MAX, and 5%) and average value and standard deviation of the last 10 propagation values (PROP) or of the last 50% of the propagation values, whichever contains the larger number of data points, from all specimens tested. Note, if less than 15 propagation values have been recorded. If a specimen is excluded from averaging, the reason for this should be noted in the report.

(s) observations from testing (e.g. deviation of the pre-crack or the delamination from the mid-plane, stick-slip, occurrence of fibre-bridging, permanent deformation after unloading, sticking of insert foil, no plateau in the R-curve) that may have affected the test procedure or the results

(t) any deviation from the prescriptions of this protocol (e.g., dimensions of specimens, fibre orientation)

(u) results from additional specimen or material characterisation (e.g. fibre volume fraction, void content) if specified

10 References

[1] Y. Wang, J.G. Williams: "Correction to Mode II Fracture Toughness Specimens of Composite Materials", Composite Science and Technology 42, 251-256 (1992).

[2] A.J. Brunner, S. Tanner, P. Davies, H. Wittich: "Interlaminar Fracture Testing of Unidirectional Fibre-Reinforced Composites: Results from ESIS-Round Robins" in: Composites Testing and Standardisation ECCM-CTS 2, (P.J. Hogg, K. Schulte, H. Wittich eds.), Woodhead Publishing, 523-532 (1994).

Figure 5: Recommended Test Result Sheet

Mixed Mode I/II Delamination Test: ADCB-Specimen

Recommended Test Result Sheet ADCB Mixed Mode I/II, Page 1 of 2 | Fibre | Resin :

Specimen label	....	**# of**	**Specimens**		
Laboratory	....	**Test personnel**	....	**Test date**	....
Laminate manufacturer	....	**Resin manufacturer**	....	**Fibre manufacturer**	....

Specimen free length	L_F [mm]	....	**Load introduction**		....
Average thickness	2h [mm]	....	**Block thickness**	H [mm]	....
Max. thickness variation	Δ2h [mm]	....	**Block length**	l_3 [mm]	....
Specimen width	B [mm]	....	**Block/hinge width**	[mm]	....
Insert material	....	....	**Surface preparation**	....	
Insert thickness	[μm]	....	**Adhesive**	....	
Insert length	A [mm]	....	**Conditioning temperature**	T_d [°]	....
Mould release agent	....	....	**Conditioning duration**	t_d [h]	....
Precrack length	a_p [mm]	....	**Test temperature**	T [°]	....
Max. cure temperature	T_{mc} [°]	....	**Relative humidity**	r.h. [%]	....
Cure duration	t_c [h]	....	**Loading/unloading rate**	[mm/min]	....
Max. delamination length	A_{max} [mm]	....	**Distances l_1/l_2**	[mm]	....
Correction from mode I	Δ_I, [mm]	....	**% Change Compliance**	[%]	....
Axial modulus	E_{11} [GPa]	....			

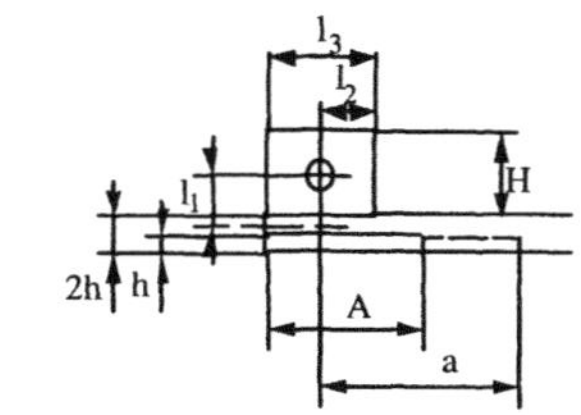

Average NL ±σ
CV NL
Averge VIS ±σ
CV VIS
Av. MAX/5 % ±σ
CV MAX/5 %

Minimum initiation value :

Text	a [mm]	Load [N]	δ [mm]
NL (insert)			
MAX/5 % (insert)			
VIS (insert)			
NL (precrack)			
MAX/5 % (precrack)			
VIS (precrack)			
PROP			
PROP			
PROP			
PROP			
PROP			
PROP			
PROP			
PROP			
PROP			
PROP			
PROP			
PROP			
PROP			
PROP			
PROP			

Specimen and Insert Location on Plate

Length .... mm

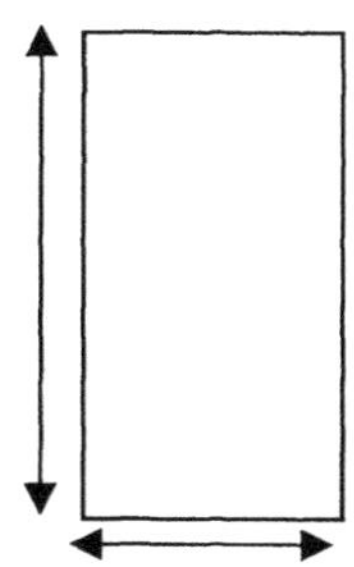

Width .... mm

G_{Ic}^{mixed} (CBT) [J/m²]	G_{IIc}^{mixed} (CBT) [J/m²]	G_{Ic}^{mixed} (ECM) [J/m²]	G_{IIc}^{mixed} (ECM) [J/m²]

Recommended Test Result Sheet ADCB Mixed Mode I/II, Page 2 of 2 | Fibre : | Resin :

Specimen label	....	**# of**	**Specimens**		
Laboratory	....	**Test personnel**	....	**Test date**	....
Laminate manufacturer	....	**Resin manufacturer**	....	**Fibre manufacturer**	....

CBT	
Δ_I=	
$\Delta_{II}=0.42\Delta_I$	
E_{11}=	
or	
ECM	
slope of fit	
Y-Intercept of fit	
m	
correlation coefficient	

Difference insert length ≤ 1 mm ?	**yes / no**
Difference precrack length ≤ 2 mm ?	**yes / no**
Precrack tip 3-5 mm from insert ?	**yes / no**
Difference total delamination ≤ 2 mm ?	**yes / no**
Specimen removed after precracking ?	**yes / no**
VIS included in linear fit ?	**yes / no**
Stable delamination growth through test ?	**yes / no**
Total delamination length > 40 mm ?	**yes / no**
Reference to ISO 291 ?	**yes / no**
Reference to ISO 4588 ?	**yes / no**
Reference to ISO 5893 ?	**yes / no**

Observations and Comments

Text	θ_1	θ_2	F	C [mm/N]	a^3 [mm^3]	$G_{I/IIC}^{mixed}$ (CBT)	$G_{I/IIC}^{mixed}$ (ECM)	δ / a	Checks ok ?
NL (insert)									
MAX/5 % (insert)									
VIS (insert)									
NL (precrack)									
MAX/5 % (precrack)									
VIS (precrack)									
PROP									
PROP									
PROP									
PROP									
PROP									
PROP									
PROP									
PROP									
PROP									
PROP									
PROP									
PROP									
PROP									
PROP									
PROP									

LIST OF SYMBOLS

A	a coefficient related to a/W
A	a constant in a power law equation
A	insert film length (as in a DCB specimen)
a	crack or notch length
a_0	initial crack length
$\tilde{a}$	average crack length
$a(n)$	crack length as a function of fatigue cycles
a_i	i^{th} value of crack length
a_{i+1}	$(i+1)^{th}$ value of crack length
a_f	final crack length
a_p	precrack length
α	ratio of crack length to width (a/W)
α	smooth fraction of fracture surface
α_{T-L}	smooth fraction of fracture surface for T-L specimen
α_{L-T}	smooth fraction of fracture surface for L-T specimen
B	specimen thickness
B_{min}	minimum value of specimen thickness
B_N	net section thickness for a side grooved specimen
b	a fitting parameter for impact analysis
b_1, b_2	particular specimen distances
β	shape factor
C	compliance
C^*	dimensionless specimen compliance
$C(a)$	compliance as a function of crack length
C_0	initial compliance or compliance when a/W=0
C_{COR}	compliance correction due to indentation and machine stiffness
C_m	machine compliance
C_i	indentation compliance
C_s	compliance of specimen
$C_{s,0}$	compliance of an unnotched specimen
$C_{0,1,2}$	coefficients
C_N	normalised compliance
C_{max}	maximum compliance
$C_{5\%}$	initial compliance increased by 5%
C_{SY}	system compliance
C_{tot}	total compliance
C_{cs}	compliance of a calibrated specimen
c	dimension of core
c_0	wave propagation velocity of impactor striker material
c_1	longitudinal wave propagation velocity of the specimen

D_{ij}	stiffness components of laminate (i,j=1,2,3)
dG/da	change in energy release rate with crack length
dK/dt	time derivative of stress intensity factor
dU	energy change
dX	change in some function X
dA	area change
da	change in crack length
dP	change in force
dC	change in compliance
dC/da	change in compliance with change in crack length
d_{LC}	distance from load cell to the end of the striker
d_{CS}	thickness of the conductive strip
d_{SG}	distance between the strain gauge and the crack tip
Δa	crack growth
ΔK	change in stress field intensity factor
ΔG	change in energy release rate
Δa_{MAX}	maximum change in crack length
Δa_{Min}	minimum change in crack length
Δ	difference in surface to centre crack length
Δ	crack length correction in delamination tests
Δ_I	delamination length correction
Δ_{II}	correction for rotation at the delamination tip
Δw_e	confidence interval for essential work of fracture
$\Delta a/\Delta t$	peel crack length
$\Delta\delta/\Delta t$	cross head speed
δ	crack opening displacement
δ	cross head displacement
δ	the difference between average through thickness crack length and the corresponding crack length measured in a fatigue test
δ_{COR}	corrected value of displacement
δ_{OFFSET}	off-set displacement
δ_{MAX}	maximum displacement
d_{MAX}	maximum horizontal displacement of the clamping arrangement
E	axial modulus
E_{STIFF}	modulus determined from stiffness
E_{FRACT}	modulus calculated from fracture data
E^1	a modulus
E^*	effective modulus
E_1	elastic modulus
E_2	plastic modulus
E_f	modulus determined from DCB test data
E_S	an independent modulus determination
E_{11}	tensile modulus parallel to fibre direction
ε_Y	yield strain

F large displacement correction
$f(\theta)$ a function in terms of the angle θ
$f(\alpha)$ a function in terms of the angle α
ϕ {or $\phi(\alpha)$ or $\phi(a/W)$} calibration factor dependent on a/W
ϕ peel angle

G energy release rate
G_I energy release rate in mode I
G_{II} energy release rate in mode II
G_{III} energy release rate in mode III
G_C critical value of energy release rate but usually known as fracture toughness; it is also referred to as fracture resistance and fracture energy
G_{IC} mode I fracture toughness
G_{IIC} mode II fracture toughness
G_{C1} plane strain value for fracture toughness
G_{MAX} maximum value of energy release rate
G_{MIN} minimum value of energy release rate
G_A interfacial fracture toughness also known as interfacial work of fracture and sometimes as adhesive strength
$G_A^{\infty E}$ energy release rate for an infinite modulus peel arm
G_A^{eb} energy release rate with corrections for elastic bending of the peel arm
G^{db} energy release rate with corrections for elastic tensile deformation and dissipated energy
in bending and tension of the peel arm
$G_{I/II}$ the sum of energy release rates from mode I and mode II in a mixed mode I/II test

$G_{I/IIC}$ fracture toughness in mixed mode I and II where G_{IC}^{mixed} and G_{IIC}^{mixed} are the respective components for mode I and mode II
G_{IC}^{mixed} mode I component of the fracture toughness during a mixed mode test
G_{IIC}^{mixed} mode II component of the fracture toughness during a mixed mode test
g acceleration due to gravity
g(u) a non dimensional energy release rate

H specimen length for DENT EWF specimen
H thickness of the load block for DCB, ADCB and ELS specimens
h specimen length for compression test
h impact drop height
h specimen gauge length for DENT EWF
h thickness of a peel arm
h thickness of each substrate or composite arm in a DCB, ADCB or ELS test
h_a thickness of adhesive layer
h_1 distance between the plane of the insert film and the top of the (DCB) specimen
h_2 distance between the plane of the insert film and the bottom of the (DCB) specimen
η_e geometry factor
η calibration factor

J fracture resistance for a non-linear solid
J_C fracture toughness for a non-linear solid
J_0 fracture resistance for a non-linear solid at crack initiation
$J_{0.2}$ fracture resistance for a non-linear solid at 0.2 mm of crack growth
J_{MAX} maximum value for fracture resistance for a non-linear solid

K stress field intensity factor
K_C critical value of stress field intensity factor, also known as fracture toughness
K_S static stress field intensity factor
K_Q a provisional fracture toughness
K_{C1} plane strain fracture toughness
K_{Id} mode I stress field intensity factor at impact velocity above 1 m/s
K_1^{dyn} mode I crack tip loading history
K_I^{qs} quasi-static mode I crack tip loading history
K_{MAX} maximum value of stress field intensity factor
K_{MIN} minimum value of stress field intensity factor
k a constant
k_d dynamic correction factor
k^{dyn} dynamic correction factor

L longitudinal direction
L specimen length
L_0 specimen gauge length
L_F free length of specimen between load line and clamp
l ligament length
l_1 distance from centre of the loading pin of the piano hinge axis to the mid-plane of the arm of the substrate beam (in a DCB, ADCB or ELS specimen)
l_2 distance between the centre of the pin-hole of the load-block and its edges measured
towards the tip of the insert (in a DCB, ABCB or ELS specimen)
l_3 total length of the load block (in a DCB, ABCB or ELS specimen)

M_W weight average molecular weight
m a constant
m mass
m a fitting parameter for impact analysis
m specimen geometry factor
m slope of a plot of C (or C/N) versus a^3
μ coefficient of friction

N a power law constant
N a load-block correction factor
N number of fatigue cycles
N_i i th cycle in fatigue
N_{i+1} (i+1) th cycle in fatigue
n slope of a plot of log C versus log a or log (C/N) versus log a

n	a constant
ν	lateral contraction ratio (Poisson's ratio)
P	force or load
P_{MAX}	maximum value of force
$\bar{P}_{max}$	mean maximum force
$P_{5\%}$	force where 5% offset compliance meets force-displacement curve
P_5	force where 5% offset compliance meets force-displacement curve
P_Q	a provisional force
$\bar{P}_Q$	mean provisional force
$P_{POP\text{-}IN}$	force at crack "pop-in"
P_H	striker force
P(t)	load as a function of time
$\bar{P}(t)$	mean load-time curve
$P^I(t)$	force as a function of time defined by the initial tangent of the force-time curve
P^I_1, P^I_2	specific values of force from the function $P^I(t)$ at the specific times of t_1 and t_2
$\bar{P}_1, \bar{P}_2$	specific values of force from the function $\bar{P}(t)$ at the specific times of t_1 and t_2
Θ	angle between force axis and the force-displacement trace in the linear region
θ	peel angle
θ	an angle measured from the crack line
R	radius
R ratio	ratio of minimum to maximum applied force or stress
r	distance from the crack tip
r_P	radius of the plastic zone
r^2	correlation coefficient of linear fits
ρ	density
S	short transverse direction
S	Span
s	skin dimension
σ	stress
σ_C	critical local stress
σ_Y	yield stress
σ_A	applied stress
σ_T	true stress
$\sigma(r)$	stress at a distance r from the crack tip
σ_{MAX}	maximum stress
σ_m	maximum stress
σ_0	a reference stress

T long transverse direction
T temperature
T stress vector at the outer side of the integration line Γ
T_g glass transition temperature
T_{mc} maximum cure temperature
T_d drying temperature
t mould thickness
t specimen thickness in DENT EWF specimen
t_f time to fracture
$t_{f\ min}$ minimum time to fracture
t_0 time at the moment of impact
t_0 a fitting parameter for impact analysis
t_i time at fracture initiation
t_c time for curing the adhesive
t_d time of drying
t_{max} a time corresponding to the maximum on an experimental load-time curve
t_Q time on the force-time curve associated with the provisional event Q
t_S sampling time
τ period of oscillation

U energy
U_B true fracture energy
U_{COR} energy correction for indentation and machine compliance
U_{dt} dissipated energy in tension
U_{db} dissipated energy in bending
U_{ext} external work
U_k kinetic energy
U_{kin} kinetic energy
U_d dissipated energy
U_{inert} area under the inertia peak of the impact force-time signal
U_Q provisional fracture energy or energy at maximum displacement
$U_{Q\ cor}$ corrected energy
U_S strain energy

U_X a function
U_t total energy

u displacement
υ notch tip radius

V test speed
V impact velocity
V crack mouth opening displacement
v_0 impact velocity

W	specimen width
w	energy density
w_e	essential work of fracture
w_P	plastic work dissipation (or non-essential work of fracture) per unit volume
W_f	total energy absorbed in fracture
w_f	total energy absorbed in fracture per unit area of ligament
w_0	reference value of plastic work dissipation per unit volume
x	co-ordinate
y	co-ordinate
ψ	geometry factor

LIST OF ABBREVIATIONS

ABS	acrylonitrile-butadiene-styrene
ADCB	asymmetric double cantilever beam
adh	adhesive
AECMA	Association Européenne des Constructeurs de Matériel Aérospatial (European Association of Aerospace Industries)
ASTM	American Society for Testing and Materials
BL	blunting line
CBEN	cantilever beam enclosed notch
CBT	corrected beam theory
CEN	Comité Européen de Normalisation (European Committee for Standardisation)
CF	carbon fibre
CFRP	carbon fibre reinforced plastic
CLS	cracked lap shear
CNF	centre notched flexure
COV (orCV)	coefficient of variation
CP	crack propagation
CS	conductive strips
CT or C(T)	compact tension
DCB	double cantilever beam
DENT	double edge notched tension
DIS	Draft International Standard (for ISO document)
DKC	dynamic key curves
ECM	experimental compliance method
ECT	edge crack torsion
EDT	edge delamination tension
EGF	European Group on Fracture (now called ESIS)
ELS	end loaded split
ENF	end notched flexure
EP	ethylene propylene
EPFM	elastic plastic fracture mechanics
ESIS	European Structural Integrity Society
EVOH	ethylene vinyl alcohol
EWF	essential work of fracture
FCP	fatigue crack propagation
4ENF	four point loaded end notched flexure
GFRP	glass fibre reinforced plastic
HDPE	high density polyethylene
IEC	International Electrotechnical Commission
IRC	impact response curve
ISO	International Organization for Standardization
JSA	Japanese Standards Association
JIS	Japanese Industrial Standards

LEFM	linear elastic fracture mechanics
LC	load cell
Liquid N_2	liquid nitrogen
LLDPE	linear low density polyethylene
L-T	specimen with applied stress in the **L** direction and crack growth in the **T** direction
MMB	mixed mode bending
MDPE	medium density polyethylene
MLL	mean load line
mPVC	modified poly(vinylchloride)
NASA	National Aeronautics and Space Administration
NL	onset of non-linearity
PA	polyamide
PE	potential energy
PE	polyethylene
PEEK	poly(ether ether ketone)
PET	poly(ethylene terephthalate)
"pop-in"	crack jumps forward by a small amount and then arrests
PMMA	poly (methylmethacrylate)
PROP	propagation
PTFE	poly(tetrafluoroethylene)
PP	polypropylene
PPO	poly(phenylene oxide)
PVC	poly(vinyl chloride)
rh	relative humidity
RS	razor slide
RT	razor tap
RTPA	rubber toughened polyamide
RTPMMA	rubber toughened poly(methylmethacrylate)
SAN	styrene-acrylonitrile
SBT	simple beam theory
SD	standard deviation
SEM	scanning electron microscopy
SENB, SE(B)	single edge notched beam
SENF	single edge notched flexure
SENF	stabilized end notched flexure
SENT	single edge notched tension
SG	strain gauge
S-N curves	stress versus log number of cycles to fracture plots
TC4	Technical committee 4 (in ESIS)
TDCB	tapered double cantilever beam
T-L	specimen with applied stress in the **T** direction and crack growth in the **L** direction
VAMAS	Versailles Agreement for Materials and Standards
VIS	visually
w/w	weight concentration

AUTHOR INDEX

AUTHOR AFFILIATIONS

Dr B.R.K. Blackman BEng, PhD
Research Lecturer
Department of Mechanical Engineering
Imperial College
Exhibition Road
London SW7 2BX
E-mail: b.blackman@ic.ac.uk

Dr Wolfgang Böhme
Fraunhofer-Institut für Werkstoffmechanik (IWM)
Wöhlerstr. 11-13
D-79108 Freiburg
E-mail: boe@iwm.fhg.de

Dr A.J. Brunner
Scientist
Polymers/Composites Department
EMPA, Swiss Federal Laboratories for Materials Testing and Research
CH-8600 Duebendorf
Switzerland
E-mail: andreas.brunner@EMPA.CH

Dr Leonardo Castellani
Senior Scientist, Polymers Characterization
Enichem
via Taliercio 14
46100 Mantova
Italy
E-mail: leonardo.castellani@enichem.it

Dr E.Q. Clutton BSc, PhD
Technical Specialist (Polymer Properties)
BP Grangemouth,
PO Box 21,
Bo'ness Road,
Grangemouth,
Scotland
FK3 9XH
E-mail: Cluttoe@bp.com

Dr Peter Davies BSc, MSc, Dipl d'Ing., PhD
Research Engineer, Materials & Structures group
IFREMER
Centre de Brest
BP70
29280 Plouzané
France
E-mail: peter.davies@ifremer.fr

Dr Geoff Hale, BMet, PhD, CEng, FIM
Internet Content Development Manager
e-Commerce Group
TWI Ltd
Granta Park
Great Abington
Cambridge
UK
CB1 6AL
E-mail: gehale@twi.co.uk

Professor A.J. Kinloch, FREng
Professor of Adhesion and Director of Postgraduate Research
University of London
Imperial College of Science, Technology and Medicine
Department of Mechanical Engineering
Exhibition Road
London, SW7 2BX, UK
E-mail: a.kinloch@ic.ac.uk

Mr Hugh J. MacGillivray
Senior Research Officer
Mechanical Engineering Department
Imperial College of Science, Technology and Medicine
Exhibition Road
London SW7 2AZ
UK
E-mail: h.macgill@ic.ac.uk

Dr D.R. Moore BSc, PhD
Business Research Associate
ICI plc
Middlesbrough
Cleveland
UK
TS90 8JE
E-mail: Roy_Moore@ici.com

Professor A. Pavan, Chem. Eng. Dr
Professor of Polymer Engineering
Dipartimento di Chimica Industriale e Ingegneria Chimica
Politecnico di Milano
Piazza Leonardo da Vinci 32
I-20133 Milano
Italy
E-mail: andrea.pavan@polimi.it

Dr. F. Ramsteiner
BASF Aktiengesellschaft
Kunststofflabor
ZKM G201
D 67056 Ludwigshafen
E-mail: falko.ramsteiner@basf-ag.de

Professor Marta Rink
Professor of Polymeric Materials
CIIC - Politecnico di Milano
Piazza Leonardo da Vinci 32
I-20133 Milano (Italy)
E-mail: marta.rink@polimi.it

Professor J.G.Williams FREng. FRS
Professor of Mechanical Engineering
Mechanical Engineering Department
Imperial College,
London
SW7 2BX,
U.K.
E-mail: g.williams@ic.ac.uk

www.ingramcontent.com/pod-product-compliance
Ingram Content Group UK Ltd.
Pitfield, Milton Keynes, MK11 3LW, UK
UKHW020232290926
15470UKWH00002B/500